A Solution to the Crises of Soil, Water, and Climate in Plant Production

A Solution to the Crises of Soil, Water, and Climate in Plant Production

Edited by

Imran

Cambridge Scholars Publishing

A Solution to the Crises of Soil, Water, and Climate in Plant Production

Edited by Imran

This book first published 2023

Cambridge Scholars Publishing

Lady Stephenson Library, Newcastle upon Tyne, NE6 2PA, UK

British Library Cataloguing in Publication Data
A catalogue record for this book is available from the British Library

ISBN (10): 1-5275-2433-7
ISBN (13): 978-1-5275-2433-0

TABLE OF CONTENTS

EDITOR BIOGRAPHY

Dr. Imran belongs to the village Sambat, Matta District Swat, working as an Agriculture Officer in the Ministry of Agriculture (Extension Wing), Govt of Khyber Pakhtunkhwa, Pakistan. He earned his PhD degree from the Department of Agronomy, and Climate Change Centre (CCC), The University of Agriculture, Peshawar, Pakistan. He was awarded a Swiss Development Cooperation scholarship (SDC) for the completion of a doctorate allied to climate change and agro-meteorology. Dr. Imran was awarded the Turk Presidential Award in 2021-22 for his postdoctoral programme and served as a visiting professor at the Rhizosphere laboratory, University of Cukurova, Faculty of Agriculture, Department of Soil Science and Plant Nutrition, Adana, Turkey. Dr. Imran is a recognized researcher and renowned scientist due to his significant and outstanding contributions to research over the past several years. His ground breaking and original research has been published in various forms like books, magazines, pamphlets, newspaper reports and journal articles. Dr. Imran published more than 95 research articles in peer reviewed journals whereas he wrote more than 40 book chapters on various topics in different books published by various publishers associated to climate change, crop production and sustainable soil management practices. His commitment to research, exploration and outcomes and its dissemination over presentations, publications, and discoveries reports has greatly facilitated the agricultural community, farmers, growers, agriculture extension workers, agriculture specialists, scientists, and policymakers. Dr. Imran has exceptional achievement in both research and academics. Beside this Dr Imran worked with many National and International organizations both as a scientist and educationist including Agriculture Research Institute Mingora Swat as a soil ecologist (Researcher of Soil, plant and microbes interaction), Lecturer of Agronomy at Hazara University Mansehra, Lecturer at Allama Iqbal Open University, Agriculture Officer in Agriculture Extension Department, District Field Agronomist in Italian Ministry of foreign affairs (IAO), Provincial Disaster Management Authority (PDMA, PaRRSA), and Agriculture Expert and Master Trainer in USAID, Sarhad Rural Support Program (SRSP). Dr Imran research interests include, Crop Nutrition, Crop Physiology, degraded Soil management, Nutrients budgeting, Crop Competition, Weed and Crop Management and Sustainable Agriculture under Stressful Environments and an additional command on climate forecasting, climate corridors and Agro-meteorology.

CHAPTER ONE

SOIL, WATER AND ENVIRONMENTAL CRISES: A PRIME FACTOR FOR HUNGER, CONFLICT AND FOOD INSECURITY

IMRAN[1, 2, 3]

[1]DEPARTMENT OF AGRONOMY, THE UNIVERSITY OF AGRICULTURE, PESHAWAR, PAKISTAN,
[2]RHIZOSPHERE LABORATORY, UNIVERSITY OF CUKUROVA, FACULTY OF AGRICULTURE, DEPARTMENT OF SOIL SCIENCE AND PLANT NUTRITION, ADANA, TURKEY,
[3]MINISTRY OF AGRICULTURE, ENTENSION WING, KHYBER PAKHTUNKHWA, PAKISTAN

Abstract

The production of more food and agricultural goods are closely connected with world peace. In a world where everyone has access to enough food and is satisfied, there will be peace and a positive outlook on life. In contrast, when food is scarce, prices are high, and agricultural products are in low, people will snip food and commit robberies, murders, and kidnappings in order to feed themselves and their children. Peace is thus closely tied to having access to more food. Climate change is exacerbating agriculture productivity by various ways including soil, water and environmental crises. Water scarcity and water-related hazards (such as floods and droughts), as rising temperatures disrupt precipitation patterns and the entire water cycle. Water and climate change are inextricably linked. Climate change affects the world's water in complex ways. From unpredictable rainfall patterns to shrinking ice sheets, rising sea levels, floods and droughts most impacts of climate change come down to water. Global warming impacts everyone's food and water security. Climate change is a direct cause of soil degradation, which limits the amount of carbon the earth is able to contain. Some 500 million people today live in areas affected by erosion, while up to 30% of food is lost or wasted as a result. Meanwhile, climate change limits the availability and quality of water for drinking and agriculture. In many regions, crops that have thrived for centuries are struggling to survive, making food security more precarious. Such impacts tend to fall primarily on the poor and vulnerable. Global warming is likely to make economic output between the world's richest and poorest countries grow wider. Disasters linked to climate and weather extremes have always been part of our Earth's system. But they are becoming more frequent and intense as the world warms. No continent is left untouched by the climate change (heat waves, droughts, typhoons, and hurricanes) causing mass destruction around the world. About 90% disasters are now classed as weather- and climate-related, costing the world economy 520 billion USD each year, while 26 million people are pushed into poverty as a result.

Keywords: Climate change, soil degradation, Nutrients depletion, precipitation, rainfall, crop yield, plant biomass, food security

Introduction

The production of more food and agricultural goods is closely correlated with world peace. In a world where everyone has access to enough food and is satisfied, there will be peace and a positive outlook on life (Imran and amanullah 2022a). In contrast, when food is scarce, prices are high, and agricultural products are in short supply, people will steal food and commit robberies, murders, and kidnappings in order to feed themselves and their children. Peace is thus closely tied to having access to more food. Years of intensive agriculture have damaged our resources (soil, water and the environment) for more food production through compaction, erosion and overuse of fertilisers and pesticides. It's estimated that one third of our agricultural soils have been lost over the past 40 years. It might be easy to dismiss the ground beneath our feet as just dirt. But without it humankind would not exist. Former US president Franklin D Roosevelt was right when he said: *"The nation that destroys its soil destroys itself."* He was speaking in the 1930s after years of severe drought and relentless winds turned states across America into dust bowls. The disaster wiped out crops and livestock and forced countless families across America to abandon their farms (Imran et al., 2022a). Today, our soils are facing chronic issues all over the world. Climate change, food insecurity, and degradation of natural resources such as water and soil are the major challenges the world is facing in the 21st century (Hall et al., 2017). Climate change causes the rainfall amount to decrease and become more erratic. Fresh water (< 500 ppm dissolved salt) is the critical resource for the survival of many lives and the most limiting factor for crop production (Albert et al., 2021). About 72% of the world's freshwater resources are utilized for agriculture production, of which irrigation takes the highest share (FAO, 2021). Irrigation is essential for sustaining crop production and achieving food security in a changing climate. According to Hoogeveen et al. (2015) the world's average irrigation efficiency is 55%, and this indicate that almost half of the water withdrawn for irrigation is lost. However, under limited moisture supply conditions, since water resources are declining, water saving through an efficient irrigation practice is required to achieve optimal crop yield (Fischer et al., 2019). Efficient and water-saving irrigation can also reduce the carbon and energy footprint of irrigation practices and help in mitigating climate change (Zou et al., 2012). As a result, water availability is a critical problem affecting many countries and is expected to be scarcer in the future (Vogelmann et al., 2013). The problems we are facing are inter-related, therefore, multi-benefit (win-win) solutions are needed. It is less likely to achieve food and water security while putting less emphasis on climate change mitigation and environmental protection, and vice versa (Mbow et al., 2019).

Food Production and Soil Crises

Soil is a mix of mineral and organic matter jumbled up with air, water, and microbes. It provides a medium in which plants grow, and a habitat for animals and microscopic life. Soils don't just help us grow food. They absorb, filter and purify our water, helping to protect against flash flooding. They also store carbon, helping to mitigate the effects of climate change. There is an estimated three times more carbon in our soils than in Earth's atmosphere. One problem facing farmers is hard or compacted soil, usually caused by heavy farm machinery. Hard soils are difficult for roots to penetrate, giving plants limited access to water and nutrients. As a result, hard soils can reduce crop yields by over 50%, causing significant losses to farmers and growers. They also become less effective at taking up water, contributing to floods. Scientists recently discovered the chemical signal that causes roots to stop growing in hard soils. Amazingly, this can be 'switched off' to allow plants to punch through compacted soil. The international research team used X-ray Computed Tomography scanners to visualise in situ how plant roots responded to compacted soil. They discovered that plants release a chemical signal when they encounter hard soils, causing roots to stop growing. If this signal is switched off, roots are able to push through compacted soil. The research could lead the way to breeding new crops that can thrive in even the most compacted soils.

Signals in the Soil

There's still a great deal we don't understand about soils, and the invisible ecosystems that are vital to their function. Nowhere is this truer than in the Arctic, where melting ice sheets and glaciers are exposing vast regions of virgin soils to the outside air. These soils haven't seen daylight in thousands of years (Imran and amanullah 2022b). But they are quickly colonised by a whole range of microbes, lichens and plants that are adapted to extreme environments. These microbes have a significant role in the carbon cycle, emitting greenhouse gases such as carbon dioxide and methane. Yet amazingly, we know almost nothing about them. One reason for this knowledge vacuum is that scientist's usually only travel to the Arctic in the summer months: when there is enough light to see by and temperatures aren't too cold. As a consequence, we don't know how these extreme ecosystems behave in the prolonged cold, dark Arctic winters. This is set to change though. Dr James Bradley, a senior lecturer in environmental science at Queen Mary, University of London is leading a project that has installed sensors in soils near a retreating glacier in Svalbard, Norway (Imran et al., 2022b). The sensors will take continuous long-term measurements of biological and chemical activity under the snow. The project is part of a wider joint initiative between UK and US scientists called Signals in the Soil (SITS), which aims to unlock the mysteries of soils using pioneering new sensor technology.

Soil Compaction, May Depleted Food Production

Numerous factors are involved depleting soil health, nutrients status, organic matter and ultimately food production. Primary factor that contributes to the degradation of soils is over ploughing. In soils, particles of clay and sand bind together with lumps of organic matter. This matter can include dead roots, fungi and secretions from earthworms. These clumps are vital for soil structure. Without them, soils can become waterlogged and wash away. Unfortunately, when farmers plough and till soil it breaks down these clumps, putting them at risk of erosion. Researchers at the universities of Sheffield, York and Leeds have found that reviving an old farming practice could help restore damage to the soil. The researchers took soil from four arable fields and three pastures at The University of Leeds farm, a commercial mixed farm in northern England. Previously the land had been used to grow cereal crops and had been subjected to decades of tillage. They looked at what impact crop rotation would have on the health of the soils. This is when fields are allowed to 'rest' by growing different plants for a season or more before replanting with the original crop. In the past, farmers commonly used crop rotation to let soils recover their nutrients. However the practice largely fell out of fashion when industrial fertilisers allowed the same crops to be grown year on year. The scientists planted a mixture of grasses and clovers in the soil. After just 19 months they found that the topsoil had recovered considerably, the soil was less compacted, earthworm numbers had increased, and the soils were much more resilient to drought and flood conditions. The research suggests that using less intensive ploughing methods, rotating crops, and allowing soils time to rest and recover could be key to improving yields.

Figure 1. Soil compaction from farm machinery is a serious soil health issue.

More People, More Food and Less Land

Climate change and growing populations mean that our soils are under more pressure than ever before. We need to feed more people using less land, and less good quality land. The truth is, our soils are degrading. Years of intensive agriculture have damaged them through compaction, erosion and overuse of fertilisers and pesticides. It's estimated that one third of our agricultural soils have been lost over the past 40 years (Imran et al., 2022c). These sick soils can lose their incredible ability to take carbon out of the atmosphere and instead become net emitters of CO2. Ironically, crops growing in degraded soils require more intensive management and fertiliser, leading to a spiral of increasing greenhouse gas emissions and soil degradation.

Restoring Peat lands

As well as providing the nourishment for plants to grow, soils play a vital role in storing carbon. Peat, which consists of partially decayed vegetation or organic matter, is especially good at this. UK peat lands currently lock away more than three billion tonnes of carbon. That's more carbon than is stored in all the world's oceans combined. However peat lands only store carbon if they are healthy. If damaged, they can release CO2 into the atmosphere. In Wales, experts at the Centre for Ecology & Hydrology (CEH) have spent nearly 20 years studying the impacts of climate change on upland heath and peat land traditionally used for hill farming. They've found that dry soils lose up to 10% more carbon. The results suggest that climate change can be a vicious circle. As global warming causes more extreme dry spells, the soil's ability to store carbon gets reduced. That in turn leads to even more carbon being released into the atmosphere. The reason is all down to the activity of bacteria and fungi breaking down the peat. When soils are water-logged, microbes and fungi are less active. This means they break down the peat more slowly and release less carbon in the form of carbon dioxide and other greenhouse gases. But when they dry out, the microbe's ramp up their activity and start to undo the centuries of carbon storage provided by peat. This research shows that by seeking out a deeper understanding of the world beneath our feet and applying it in the right ways there's hope yet for a future where we can all benefit from healthy soil.

Figure 2. Peat bogs can be a great asset in fighting climate change but it all depends on their health.

Food Security in Relation to Water Security

Food and Water Insecurity: A Catalyst for Conflict

Climate change is a major threat to international peace and security. The effects of climate change heighten competition for resources such as land, food, and water, fuelling socioeconomic tensions and, increasingly often, leading to mass displacement. Climate is a risk multiplier that makes worse already existing challenges. Droughts in Africa and Latin America directly feed into political unrest and violence. The World Bank estimates that, in the absence of action, more than 140 million people in Sub-Saharan Africa, Latin America, and South Asia will be forced to migrate within their regions by 2050 (Imran et al., 2022d). Global warming impacts everyone's food and water security. Climate change is a direct cause of soil degradation, which limits the amount of carbon the earth is able to contain. Some 500 million people today live in areas affected by erosion, while up to 30% of food is lost or wasted as a result. Meanwhile, climate change limits the availability and quality of water for drinking and agriculture. In many regions, crops that have thrived for centuries are struggling to survive, making food security more precarious. Such impacts tend to fall primarily on the poor and vulnerable. Global warming is likely to make economic output between the world's richest and poorest countries grow wider. While science tells us that climate change is irrefutable, it also tells us that it is not too late to stem the tide. This will require fundamental transformations in all aspects of society; how we grow food, use land, transport goods, and power our economies. While technology has contributed to climate change, new and efficient technologies can help us reduce net emissions and create a cleaner world (Imran et al., 2022d). Readily-available technological solutions already exist for more than 70% of today's emissions. In many places renewable energy is now the cheapest energy source and electric cars are poised to become mainstream. In the meantime, nature-based solutions provide 'breathing room' while we tackle the decarbonisation of our economy. These solutions allow us to mitigate a portion of our carbon footprint while also supporting vital ecosystem services, biodiversity, and access to fresh water, improved livelihoods, healthy diets, and food security. Nature-based solutions include improved agricultural practices, land restoration, conservation, and the greening of food supply chains.

Water Scarcity

About two billion people worldwide don't have access to safe drinking water today (SDG Report 2022), and roughly half of the world's population is experiencing severe water scarcity for at least part of the year (IPCC). These numbers are expected to increase, exacerbated by climate change and population growth. Only 0.5 per cent of water on Earth is useable and available freshwater and climate change is dangerously affecting that supply (Imran 2022b). Over the past twenty years, terrestrial water storage including soil moisture, snow and ice has dropped at a rate of 1 cm per year, with major ramifications for water security (WMO). Water supplies stored in glaciers and snow cover are projected to further decline over the course of the century, thus reducing water availability during warm and dry periods in regions supplied by melt water from major mountain ranges, where more than one-sixth of the world's population currently live. Sea-level rise is projected to extend salinization of groundwater, decreasing freshwater availability for humans and ecosystems in coastal areas. Limiting global warming to 1.5°C compared to 2°C would approximately halve the proportion of the world population expected to suffer water scarcity, although there is considerable variability between regions (Imran 2022c). Water quality is also affected by climate change, as higher water temperatures and more frequent floods and droughts are projected to exacerbate many forms of water pollution – from sediments to pathogens and pesticides. Climate change, population growth and increasing water scarcity will put pressure on food supply as most of the freshwater used, about 70 per cent on average, is used for agriculture (it takes between 2000 and 5000 litters of water to produce a person's daily food) (FAO).

Water-Related Hazards

Climate change has made extreme weather events such as floods and droughts more likely and more severe. Rising global temperatures increase the moisture the atmosphere can hold, resulting in more storms and heavy rains, but paradoxically also more intense dry spells as more water evaporates from the land and global weather patterns change. Drought and flood risks, and associated societal damages, are projected to further increase with every degree of global warming. The frequency of heavy precipitation events will very likely increase over most areas during the 21st century, with more rain-generated floods. At the same time, the proportion of land in extreme drought at any one time is also projected to increase. Water-related disasters have dominated the list of disasters over the past 50 years and account for 70 per cent of all deaths related to natural disasters. Since 2000, flood-related disasters have risen by 134 per cent compared with the two previous decades. Most of the flood-related deaths and economic losses were recorded in Asia. The number and duration of droughts also increased by 29 per cent over this same period. Most drought-related deaths occurred in Africa.

Saving Precious Water

Another SITS project is led by Paul Hallett, a researcher at the University of Aberdeen. Hallett and his team are developing a network of underground sensors that can determine the health of soils by measuring soil moisture, temperature and salinity. Amazingly, the sensors don't contain batteries or have any direct source of power. Instead, they 'download' energy from autonomous robots and drones above ground. At first the team aim to use the network of sensors to investigate one of biggest problems facing the food industry: how to feed over 9 billion people by 2050 without outstripping our soil's capabilities (Imran 2021). To boost productivity, farmers use irrigation: the process of applying controlled amounts of water to land to assist in the production of crops. However irrigation consumes over 70% of freshwater reserves in many regions of the world. Alternative sources that could be used include water reclaimed from waste sources, or even seawater. But these can contain contaminants or excess nutrients and salt that are harmful to the environment. Hallett's team will use the network of sensors to see what impact water from different sources has on soil health.

Figure 3. Glacial melt is exposing soils that have been trapped under ice for thousands of years.

Solution of Soil, Water and Climate Crises

Healthy aquatic ecosystems and improved water management can lower greenhouse gas emissions and provide protection against climate hazards (Water and Climate Coalition). Wetlands such as mangroves, sea grasses, marshes and swamps are highly effective carbon sinks that absorb and store CO2, helping to reduce greenhouse gas emissions. Wetlands also serve as a buffer against extreme weather events. They provide a natural shield against storm surges and absorb excess water and precipitation. Through the plants and microorganisms that they house, wetlands also provide water storage and purification. Early warning systems for floods, droughts and other water-related hazards provide a more than tenfold return on investment and can significantly reduce disaster risk: a 24-hour warning of a coming storm can cut the ensuing damage by 30%. Water supply and sanitation systems that can withstand climate change could save the lives of more than 360,000 infants every year (New Climate Economy report). Climate-smart agriculture using drip irrigation and other means of using water more efficiently can help reduce demand on freshwater supplies (Imran and Amanullah 2021a).

A Climate Crisis Solution: A Race We Can Win

Climate change is the defining crisis of our time and it is happening even more quickly than we feared. But we are far from powerless in the face of this global threat. As Secretary-General Antonio Gutierrez pointed out in September, "the climate emergency is a race we are losing, but it is a race we can win". No corner of the globe is immune from the devastating consequences of climate change. Rising temperatures are fuelling environmental degradation, natural disasters, weather extremes, food and water insecurity, economic disruption, conflict, and terrorism. Sea levels are rising, the Arctic is melting, coral reefs are dying, oceans are acidifying, and forests are burning. It is clear that business as usual is not good enough. As the infinite cost of climate change reaches irreversible highs, now is the time for bold collective action. Billions of tons of CO2 are released into the atmosphere every year as a result of coal, oil, and gas production. Human activity is producing greenhouse gas emissions at a record high, with no signs of slowing down (Imran and Amanullah 2021b). According to a ten-year summary of UNEP Emission Gap reports, we are on track to maintain a "business as usual" trajectory. The last four years were the four hottest on record. According to a September 2019 World Meteorological Organization (WMO) report, we are at least one degree Celsius above preindustrial levels and close to what scientists warn would be "an unacceptable risk". The 2015 Paris Agreement on climate change calls for holding eventual warming "well below" two degrees Celsius, and for the pursuit of efforts to limit the increase even further, to 1.5 degrees. But if we don't slow global emissions, temperatures could rise to above three degrees Celsius by 2100, causing further irreversible damage to our ecosystems. Glaciers and ice sheets in polar and mountain regions are already melting faster than ever, causing sea levels to rise. Almost two-thirds of the world's cities with populations of over five million are located in areas at risk of sea level rise and almost 40 per cent of the world's population live within 100 km of a coast. If no action is taken, entire districts of New York, Shanghai, Abu Dhabi, Osaka, Rio de Janeiro, and many other cities could find themselves underwater within our lifetimes, displacing millions of people (Imran and Amanullah 2021c).

Scalable new technologies and nature-based solutions will enable us all to leapfrog to a cleaner, more resilient world. If governments, businesses, civil society, youth, and academia work together, we can create a green future where suffering is diminished, justice is upheld, and harmony is restored between people and planet. Climate change is a global challenge and a threat to food security. Agriculture is expected to produce food for the world's population, estimated to reach 9.1 billion in 2050 and more than 10bn by the end of the century. Weather patterns/climate is one of the main factors influencing the productivity of agriculture. According to researchers, to improve the productivity and stability of smallholder agricultural production in the wake of climate change, agricultural systems must be transformed. Significant effects of this change have already been seen on water resources, human health, and food security. Rising temperatures, changes in precipitation, floods and famines affect crop production and negatively affect land and water resources. Fluctuations in climatic factors such as rainfall and temperature and elevated level of carbon dioxide have increased the frequencies of climatic disasters like floods, droughts and cyclones all over the Indian subcontinent. Improved irrigation facilities, highly genetically modified seeds, and increased usage of fertilisers and pesticides have restricted negative trends. But, the imbalance in the use of fertilisers and pesticides is also responsible for declining soil fertility. Climate-smart agriculture (CSA) refers to farming methods that tackle both the problems of climate change and global food security. It is becoming an increasingly crucial concept to ensure future food security and sustainability as the global climate becomes more hostile to conventional agricultural practices. The Food and Agriculture Organisation (FAO) has encouraged a sustainable agricultural production system due to the social, environmental, and economic issues caused by climate change and traditional agriculture. According to FAO, CSA practises are viewed as a way to improve resilience while slowing down environmental degradation. Consequently, its adoption will boost yield, enhance resource use efficiency, increase agricultural income, and mitigate the adverse effects of climate change on its production. CSA includes cutting back on chemical inputs, improving soil fertility, and managing water resources better. Crop diversification, changes in the cropping pattern and planting dates are other farming techniques that can lower the risk of crop loss due to climate change. The National Agriculture Extension Policy encourages farmers to adopt CSA practices such as water harvesting, efficient irrigation systems, improved crop varieties, and agroforestry. It also promotes the use of climate-smart farm inputs such as organic fertilisers and soil amendments.

Additionally, CSA focuses on enhancing the resistance of crops to extreme weather events. This includes the use of drought-resistant varieties, mulch or cover crops to prevent soil erosion, terracing and other soil conservation techniques. Farmers can also practise conservation agriculture, which reduces the amount of tilling and maintains a biological mulch cover on the soil. Precision farming and remote sensing are two examples of cutting-edge technologies that climate-smart agriculture encourages farmers to adopt. While remote sensing can be used to monitor crop health and lower the risk of crop losses because of changes in weather patterns, precision farming helps farmers apply inputs like fertiliser and irrigation water more efficiently. The policy implication is that promoting and scaling up the adoption of CSA practices can help farmers raise their incomes by improving agricultural productivity and profitability. The Pakistan Agricultural Research Council studied farmers' knowledge and the intensity of their adoption of CSA practices. According to this research, most farmers were only moderately knowledgeable about the various indicators of climate-smart agriculture. The overall knowledge level showed that the sampled farmers knew more about crop rotation and varieties but less about pesticide application precautionary measures, crops recommended fertiliser application per acre and laser leveller. The sampled farmers had the lowest knowledge level in early maturity (short duration) crop varieties, integrated pest management techniques, wheat on ridges/seed beds, heat and drought tolerant varieties, and organic farming. These results indicate a significant knowledge gap and low practice of CSA, in spite of the positive outlook to climate-smart practices attributed to the collapse of extension services over the years. Insufficient understanding of climate-smart practices, limited access to weather and climate information, low financial capacity, and a lacklustre policy framework were the main barriers to adopting these practices. In general, the government is moving in the right direction to support CSA in the country. Nevertheless, there is still a need to raise knowledge among farmers. Using modern technology to connect with farmers and conduct targeted communication efforts can accomplish this. The government should also focus on providing technical assistance and access to finance to farmers so they can adopt CSA practices. This could include providing financial incentives to farmers to purchase climate-smart inputs and equipment and providing training and capacity-building programmes. By promoting CSA in Pakistan, farmers can improve their productivity and profitability while adapting to the effects of climate change. This will be essential for ensuring food security in the country and for building the resilience of its agricultural sector.

Conclusion

Peace can be ensure around the globe with food security. If everyone has access to enough food and to quality produce, will have smooth life while in contrast, when food is scarce, the life will be hard and full of conflicts. Peace is thus closely tied to having access to more food. Global peace can be ensure to produce more food with the objectives to protect the soil, water and environment. Crops and animals should be raised not on the cost of future generation but by owing the principles of climate, soil and water smart farming to survive with peace and health.

References

1. Imran & Amanullah (2022a): Soybean quality and profitability improved with peach (Prunus persica L) remnants, phosphorus and beneficial microbes, *Journal of Plant Nutrition*, DOI: 10.1080/01904167.2022.2068438
2. Imran, Amanullah & Abdel Rahman Al Tawaha (2022a): Indigenous organic resources utilization, application methods and sowing time replenish soil nitrogen and increase maize yield and total dry biomass, *Journal of Plant Nutrition*, DOI: 10.1080/01904167.2022.2067055.
3. Hall, Dawson, Macdiarmid, Matthews, & Smith. (2017). The impact of population growth and climate change on food security in Africa: looking ahead to 2050. *2017/03/04, 15*(2), 124-135. doi:https://doi.org/10.1080/14735903.2017.1293929
4. Hoogeveen, Faurès, Peiser, Burke, & van de Giesen. (2015). GlobWat – a global water balance model to assess water use in irrigated agriculture. *Hydrol. Earth Syst. Sci., 19*(9), 3829-3844. doi:https://doi.org/10.5194/hess-19-3829-2015
5. Albert, Destouni, Duke-Sylvester, Magurran, Oberdorff, Reis, . . . Ripple. (2021). Scientists' warning to humanity on the freshwater biodiversity crisis. *Ambio, 50*(1), 85-94. doi:https://doi.org/10.1007/s13280-020-01318-8
6. FAO. (2021). AQUASTAT Database. Global Information System on Water and Agriculture. Retrieved from https://www.fao.org/aquastat/en/. Retrieved 28 May 2022, from Food and Agriculture Organization of the United Nations https://www.fao.org/aquastat/en/
7. Fischer, Manzoni, Morillas, Garcia, Johnson, & Lyon. (2019). Improving agricultural water use efficiency with biochar – A synthesis of biochar effects on water storage and fluxes across scales. *2019/01/10/, 657*, 853-862. doi:https://doi.org/10.1016/j.scitotenv.2018.11.312
8. Zou, Li, Gao, & Wan. (2012). How water saving irrigation contributes to climate change resilience—a case study of practices in China. *Mitigation and Adaptation Strategies for Global Change, 17*(2), 111-132. doi:10.1007/s11027-011-9316-8

9. Vogelmann, Reichert, Prevedello, & Awe. (2013). Hydro-physical processes and soil properties correlated with origin of soil hydrophobicity. *Ciência Rural, 43*, 1582-1589. doi:https://doi.org/10.1590/S0103-84782013005000107
10. Mbow, Rosenzweig, Barioni, Benton, Herrero, Krishnapillai, . . . Sapkota. (2019). Food security. In *Climate Change and Land: an IPCC special report on climate change, desertification, land degradation, sustainable land management, food security, and greenhouse gas fluxes in terrestrial ecosystems* (pp. 437-550). Retrieved from https://www.ipcc.ch/srccl/chapter/chapter-5/
11. Imran, Amanullah and Ibrahim Ortash. 2022b. Agronomic Practices Improved Cucumber Productivity, Nutrients Uptake and Quality. *Gesunde Pflanzen*. 10.1007/s10343-022-00634-1.
12. Imran, Amanullah, and Abdel Rahman Al Tawaha. 2022c. Regenerating potential of dual purpose rapeseed (*Brassica napus* L.) as influenced by decapitation stress and variable rates of phosphorous. *Communication in soil science and plant analysis. Doi.org/10.1080/00103624.2022.2118297*
13. Imran and Amanullah. (2022b). Integration of peach (*Prunus persica* L) remnants in combination with beneficial microbes and phosphorus differ phosphorus use efficiency, agronomic efficiency and partial factor productivity in soybean Vs maize crops. *Journal of Plant Nutrition.* doi.org/10.1080/01904167.2022.2099890
14. Imran, & Abdel Rahman Al Tawaha (2022): Humic Acid and Sulfur Integration Enhances Growth and Yield Related Traits of *Brassica Napus L. Journal of Plant Nutrition.* *d*oi.org/10.1080/01904167.2022.212882
15. Imran. 2022a. Growing of Off-Season Tomato in High Tunnel and Its Nutritional Value Augmentation with Integrated Nutrients Management. *Journal of Plant Nutrition.* Doi: 10.1080/01904167.2022.2046062
16. Imran and Amanullah. 2021a. Assessment of chemical and manual weed control approaches for effective weed suppression and maize productivity enhancement under maize-wheat cropping system. *Gesunde Pflanzen*, DOI: 10.1007/s10343-021-00599-7.
17. Imran, Amanullah and Abdel Rehman Altawaha. 2022c. Carbon Assimilation and Dry Matter Partitioning in Soybean Ameliorates with the Integration of Nano-Black Carbon, Along with Beneficial Microbes and Phosphorus Fertilization. *Journal of Plant Nutrition, DOI:10.1080/01904167.2022.2035753*

18. Imran. 2022b. (Review) Phosphorus availability enhanced with combine application of organic amendments and beneficial microbes under soybean-wheat cropping system. *Communication in soil science and plant analysis.* doi.org/10.1080/00103624.2022.2034848
19. Imran.2022c. Management of Abiotic Stresses with Nano-black carbon is a tool for Crop Production. *Journal of Plant Nutrition.* doi.org/10.1080/01904167.2022.2046085
20. Imran and Amanullah. 2021b. Phosphorus biofortification and uptake in maize enhanced with integrated phosphorus management. *Phosphorus, Sulfur, and Silicon and the Related Elements.* DOI: 10.1080/10426507.2021.2022677.
21. Imran. 2021. The bioavailability of Phosphorus in composite vs. Hybrid maize differ with phosphorus and boron fertilization. *Phosphorus, Sulfur, and Silicon and the Related Elements. https://doi.org/10.1080/10426507.2021.1920588.*
22. Imran, Amanullah, Asad Ali khan, Tariq Mahmood, Abdel Rahman Al Tawaha and Samia Khanum. 2021a. Adequate Fertilization, Application Method and Sowing Techniques Improve Maize yield and related traits. *Communication in soil science and plant analysis*. DOI:10.1080/00103624.2021.1925688
23. Imran, Amanullah and Abdel Rahman M. Al Tawaha. 2021b. Management of Nano-black carbon, Phosphorous and Bio Fertilizer improve Soil organic carbon and ensilage biomass of soybean and maize. *Communication in Soil Science and Plant Analysis.* DOI.10.1080/00103624.2021.1966439
24. Imran and Amanullah· 2021c. Phosphorus and Boron Application Optimizing Biofortification Of phosphorus and Productivity of French Bean (*Phaseolus Vulgaris* L.). *Communication In Soil Science and Plant Analysis.* DOI:10.1080/00103624.2021.1971689
25. Imran, Amanullah and Abdel Rahman Al-Tawaha. 2021c. Carbon Sources Application Increase Wheat Yield And Soil Fertility. *Communication in soil science and plant analysis DOI: 10.1080/00103624.2020.1865397.*
26. Imran, Amanullah and Abdel Rehman Al-Tawaha.2020a. The Productivity of subsequent wheat enhanced with residual carbon sources and phosphorus under improved irrigation system. *Communication in soil science and plant analysis*, DOI:10.1080/00103624.2020.1763387
27. Imran, Amanullah, Muhammad Arif, Zahir Shah and Abdul Bari. 2020b. Soil application of *Trichoderma* and Peach (*Prunus persica* L.) residues possesses biocontrol potential for weeds and enhances growth and

profitability of soybean (Glycine max). *Sarhad Journal of Agriculture* 36 (1) 10-20. DOI | http://dx.doi.org/10.17582/journal.sja/2020/36.1.10.20

28. Izhar Ali, Saif Ullah, Imran, Zixiong Guo 2022. Changes in Soil Properties and Microbial Communities with Biochar Amendment in Combination with Nitrogen Fertilizer in a paddy field. Frontiers Microbiology. DOI: 10.3389/fmicb.2022.834751
29. Ali M. Qaisi, Abdel Rahman Al Tawaha, Imran, Moh, d Al-Rifaee.2022. Effects of chitosan and Biochar-Mended soil on growth, yield and yield components and mineral composition of Fenugreek. *Gesunde Pflanzen. doi.org/10.1007/s10343-022-00727-x*
30. Izhar Ali, Saif Ullah, Imran, Zixiong Guo1, Shangqing Wei1 and Ligeng Jiang. 2021. Combined application of biochar and nitrogen fertilizer promotes the activity of starch metabolism enzymes and the expression of related genes in rice in a dual cropping system. BMC Plant Biology *(2021) 21:600* https://doi.org/10.1186/s12870-021-03384-w
31. Imran, Amanullah & Abdel Rahman Al Tawaha (2022d): Indigenous organic resources utilization, application methods and sowing time replenish soil nitrogen and increase maize yield and total dry biomass, *Journal of Plant Nutrition*, DOI: 10.1080/01904167.2022.2067055.

CHAPTER TWO

A SOLUTION TO SOIL, WATER AND ENVIRONMENTAL CRISES

ABID KAMAL[1], IMRAN[2,3], ASAD ALI KHAN[2] AND IBRAHIM ORTAS[3]

[1]DEPARTMENT OF SOIL AND ENVIRONMENTAL SCIENCES, THE UNIVERSITY OF AGRICULTURE PESHAWAR-PAKISTAN

[2] DEPARTMENT OF AGRONOMY, THE UNIVERSITY OF AGRICULTURE PESHAWAR-PAKISTAN

[3]DEPARTMENT OF SOIL AND PLANT NUTRITION, FACULTY OF AGRIULTURE, CUKUROVA UNIVERSITY ADANA TURKEY

Abstract

The loss of soil quality and fertility, which is a global concern, is referred to as the "soil crisis." Soil quality degradation has a substantial economic impact, as it lowers crop production and raises the cost of food production. Soil crises is caused by several factors, including intensive agriculture, deforestation, and urbanization, overuse of fertilizer and pesticides, and climate change. As a results of these factors, soil erosion, desertification, salinization, and nutrients depletion have become major issues, which threaten food security, biodiversity, and environmental sustainability. The soil crisis requires a range of intervention, including the adoption of sustainable land use practice, conservation of soil resources, and promotion of soil restoration techniques such as agroforestry, conservation agriculture and composting. It also requires the cooperation and coordination of stakeholder at all levels, form farmers to policymakers, to ensure the long-term sustainability of our soil resources. The lack of freshwater resources to meet environmental and human demands is what is known as the water crisis. The first issue for global growth is the lack of clean drinking water, hence the water crisis is closely tied to human rights. Some of the major causes of the water crisis include population growth, climate change, overconsumption, and pollution. As the world's population continues to grow, demand for water also increases, and many regions are already struggling to meet the water needs of their populations. Climate change is also exacerbating the water crisis, as it alters weather patterns and causes more extreme weather events such as droughts and floods. The environmental crisis has significant impacts on human health, well-being, and the economy, and it threatens the sustainability of the planet. Rising temperatures are fuelling environmental degradation, natural disasters, weather extremes, food and water insecurity, economic disruption, conflict, and terrorism. Sea levels are rising, the Arctic is melting, coral reefs are dying, oceans are acidifying, and forests are burning.

Keywords: Soil contamination, Nutrients depletion, Food shortage, water shortage, land degradation, pollution, crises.

Introduction

1. Soil Crises

Planet Earth is currently going through an unparalleled crisis of soil deterioration, desertification, and erosive loss that is adversely affecting the services it offers as a result of anthropogenic abuses, natural disasters, and extreme weather events encouraged by global warming (Imran and amanullah 2022a). The soil crisis refers to the degradation of soil quality and fertility, which is a global concern and the degradation of soil; quality also has significant economic impact, as it reduces crop yield and increase the cost of food production. The problem is caused by several factors, including intensive agriculture, deforestation, and urbanization, overuse of fertilizer and pesticides, and climate change. As a results of these factors, soil erosion, desertification, salinization, and nutrients depletion have become major issues, which threaten food security, biodiversity, and environmental sustainability (Imran et al., 2022a). The soil crisis requires a range of intervention, including the adoption of sustainable land use practice, conservation of soil resources, and promotion of soil restoration techniques such as agroforestry, conservation agriculture and composting. It also requires the cooperation and coordination of stakeholder at all levels, form farmers to policymakers, to ensure the long-term sustainability of our soil resources (Imran and amanullah 2022b).

A mix of intensive farming methods and natural processes is causing serious soil degradation. It becomes more challenging to raise food crops as the topsoil layer that is fertile becomes thinner. The global food supply begins to appear vulnerable if farming methods are not changed and urgent measures are not taken to keep soil. Most natural resource commodities are currently undergoing increased demand in the twenty-first century. Natural resource governance is becoming increasingly complex, especially as the interdependence and linkages between different resources are becoming more apparent are considered when making decisions (Andrews-Speed et al. 2012). Erosion, soil fertility decline, and other factors frequently limit the productivity of available land. Pollution, salinization, and flooding (Gupta et al. 2009). Loessy soil substrates have a favourable root structure, but they are prone to hydro collapse and water erosion (Smalley et al. 2006). Cropland degradation in rainfed agriculture is primarily caused by soil erosion caused by both wind and water (Suleimenov et al. 2012). Water erosion caused by furrow irrigation is also a significant issue (Saparov et al. 2013).

According to Vazhenin (1987), the following levels of heavy metal contamination were assessed: weak, moderate, medium, increased, high, and very high. The research findings revealed that phosphorus fertilizers had the greatest amount of heavy metals penetrating the soil in terms of both accumulation and concentration of the impurity. Potash fertilizer figures are slightly lower, and nitrogen fertilizer penetration of heavy metals in the soil is negligible. The calculation confirms that using scientifically justified fertilizer quantities does not result in heavy metal contamination of soil (Imran et al., 2022b). There are numerous solutions to these issues. A unique approach must be developed for each region, taking into account the natural climatic and ecological conditions. To halt and reduce land degradation and desertification, salinization, and pollution, scientifically justified land cultivation systems, as well as complex organizational, economic, agro technical, and hydro-technical activities, forest reclamation schemes, and so on, must be implemented (Imran et al., 2022c).

- A thorough inventory of arable lands, taking into account their existing level of degradation and desertification, as well as their vulnerability to erosion, salinization, pollution, and other adverse conditions;
- land cultivation systems, industrial and creative technologies, pasture rotation, surface and radical improvement of soil reclamation and environmental conditions on unproductive degraded agricultural lands, rangelands and their irrigation; integrated soil resource management and their protection;

- Efficient use of water resources; stepping up efforts to stop water loss; implementing contemporary irrigation techniques;
- Establishing environmental monitoring programs and enhancing ecological conditions through the application of environmental certification and standards in agriculture;
- Creation of incredibly efficient hydrocarbon extraction technology.
- Preserving inundated oil wells

One of the most important goals of our day is to create fresh, highly applicable approaches to managing soil resources. Other major tasks include the rational use of the soil landscape, enhancement of soil reclamation, and environmental status of land resources. In this sense, the importance and relevance of the problem are clear, and it calls for quick action and a solution (Imran et al., 2022d). It is required to create specific programs for researching contaminated and disturbed lands, create legislation, and provide pertinent advice for re-cultivation and land restoration in order to address these issues (Imran 2022b; Imran 2022c). The agreed measures, which are intended to implement a single development policy in the agricultural sector, will form the framework for the sustainable and intense growth of agro-industrial complexes in the Republic and guarantee food security for the nation.

Soil or land Contamination

Soil or land contamination refers to the presence of harmful substances or pollutants in the soil, which can affect the quality of the soil and have adverse effects on human health and the environment. The contaminants can come from a variety of source, including industrial activities, agriculture, mining, and improper disposal of waste. Common example of soil contamination include heavy metal such as lead, mercury, and cadmium; pesticide and herbicides; petroleum product; and industrial chemical. These contaminants can enter the soil through direct application, leaching from contaminated ground water or surface water, atmospheric deposition, or through spills and leaks. Soil contamination can significant impact on the environment, including the loss of biodiversity, soil degradation, and water pollution. It can also pose health risks to humans, especially those who work or live in contaminated areas, and can lead to respiratory problems, skin irritation, and even cancer. To address soil contamination, various measure can be taken, including site assessments in the soil. Prevention measures, such as proper waste management and the use of environmentally friendly farming practices, can be also help reduce the risks soil contamination. Health hazards, direct contact with contaminated soil, vapour from toxins, or secondary pollution of water supplies inside and beneath the soil are the main causes of worry with soil contamination. (US EPA) Environmental contamination mapping and clean-ups are time-consuming and expensive processes that call for knowledge of geology, hydrology, chemistry, computer modelling, and GIS, as well as an understanding of the development of industrial chemistry. (George et al., 2016)

A few factors can destroy the soil:

- Fertilizers, pesticides, and weed killers.
- Mining
- Fuel and oil spills
- Coal ash disposal
- Landfills' leaching
- Surface water contamination that seeps into the ground
- Excreting excrement and urine in the open
- Electronic garbage

Contaminants' potential influence on soil functions

The activity, species composition, and abundance of soil microorganisms can be negatively impacted by heavy metals and other soil pollutants, endangering soil functions such the biogeochemical cycling of carbon and nitrogen. (Rijk et al., 2020) Yet, with time, soil pollutants may become less bioavailable, and ecosystems and microbes may adapt to new environmental conditions. In polluted soils, soil characteristics including pH, organic matter concentration, and texture affect the mobility, bioavailability, and toxicity of contaminants. (Alloway 2012) The same quantity of a pollutant may be completely safe in one soil and harmful in another. This highlights the requirement for risk assessment and mitigation particular to soil.

Nutrients Depletion

Nutrient depletion refers to a condition in which the body lacks essential nutrients due to inadequate intake or poor absorption. Nutrient depletion can result from a variety of factors such as poor diet, mal-absorption, medication use, and certain medical conditions. Inadequate intake of essential nutrients such as vitamins, minerals, and protein can lead to nutrient depletion. Poor absorption can occur due to conditions such as celiac disease, inflammatory bowel disease, and certain medications that interfere with nutrient absorption. Nutrient depletion can lead to a variety of health problems, including fatigue, weakness, impaired immune function, and increased risk of infections. Long-term nutrient depletion can also increase the risk of chronic diseases such as osteoporosis, anaemia, and cardiovascular disease. To prevent nutrient depletion, it's important to eat a balanced diet that includes a variety of nutrient-dense foods. If you have a medical condition that affects nutrient absorption or are taking medications that interfere with nutrient absorption, speak with your healthcare provider to decide if you need to take supplements or make changes to your diet. The depletion of nutrients and micronutrients in an ecosystem or areas of the biosphere, most frequently the soil, is referred to as "nutrient depletion," which is a type of resource depletion (soil depletion, soil degradation). (Tan et al., 2005) Nutrient depletion can also result from the loss of the nutritional substrate at the level of an entire ecological niche or ecosystem (soil loss, wetland loss, etc.). The loss of nutrients in a habitat will have an impact on nutrient availability and ultimately the entire food chain because nutrients are typically the first link in the food chain. (Dalal and probert 2021, Lal 2009). Shifts in the relative nutrient composition and essential nutritional quantity are indications of nutrient depletion. Large-scale changes brought about by human activity have had a harmful impact on both the natural environment and wildlife, particularly on flora and fauna. (Whitcomb et al., 2014) Eutrophication, often referred to as nutrient contamination, is the opposite consequence. (Birnie et al., 2017). Modifications in biodiversity and species abundance are brought on by both depletion and eutrophication (WHO 2020). The effects are reciprocal, meaning that a change in the species composition of a habitat may also result in a change in the composition of the nutrients. (Sherrard et al., 2019)

1.1.1. Organic matter Depletion

Organic matter depletion is a serious issue that can lead to a range of environmental and agricultural crises. Organic matter depletion refers to the reduction in the quantity and quality of organic matter in soil, which can be caused by a range of factors including intensive agricultural practices, deforestation, and land-use change. One of the main consequences of organic matter depletion is soil degradation. When soil loses organic matter, it becomes less fertile and less able to support plant growth. This can lead to decreased crop yields, which can have serious economic and social consequences, especially in developing countries where agriculture is a major source of income and food security. Organic matter depletion can also contribute to climate change. Soil organic matter contains carbon, and when it is depleted, carbon is released into the atmosphere in the form of carbon dioxide. This contributes to the build-up of greenhouse gases in the atmosphere, which can lead to global warming and climate change (Imran and Amanullah 2021a,b). Furthermore, organic matter depletion can reduce soil water-holding capacity, which can lead to increased erosion and water pollution. In addition, the loss of organic matter can lead to decreased biodiversity in soil, which can have negative impacts on soil health and ecosystem functioning. To address organic matter depletion, sustainable agricultural practices should be implemented, such as conservation tillage, cover cropping, and crop rotation. These practices can help to maintain or increase soil organic matter levels, improve soil health, and reduce greenhouse gas emissions. Additionally, efforts to reduce deforestation and land-use change can help to preserve and protect organic matter-rich soils (Imran and Amanullah 2021c).

What causes organic matter decline?

Soil organic carbon content is affected mostly by climate, texture, hydrology, land use and vegetation.

Climate: Organic matter decays more rapidly at higher temperatures, so soils in warmer climates tend to have less organic matter than those in cooler climates.

Soil texture: Fine-textured soils tend to have more organic matter than coarse soils; they hold nutrients and water better, thus supplying good conditions for plant growth. Coarse soils are better aerated, and the presence of oxygen results in a more rapid decay of organic matter.

Soil hydrology: (drainage) the wetter a soil is, the less oxygen is available for organic matter to decay, so that it accumulates.

Land use: (tillage) Tillage mixes oxygen into the soil and raises its average temperature, thereby contributing to an increased rate of organic matter decay. Loss of organic matter also occurs because erosion washes away topsoil and humus. Overall, cropping returns less organic matter to the soil than does native vegetation.

Vegetation: Roots are a great contributor to soil organic matter. Grassland produces deep roots that decay deep in the soil. Forest soils in contrast rely primarily on the decay of surface litter for organic matter input. Crops produce more above-ground biomass than roots. The organic matter input on cropland depends on the land management practices, including whether crop residues are removed or left behind.

Soil Microbiota Depletion

Soil microbiota depletion refers to the loss of diversity and abundance of microorganisms in the soil. Soil microorganisms play critical roles in soil health and ecosystem functioning, including nutrient cycling, organic matter decomposition, and plant growth promotion (Imran 2021). The depletion of soil microbiota can occur due to various factors, such as intensive agriculture, use of synthetic fertilizers and pesticides, land-use change, and urbanization. These factors can disrupt the balance of the soil microbial community, leading to a reduction in beneficial microbes and an increase in harmful ones. The consequences of soil microbiota depletion can be severe and long-lasting. It can lead to decreased soil fertility, reduced crop yields, increased susceptibility to soil-borne diseases, and decreased resistance to environmental stressors such as drought and extreme temperatures. To address soil microbiota depletion, sustainable soil management practices such as crop rotation, reduced tillage, and organic farming can be employed. These practices help to promote soil microbial diversity and function, leading to improved soil health and productivity. One gram of soil contains between 108 and 1010 microbial cells, with thousands, if not millions, of distinct species (Zhang et al., 2017, and Ruesch et al., 2007) A "plant microbiome" that lives in both soil and plant tissues is created as a result of plant roots interacting with the local soil flora (Lebeis et al., 2012). The plant microbiota is typically divided into four fractions: bulk soil (microorganisms living more than a few millimetres from the plant), rhizosphere (microorganisms living within a few millimetres of the roots), rhizoplane (microorganisms tightly attached to roots), and root associated (endophytic) microorganisms. Legumes also have a root nodule microbiome in addition to the leaf bacteria living in the phyllo sphere. The earliest ancient bacteria and microbes are thought to have originated in the waters of Earth between two and four billion years ago. These bacteria had the ability to fix nitrogen, and as they grew over time, they released oxygen into the environment. (Farqunar et al., 2000 and Lanfield 2004). This resulted in more sophisticated microorganisms, (Falkowski 2015 and Jelen et al., 2016) which are crucial because they impact soil fertility and structure. The several types of soil microorganisms include bacteria, actinomycetes, fungus, algae, and protozoa. Each of these groupings has traits that characterise them and define their roles in the soil. (Rao 1999 and Islam et al., 2021). Each gram of soil around and surrounding plant roots, or the rhizosphere, contains up to 10 billion bacteria. In the roots of sugar beets, a researcher found more than 33,000 different bacterial and archaeal species in 2011. (Vrieze 2015). Either because their small size and close relationship with soil particles and plant roots, soil biota are capable of experiencing fast changes in growth. They carry out or support a variety of essential ecological processes, such as nutrient uptake of N and P and their consequences on the biotic communities of the soil, where some taxa experience an increase in diversity and abundance while others experience a decrease in response to N enrichment. Yet, numerous studies on mycorrhizal fungi show a drop in the reproductive production of species adapted to N-poor habitats and the consequent extinction of those species. This is frequently accompanied by an increase in parasitic fungal species, a decrease in the diversity of microorganisms, and/or a change in the soil community towards a greater bacterial predominance. These patterns might be more common in systems with phosphorus-rich soils (Johnson et al.,

2003). The term "soil biota" refers to the variety of living organisms that interact with the soil's plant life, including bacteria, fungi, and other microbes (Balestrini et al., 2015). The ability of soil microbes to withstand drought stress is a well-known and well researched phenomenon. Plants may tolerate water stress thanks to soil microorganisms and arbuscular mycorrhizae (AM). In comparison to fungi-inoculated plants, which have a higher tolerance to drought stress, plants without symbiotic relationship with AM had higher stomatal conductance, superoxide dismutase (SOD) activity, but lower tissue water levels (Ortiz et al., 2015). Unfortunately, there aren't many generalizations that can be drawn about how soil biota respond given the need for more research.

Soil threats due to climate change

Climate change poses several threats to soil health and productivity. Some of the most significant soil threats due to climate change are:

- Soil erosion: Increased frequency and intensity of extreme weather events such as heavy rainfall and droughts can lead to soil erosion. Soil erosion can cause loss of topsoil, decrease soil fertility, and reduce crop yields.
- Soil compaction: Increased rainfall intensity and frequency, coupled with heavy agricultural machinery, can lead to soil compaction, reducing soil porosity and water infiltration rates. This can decrease soil fertility and limit plant growth.
- Soil degradation: Changes in temperature and precipitation patterns can cause soil degradation, leading to reduced soil quality and productivity. Soil degradation can also lead to increased soil erosion, nutrient depletion, and loss of soil organic matter.
- Salinization: Climate change can cause sea-level rise and increased coastal flooding, leading to soil salinization. Salinization can cause soil degradation, reduce crop yields, and limit plant growth.
- Changes in soil microbial communities: Changes in temperature and precipitation patterns can alter the composition and activity of soil microbial communities, which can have significant impacts on soil health and ecosystem functioning.

To address these threats, sustainable soil management practices such as conservation agriculture, agroforestry, and organic farming can be employed. These practices help to improve soil health, reduce soil erosion, and increase soil fertility, making soil more resilient to the impacts of climate change.

Pakistan is particularly susceptible to the negative effects of climate change. According to the study, Pakistan would see an extraordinary heat wave and drought in the summers of 2022, followed by unusual rain and flooding during the monsoon season. Pakistan's economy, which has been severely damaged by both drought and floods, is based primarily on agriculture. Even though the flood waters are progressively receding, the stagnant, filthy water is endangering the residents' health. According to this study, Pakistan's national security is currently facing a challenge related to water security. The article looks into the water availability situation in relation to the expanding population, agriculture, and other water needs. Adverse effects of excessive glacier melting, lack of dams to store rainwater, and lack of smart means for agriculture water have been examined to empirically validate the arguments. These days, "climate change is no more an obscure term, which can be understood through personal knowledge, experience, and interactions" (Deng et al., 2012). It has become a global buzzword. Most human-induced activities, more notably the atmospheric emissions of greenhouse gases, are to blame for the phenomenon of climate change (Mcmicheal et al., 2003). As a result, the effects of climate change are creating a variety of risks on the surface of the Earth, such as droughts, heavy precipitation, heat waves, and previously unheard-of tropical cyclones (Woodruff et al., 2006).

Climate is a crucial factor for agriculture and a driver of rural development. Central Asia is a dryland region and has a continental climate with hot summers and cold winters. 90 % of this area has less than 400 mm of precipitation per year, and the average over the total area is 266 mm (de Pauw 2007, in Gupta et al. 2009). Due to Pakistan's susceptibility to the detrimental effects of climate change, new weather patterns may arise, which may present strategic difficulties (Bacha et al., 2021). In the north, the glaciers are melting quickly due to rising temperatures, and this year's unique monsoon brought about record floods in Pakistani history during the past 30 years. According to Antonio Guterres, secretary general of the UN, Pakistan is suffering from "the persistent effects of epochal amounts of rain and flooding" (Guterres et al., 2022). Pakistan is likely to have a water crisis, food shortages, and major consequences for human security as a result of the tremendous material and human losses. Mr. Miftah Ismail, Pakistan's finance minister, responded to the recent flood scenario by saying that "Pakistan is coping with the worst effects of the climate change, which has caused over US$ 30 billion loss to Pakistan's economy and displacing 33 million people" (Sultan 2022). The effects of climate change on the availability, use, and storage of water in Pakistan are examined in this study. In order to examine the effects of climate change throughout Pakistan by sampling both rural and urban areas, the drought and floods of the year 2022 have been used as a case study.

Pakistan is most at risk from the harmful effects of climate change, which is an emerging worldwide concern. Extreme drought on the one hand and exceptional floods over a brief period of 2-3 months were both seen in 2022. Alarm bells are ringing for Pakistan to take an all-encompassing inventory of the situation, declare a climate emergency, and adopt "whole of nation" and "whole of government" approaches for a comprehensive response. This will ensure strong interagency cooperation and capitalize on the synergetic application of all Elements of National Power (EoNP) for the best outcomes. For effective operation, coordination, and execution, the relevant departments must be integrated under the aegis of national and provincial disaster response organizations. To "conserve, maintain, and proportionally distribute current water resources" (Ishaque et al., 2015) there is an urgent need to develop a strong national realization. Also, as these techniques have been implemented by the majority of developed countries, smart ways to use water for agriculture and recycling water for purposes other than drinking would be beneficial. The present flooding in Pakistan is evidence that larger water storage facilities need to be built. In order to handle Pakistan's water security challenges on a long-term basis, it is believed that this study will be helpful to the appropriate government ministries as an academic policy contribution. Soils will be affected by climate change, which will impact soil erosivity, organic carbon, nutrients, and alkalinity. The accounting of carbon emissions from the land, a crucial step in helping NSW reach its goal of having net zero emissions by 2050, is affected by the declining soil carbon as a result of climate change.

1.1.2. Soil health and climate extreme

Soil health plays a critical role in mitigating and adapting to climate extremes. Healthy soil is better able to withstand and recover from the impacts of climate extremes such as droughts, floods, and heat waves. Changes in land use and soil can either speed up or slow down climate change, which has a significant impact on both. We cannot address the climate catastrophe, produce enough food, or prepare for a changing climate without better soils and sustainable land and soil management. Some of the ways that soil health can help to address climate extremes are:

- Water retention: Healthy soil with good structure and organic matter content can retain more water, reducing the risk of drought and water stress during dry periods.
- Flood management: Healthy soil with good structure and organic matter content can also help to absorb and store excess water during wet periods, reducing the risk of flooding and erosion.
- Carbon sequestration: Healthy soil can act as a carbon sink, sequestering carbon from the atmosphere and helping to mitigate climate change. This is because soil organic matter contains carbon that is stored in the soil for long periods.
- Temperature regulation: Healthy soil can help to regulate soil temperature, reducing the risk of soil overheating during heat waves.

- Biodiversity conservation: Healthy soil supports diverse microbial communities and promotes the growth of diverse plant species, which can help to maintain ecosystem functioning and resilience to climate extremes.

To promote soil health and resilience to climate extremes, sustainable soil management practices such as conservation agriculture, agroforestry, and organic farming can be employed. These practices help to improve soil health, reduce soil erosion, and increase soil fertility, making soil more resilient to the impacts of climate extremes. In a recent article published in Agriculture and Environmental Letters (Sprunger et al. 2020), we report the findings of a study that compared a number of practices that promote soil health, including annuals grown in rotation with a cover crop, perennial monoculture systems, and perennial polyculture systems. The ability of the soil to carry out a variety of agronomic and ecosystem functions, thereby sustaining biological productivity, maintaining environmental quality, and promoting plant and animal health, is referred to as soil health (Kibble white et al., 2008; Lal, 2011). Emissions of greenhouse gases (GHGs) can come from the soil both as a source and as a sink. Carbon dioxide (CO2) is released as a result of the microbial degradation of SOM. Climate change-related increases in soil temperature can speed up SOM decomposition. (Singh et al. 2017). So Climate change has a variety of effects on our daily lives around the world, some of which are mild and some of which are severe. One of the significant effects of climate change is that it interferes with many ecosystem processes, resulting in a decrease in the quantity and quality of fresh produce. Water, species extinction, declining biodiversity, and landscape deterioration (McCarty, 2000). The plant and animal life on and in the soil is greatly influenced by the climate. Yet, unsustainable land use and management practises, such as converting grasslands or forests to crops or using intensive tillage in cropland, have been endangering the health of the local and global environment as well as the soil. These outcomes prevent soil from acting as a carbon sink and instead make it a source of GHG emissions, lowering the physical, chemical, and biological aspects of soil quality and health. This chapter provides an overview of sustainable management techniques that help build SOM and increase soil resilience to management and environmental changes by adhering to the principles of soil health. This chapter views sustainable management practises holistically within the context of climate change and agricultural systems control, and the thoughtful integration of livestock into farming systems.

Solution to soil Crises

There are several solutions to address soil crises and promote sustainable soil management practices. Some of the key solutions are:

- Conservation agriculture: Conservation agriculture is a sustainable farming approach that involves minimal soil disturbance, crop rotation, and the use of cover crops to improve soil health and fertility. This practice can help to reduce soil erosion, increase soil organic matter, and improve water retention.
- Agroforestry: Agroforestry involves integrating trees with crops or livestock in a farming system. This practice can improve soil health by increasing soil organic matter, improving soil structure, and enhancing nutrient cycling.
- Organic farming: Organic farming practices avoid the use of synthetic fertilizers and pesticides and focus on improving soil health through the use of compost, crop rotation, and cover crops. Organic farming can help to improve soil fertility, reduce soil erosion, and enhance biodiversity.
- Soil conservation: Soil conservation practices such as terracing, contour farming, and strip cropping can help to reduce soil erosion, improve soil structure, and increase water retention.
- Land-use planning: Land-use planning can help to ensure that land use is optimized for sustainable soil management practices, taking into account soil characteristics, climate, and other factors.
- Soil testing and analysis: Soil testing and analysis can help farmers to determine soil fertility levels and nutrient deficiencies, allowing them to tailor their fertilizer application and other soil management practices to the specific needs of their soil.
- Education and outreach: Education and outreach programs can help to increase awareness of sustainable soil management practices and encourage farmers and other stakeholders to adopt these practices.

Overall, promoting sustainable soil management practices is key to addressing soil crises and ensuring the long-term health and productivity of our soils. By implementing these practices, we can improve soil health, reduce soil erosion, enhance biodiversity, and promote resilient and sustainable farming systems.

A complete preventive monitoring of all agricultural land would be an unattainable goal as well. Measuring and evaluating soil quality in terms of its societal functions is more useful and common. The specific role of soil and land in producing plant biomass for humans, for example (productivity function, Mueller et al. 2010), remains critical. As a result, higher soil quality indicates that the land has a higher crop yield potential. Soil can be thought of as a resource in the context of agriculture. The process of eroding away the topsoil, or surface layer, through the mechanical movement of surface particles caused by physical forces like wind and water, is known as soil erosion. (Agroconection 2021). By endangering the viability of their gardens and agricultural land, it can lead to a variety of issues for homeowners and farmers. Soil erosion essentially depletes a plot of land of its rich natural resources. (Sinha, R. 2021) The adversary of agriculture is soil erosion, which poses a serious environmental risk to productivity and sustainability with downstream consequences for the global warming catastrophe and food security. Sustainable land management can tackle a wide range of issues by preventing soil erosion. (Sulaeman, D. & Westhoff, T. 2020) The process of soil eroding is ongoing and can happen slowly or rapidly. It produces in a continuous loss of topsoil, ecological degradation, soil collapse, (Byjus 2021). To improve soil structure and drainage, increase the soil organic matter level with compost, organic mulches, living plants, and cover crops (vegetable gardens). More information can be found on this page Organic Matter and Soil Amendments.

1.1.3. Solution of Soil or land Contamination

Soil or land contamination can have a significant impact on the environment and human health. Therefore, it is essential to address and remediate contaminated soil to reduce the potential risks. Reduce, reuse, and recycle to lower land emissions. Reforestation and afforestation techniques must be used. Farmers can utilize a variety of techniques, including crop rotation, organic fertilizers, and integrated pest management.

The solution to soil or land contamination depends on the type and extent of contamination. Once land is destroyed that could be directly or indirectly used in a positive way by human activity, it is said to have been contaminated. Now, harmful substances like lead and arsenic are considered to have poisoned 300,000 acres of UK soil. The other developed countries, which are the hardest afflicted, are similarly affected. The developing nations are likewise making slow but sure progress in this regard. Given the severe harm that soil pollution causes to the environment, this should raise serious concerns. There are several possible solutions to land pollution, including conservation. Conservation focuses on preserving natural resources, such as soil and plants. The efforts to conserve resources can start with utilizing sustainable practices. For instance, leaving some of the trees in a forest to naturally die and decay. This not only leaves the cover needed for the soil and other vegetation, but it helps to provide the nutrients that the soil needs to remain fertile. Here are some common remediation techniques:

- Excavation and removal: The contaminated soil is excavated and removed from the site to be treated or disposed of at a licensed facility.
- Soil vapour extraction: This technique involves extracting volatile organic compounds (VOCs) from the soil by applying a vacuum to the soil.

- Bioremediation: Microorganisms are used to break down and degrade contaminants in the soil, converting them into harmless substances. Bioremediation can be done naturally or by adding microorganisms to the soil.
- Soil washing: The contaminated soil is washed with water or a chemical solution to remove contaminants. The soil is then treated or disposed of accordingly.
- Chemical oxidation: This process involves adding chemicals to the contaminated soil to oxidize and break down the contaminants.
- Phytoremediation: This method uses plants to absorb and break down contaminants in the soil. Plants can take up contaminants through their roots and store them in their tissues or break them down through natural metabolic processes. Phytoremediation can be effective for removing low to moderate levels of contaminants but may not work well for highly contaminated sites.
- Chemical treatment: This technique involves adding chemicals to the soil to neutralize or break down the contaminants. This method can be effective for removing specific types of contaminants but may be expensive and can produce secondary waste.
- Thermal treatment: This approach involves heating the soil to high temperatures to volatilize or break down the contaminants. This method is effective for removing certain types of contaminants but can be expensive and energy-intensive.
- Solidification/stabilization: This technique involves adding materials to the contaminated soil to reduce the mobility and toxicity of contaminants.

It is important to note that each remediation technique has its advantages and disadvantages, and the most proper method will depend on the specific contamination and site conditions. Additionally, remediation efforts must follow relevant regulations and guidelines. About half of Earth's population, or 4.2 billion people, live in urban areas, and by 2050, that figure is projected to rise by 2.5 billion. (UN 2019) Groundwater monitoring needs to be done more often during seasons of precipitation and drought for brownfield sites with residual contaminants and post-remediation management because these processes directly alter contamination concentrations (Libera, A. et al 2019). The character of urban economic activity is changing in tandem with this trend, and industrial sites that were formerly in the centre of industrialized urban centres are being abandoned at an increasing rate. (US EPA 2022) Due to land-use constraints or urban planning controls over hazardous material contamination of the soil and groundwater, many of these formerly constructed properties remain abandoned or underutilized (Adams et al., 2010). We discuss long-term BRR tactics in this review. The primary, secondary, and tertiary effects of conventional techniques over the course of remediation are discussed first. Then, we list some of the most effective in-situ soil and groundwater remediation technologies as well as redevelopment-integrated remediation strategies. We conclude by outlining problems and potential research directions. The negative consequences of land pollution can be greatly reduced with the cooperation of everyone. By making a conscious effort to contribute to a safer environment, the health and well-being of all can be protected.

Solution of Soil Nutrients Depletion

Soil nutrient depletion can occur due to various reasons, such as overuse of fertilizers, erosion, and intensive farming practices. Some areas with light erosion problems, replanting with vegetation and covering with mulch are good solutions. For erosion along footpaths, covering with mulch or stone is the best option. For heavy erosion in areas of concentrated flow, the most effective solutions are check dams or terraces. In an agricultural setting, nitrogen deficiency can be combated by the addition of nitrogen-rich fertilizers to increase the availability of nutrients and thereby increase crop yield. For this reason, protecting a soil from these losses and replenishing the nutrients lost by adding crop residues (straw, cornstalks, other stems, and roots), manure, and fertilizer materials (ash, phosphate rock, bone, chemical fertilizers), are crucial strategies used by food producers to maintain production. To address soil nutrient depletion, some solutions are:

- Crop rotation: Crop rotation involves growing different crops on the same land over a period of time. This practice helps to restore soil fertility by alternating nutrient-depleting crops with those that replenish the soil with nutrients. Crop rotation can also help to reduce soil erosion and improve soil structure.
- Cover cropping: Cover cropping involves planting crops that are specifically grown to protect and improve the soil between periods of main crop cultivation. Cover crops help to prevent erosion, fix nitrogen in the soil, and add organic matter, which improves soil structure and fertility.
- Composting: Composting is a natural process that turns organic waste into a nutrient-rich soil amendment. Composting helps to improve soil structure and fertility by adding organic matter, which promotes the growth of beneficial microorganisms and increases soil nutrient levels.

- Reduced tillage: Reduced tillage involves minimizing soil disturbance during farming practices. This helps to reduce soil erosion and improve soil structure, which can help to retain nutrients in the soil.
- Fertilizer management: Proper management of fertilizer application can help to minimize nutrient depletion. This involves applying fertilizers at the right time and in the right amounts, as well as using slow-release fertilizers to improve nutrient uptake and reduce nutrient leaching.
- Soil testing: Soil testing helps to identify nutrient deficiencies and imbalances in the soil. This information can then be used to develop a targeted nutrient management plan, which can help to optimize nutrient use and reduce depletion.
- Nutrient recycling: Nutrient recycling involves recycling organic waste, such as manure and crop residues, to provide nutrients to the soil. This practice helps to reduce waste and increase soil fertility.

Implementing these solutions can help to address soil nutrient depletion and promote sustainable agriculture practices. Researchers recommend that farmers rotate through perennial crops, forages, and legumes to fix this and lessen nutrient depletion. This will consistently deliver needed nutrients and stop erosion. By increasing soil nitrogen levels through bacterial conversions, introducing legumes to the farm, for instance, can lessen the demand for synthetic nitrogen fertilizer (Teague et al., 2016). Synthetic fertilizers are primarily used for large-scale crop growth on industrial-grade farms. Inequities in nutrient levels and soil homeostasis result from the killing of soil microorganisms by synthetic fertilizers (Teague et al., 2016). In addition, the production of synthetic fertilizers' uses a lot of energy and fossil fuels, which has an adverse influence on the ecosystem on a much larger scale than merely the soil (Mulvaney et al., 2009). Tilling causes soil to be flattened, topsoil to be mixed, crop residues to be removed, microbial populations to be reduced, soil erosion to be accelerated, and the release of harmful greenhouse gases. Most croplands on the planet use tilling techniques that dramatically diminish soil nutrients (Jerry L. Hatfield and Charles L. Walthall 2015). Grass is consumed by ruminants and cows, who then use their excrement to turn it into nutrient-rich soil. Plants have time to develop and replenish when animals roam fields often. Monotonous grazing causes soil erosion and lowers soil fertility. Almost 20% of the existing grasslands worldwide have been lost due to overgrazing (Mottet et al., 2018). Global farmlands may shortly run out of the fertile soil needed to produce nutrient-rich, healthful food if humans continue to produce food at the same rate and using the same techniques. By reversing the effects, such calamities can only be avoided. Agricultural industries should:

- Amplify crop variety.
- Add added ruminants, or grazing animals, such as cows.
- Rotate the pastures.
- fewer species that have been mechanically enhanced
- Choose natural and organic fertilizers' and insecticides.

A serious issue that affects all people, not just Americans, is the depletion of nutrients in American farmlands and soil. Unsafe agricultural and industrial practices damage the ecosystem and make it harder for everyone to acquire the nutrients they need to survive. Before the few remaining resources are completely depleted, farmlands must make a change.

1.1.4. Solution of Soil Organic matter Depletion

Soil nutrient depletion can occur due to various reasons, such as overuse of fertilizers, erosion, and intensive farming practices. One of the greatest methods for increasing soil quality and organic matter levels is to use cover crops. Growing cover crops has several advantages, such as: By securing the soil and reducing the force of rains, they prevent erosion. They incorporate plant material into the soil to replace organic materials. To address soil nutrient depletion, some solutions are:

- Crop rotation: Crop rotation involves growing different crops on the same land over a period of time. This practice helps to restore soil fertility by alternating nutrient-depleting crops with those that replenish the soil with nutrients. Crop rotation can also help to reduce soil erosion and improve soil structure.
- Cover cropping: Cover cropping involves planting crops that are specifically grown to protect and improve the soil between periods of main crop cultivation. Cover crops help to prevent erosion, fix nitrogen in the soil, and add organic matter, which improves soil structure and fertility.
- Composting: Composting is a natural process that turns organic waste into a nutrient-rich soil amendment. Composting helps to improve soil structure and fertility by adding organic matter, which promotes the growth of beneficial microorganisms and increases soil nutrient levels.
- Reduced tillage: Reduced tillage involves minimizing soil disturbance during farming practices. This helps to reduce soil erosion and improve soil structure, which can help to retain nutrients in the soil.
- Fertilizer management: Proper management of fertilizer application can help to minimize nutrient depletion. This involves applying fertilizers at the right time and in the right amounts, as well as using slow-release fertilizers to improve nutrient uptake and reduce nutrient leaching.
- Soil testing: Soil testing helps to identify nutrient deficiencies and imbalances in the soil. This information can then be used to develop a targeted nutrient management plan, which can help to optimize nutrient use and reduce depletion.
- Nutrient recycling: Nutrient recycling involves recycling organic waste, such as manure and crop residues, to provide nutrients to the soil. This practice helps to reduce waste and increase soil fertility.

Implementing these solutions can help to address soil nutrient depletion and promote sustainable agriculture practices. When high-quality and low-quality residues are managed using fertilizer, the production potential of many crops can be maximized (Snapp, Nyiraneza, Otto, & Kirk, 2003). Manure's quality might vary, but it often encourages a diverse population of soil critters and living organic materials. Reduced tillage, however, is most efficient at enhancing SOM when combined with rotations or cover crops that comprise deep-rooted, high-residue-producing plants like alfalfa, according to long-term trials (Franzluebbers, 2010). Selling some wastes while replacing them with compost or a cover crop to protect the soil is one win-win strategy. Even with the total removal of corn stover, manure and compost inputs are efficient ways to increase soil organic matter (Thelen, Fronning, Kravchenko, Min, & Robertson, 2010). There are several pools of soil organic matter, from a short-term, active SOM pool to a long-term, stable SOM pool. There is mounting proof that a mixture of residue qualities can improve SOM pools by fostering a diversified soil food web. For specific soil types and crop rotations, it is challenging to forecast SOM reactions. Responses of soil organic matter take place over a lengthy period, sometimes a decade or more. It is crucial for farmers to verify any soil quality promotion claims for an organic amendment, kind of manure, or residue management strategy by testing on an economically small scale.

Solution of Soil Microbiota Depletion

Soil microbiota depletion refers to the reduction of microorganisms in soil due to various factors, such as excessive use of chemical fertilizers, pesticides, and intensive farming practices. Soil microbiota plays a crucial role in keeping soil health, nutrient cycling, and plant growth. Therefore, the depletion of soil microbiota can have adverse effects on soil quality and crop productivity. The inherent suppressiveness of soil is improved by the soil microbiota, which is essential in the suppression of soil-borne plant pathogens. The network of soil microbial communities is changed and shaped by microbiome disruption to increase suppression against phytopathogens and related illnesses. By agronomical approaches based on soil pre-fumigation, organic amendment, crop rotation, and intercropping, very significant goals have been achieved in the manipulation of soil microbiota. Therefore, a thorough understanding of the microbiome alterations during its manipulation is required in light of the microbiome-assisted solutions to limit inconsistencies, disadvantages, and failures associated to soil microbiota disruption.

Here are some solutions to prevent or mitigate soil microbiota depletion:

- Reduce the use of chemical inputs: Chemical fertilizers and pesticides can have negative impacts on soil microbiota, so reducing their use can help to maintain soil health. Instead, farmers can use organic fertilizers and pest control methods, such as compost, crop rotation, and integrated pest management (IPM).

- Increase soil organic matter: Soil organic matter is an essential component of soil microbiota, providing food and habitat for microorganisms. Increasing soil organic matter through the use of cover crops, green manure, and compost can enhance soil microbiota and improve soil health.
- Practice conservation agriculture: Conservation agriculture involves practices such as reduced tillage, cover cropping, and crop rotation, which help to maintain soil structure and reduce erosion. These practices can also support soil microbiota by reducing soil disturbance and increasing organic matter.
- Utilize microbial inoculants: Microbial inoculants are formulations of beneficial microorganisms that can be added to soil to enhance soil health and plant growth. These inoculants can help to restore soil microbiota and improve nutrient cycling.
- Foster biodiversity: Increasing biodiversity in the farm ecosystem can also support soil microbiota. For example, planting hedgerows, maintaining wildflower strips, and providing habitat for beneficial insects and other organisms can enhance soil health and promote soil microbiota.

Despite the presence of virulent pathogens and susceptible hosts, soil biotic and abiotic factors can successfully suppress plant illnesses brought on by soil-borne plant pathogens (filamentous fungus, oomycetes, and bacteria) (Baker 1974). In agro ecosystems, the dynamics of soil organic matter (SOM) and plant nutrient availability are controlled by the microbiota (Kibblewhite 2008, and O'Donnell et al., 2005) The success or failure of a defensive plan based, for instance, o The key factors determining the success (or failure) of disease-suppressive compost-amended soils are the intricate interactions among beneficial microorganisms, pathogens, SOM, and physicochemical qualities (such as pH, electrical conductivity, macro-micronutrient content) (Haggag and Saber 2000). In the addition of various degrees of municipal waste compost under greenhouse conditions for preventing soil-borne diseases (Morra et al., 2010) is largely dependent on the microbiota. The manipulation of soil microbiota has been used in numerous research relating to the eradication of plant diseases and the adoption of acceptable agronomic techniques for agroecosystems that are sustainable. Under the guidance of microbiome-assisted techniques, novel organic farming systems based on the reduced use (or absence) of synthetic pesticides (fungicides and fumigants) have been developed (Yamada et al., 2016). By altering the microbial community composition of the soil along the rhizosphere profile, crop rotation can reduce the incidence and severity of disease in various agricultural systems (Xu LH. et al., 2012). Overall, promoting soil microbiota is critical to maintaining soil health and productivity. By adopting these solutions, farmers and land managers can mitigate soil microbiota depletion and promote sustainable agriculture. As soil characteristics and crop yields are gradually deteriorating, microbiota can boost the natural soil suppressiveness against soil-borne pathogens. The main causes of soil depletion include intensive farming systems, which are characterised by increased inputs of synthetic pesticides, reduced SOM accumulation, sparse humification degrees, and frequent soil tillage. Further research is needed on how OAs and specially formulated microbe-based formulations affect the soil microbiota, which is a significant component in plant health. Alternatively, if exogenous BCAs of unknown origin are added, very little focus has been placed on the ecological impact of beneficial microbial consortia recruited from compost. In order to better comprehend the microbial community network inhabiting soil, the ecological responsibilities of soil microbiota should be more thoroughly explained.

Solution of Soil threats due to climate change

Soil is a vital natural resource that is essential for food production, supporting biodiversity, and mitigating climate change. However, climate change poses significant threats to soil health and function, such as increased erosion, nutrient depletion, and soil compaction. Here are some potential solutions to address soil threats due to climate change:

- Conservation Agriculture: This farming practice involves minimal soil disturbance, permanent soil cover, and crop rotation to enhance soil health, reduce erosion, and increase carbon sequestration.

Agroforestry: This integrated farming system combines trees with crops and/or livestock to improve soil fertility, increase carbon sequestration, and reduce soil erosion.

Cover Crops: Growing cover crops during fallow periods can help reduce soil erosion, enhance soil organic matter content, and increase nutrient availability.

Soil Organic Matter Management: Adding organic matter to the soil through composting, manure application, or using crop residues can improve soil structure, water-holding capacity, and nutrient cycling.

Nutrient Management: Proper nutrient management practices, such as precision fertilizer application and using organic amendments, can reduce nutrient depletion and increase soil fertility.

Soil Conservation Practices: Implementation of soil conservation practices, such as terracing, contour farming, and grassed waterways, can reduce soil erosion and increase soil water retention.

Land Use Management: The management of land use, such as reducing deforestation and promoting reforestation, can help to protect soil health and function by reducing soil erosion and increasing carbon sequestration.

In conclusion, a combination of these solutions could be implemented to address soil threats due to climate change. By implementing sustainable agricultural practices, protecting land use, and managing soil organic matter and nutrients, we can mitigate the impact of climate change on soil health and ensure long-term food security and environmental sustainability.

Solution of Soil health declining under climate extreme

The health of soil is intimately linked to climate extremes, as both are intricately interconnected. Climate change can cause extreme weather conditions such as droughts, floods, and heatwaves, which can adversely affect soil health. On the other hand, healthy soil can play a significant role in mitigating climate change by sequestering carbon and reducing greenhouse gas emissions.

To address this issue, there are several solutions that can be implemented to improve soil health and mitigate the impacts of climate extremes:

- **Conservation Agriculture**: Conservation agriculture is a farming technique that involves minimal soil disturbance, permanent soil cover, and crop rotation. This technique helps to maintain soil health by reducing erosion, increasing water infiltration, and promoting soil organic matter.
- **Agroforestry**: Agroforestry is the practice of integrating trees into farming systems. Trees provide multiple benefits, including carbon sequestration, soil stabilization, and water retention. Agroforestry can also help to buffer crops against extreme weather events such as droughts and floods.
- **Cover crops**: Cover crops are planted during fallow periods to protect the soil from erosion and nutrient depletion. They can also

help to build soil organic matter and increase soil water-holding capacity.

- **Crop diversity**: Crop diversity can help to maintain soil health by improving soil structure, reducing erosion, and increasing soil organic matter. Crop diversity can also help to buffer crops against extreme weather events.
- **Soil carbon sequestration**: Soil carbon sequestration is the process of capturing atmospheric carbon dioxide and storing it in the soil. This can be done through techniques such as reduced tillage, cover cropping, and adding organic matter to the soil.

2. Water Crisis

The term "water crisis" refers to a situation where there is a shortage of fresh water supply, or a lack of access to safe and clean drinking water. The world is facing a growing water crisis, as many regions are experiencing water scarcity, pollution, and mismanagement. The lack of freshwater resources to meet environmental and human demands is what is known as the water crisis. The first issue for global growth is the lack of clean drinking water, hence the water crisis is closely tied to human rights. Some of the major causes of the water crisis include population growth, climate change, overconsumption, and pollution. As the world's population continues to grow, demand for water also increases, and many regions are already struggling to meet the water needs of their populations. Climate change is also exacerbating the water crisis, as it alters weather patterns and causes more extreme weather events such as droughts and floods.

Overconsumption of water by industries, agriculture, and individuals is another major cause of the water crisis. Many regions are using water resources faster than they can be replenished, leading to depletion of aquifers and other water sources. Pollution from industrial and agricultural activities is also a major problem, as it contaminates freshwater sources and makes them unsafe for human consumption. A water crisis occurs when the amount of potable, clean water that is available in a region is insufficient to meet the region's demand. Climate change, which causes droughts, floods, and pollution, as well as abuse of water, increased human demand due to population growth, and intense agricultural use, are some of the causes of these water scarcity challenges (Alam, 2015). Moreover, the human, agricultural, and industrial pollutants have prevented the replenishment of subsurface waters' fresh source (World Bank, 2018). A country or region is declared to be under "water stress" according to the Falkenmark Water

Stress Indicator described in the climate change (2001) report when annual water resources fall below 1,700 cubic metres per person per year.

To address the water crisis, it is important to implement policies and practices that promote sustainable use of water resources, including conservation measures, better water management, and investment in infrastructure to improve water supply and treatment. Education and awareness campaigns can also help to raise awareness about the importance of water conservation and the impacts of the water crisis on human health, ecosystems, and the economy.

Water crisis causes Soil or land Contamination

Soil or land contamination is a major cause of the water crisis, as it can lead to pollution of groundwater and surface water sources. Contaminated soil can release harmful chemicals and pollutants into the water, making it unsafe for human consumption. Contaminants such as heavy metals, pesticides, and industrial chemicals can leach into the soil from sources such as agricultural practices, industrial activities, and waste disposal. These contaminants can then enter water sources through runoff or infiltration, leading to a wide range of health and environmental problems. The effects of soil or land contamination on water resources can be long-lasting and difficult to reverse. Once pollutants have entered the groundwater or surface water sources, it can be challenging to remove them completely. To address the water crisis caused by soil or land contamination, it is important to implement measures to prevent contamination from occurring in the first place. This can include better management of industrial and agricultural waste, improved waste disposal practices, and stricter regulations on the use of hazardous chemicals.

In addition, remediation measures can be used to clean up contaminated sites and prevent further contamination of water resources. These can include technologies such as bioremediation, phytoremediation, and chemical treatments to remove pollutants from the soil.

Water crisis causes Nutrients Depletion

Water crisis and nutrient depletion are two separate but related issues that can have serious impacts on human health and the environment. Water crisis refers to a situation where there is not enough water to meet the needs of a population or ecosystem. This can be caused by a variety of factors such as drought, overuse of water resources, pollution, climate change, and inefficient use of water. In areas where water is scarce, people may struggle to access safe and clean drinking water, leading to health problems such as dehydration, water-borne diseases, and malnutrition.

Nutrient depletion, on the other hand, refers to the loss of essential nutrients in the soil or water that can lead to poor crop yields, malnutrition, and ecosystem degradation. This can be caused by factors such as overuse of fertilizers, soil erosion, deforestation, and pollution. In areas where nutrient depletion is prevalent, people may struggle to access a diverse and nutritious diet, leading to health problems such as stunted growth, anemia, and other nutrient deficiencies.

Both water crisis and nutrient depletion are complex issues that require a multi-faceted approach to address. Strategies such as sustainable water management, efficient use of fertilizers, reforestation, and pollution control can help to mitigate these issues and promote a healthier and more sustainable future for all.

Water crisis causes organic matter Depletion

Water crisis and organic matter depletion are two distinct environmental issues, but they can be interconnected in some ways.

- Water crisis refers to a situation where there is a scarcity of water resources or a lack of access to safe and clean water for human consumption and other uses. This can occur due to various factors such as climate change, drought, and overpopulation, overuse of water resources, pollution, and poor water management.
- On the other hand, organic matter depletion refers to a decline in the quantity and quality of organic matter in soil, water, and other ecosystems. Organic matter is essential for the health of soil and the growth of plants, and it also plays a crucial role in the water cycle by promoting water infiltration and retention in soil.
- The connection between water crisis and organic matter depletion lies in the fact that both issues can be influenced by human activities and can have detrimental impacts on ecosystems and human well-being. For example, overuse of water resources can lead to soil degradation and reduced organic matter content, which can further exacerbate water scarcity and reduce agricultural productivity.
- Conversely, organic matter depletion can also contribute to water pollution by reducing the ability of soil and aquatic ecosystems to

filter and absorb pollutants, leading to contamination of water sources.

Addressing these environmental issues requires a holistic approach that considers the interconnections between water resources, soil health, and other ecosystem components. This can involve implementing sustainable water management practices, reducing the use of harmful chemicals and fertilizers, and promoting conservation and restoration of ecosystems.

Water crisis causes Soil Microbiota Depletion

The water crisis and soil microbiota depletion are two interconnected environmental issues that pose a significant threat to our planet's health and well-being.

Water crisis refers to the global scarcity of clean and safe water, which is exacerbated by overconsumption, pollution, and climate change. It affects more than 2 billion people worldwide and poses a severe threat to public health, economic growth, and ecosystem stability.

Soil microbiota depletion refers to the loss of beneficial microorganisms in the soil, including bacteria, fungi, and other microbes. These microorganisms play a crucial role in maintaining soil fertility, nutrient cycling, and ecosystem health. However, agricultural practices such as monoculture farming, the use of chemical fertilizers and pesticides, and land degradation have led to a significant reduction in soil microbiota diversity and abundance.

The depletion of soil microbiota can further exacerbate the water crisis as it reduces soil's ability to retain water and nutrients, leading to soil erosion, desertification, and increased runoff. In turn, this can cause water pollution, flooding, and the loss of biodiversity.

To address these interconnected issues, it is essential to adopt sustainable agricultural practices that prioritize soil health and water conservation. This can include practices such as crop rotation, conservation tillage, the use of organic fertilizers and compost, and the restoration of degraded lands. Additionally, implementing water conservation strategies such as rainwater harvesting, drip irrigation, and efficient water management can help mitigate the impact of the water crisis.

Overall, the water crisis and soil microbiota depletion are complex environmental issues that require a multi-faceted approach to address. By prioritizing sustainable practices and conservation efforts, we can help ensure a healthy and thriving planet for generations to come.

Water crisis and Soil threats due to climate change

Climate change is having a significant impact on the availability and quality of water resources, as well as the health of soil. These impacts are contributing to a water crisis and soil threats that pose serious challenges to sustainable development and human well-being.

- One of the most visible impacts of climate change is changes in precipitation patterns. In some regions, this can lead to more frequent and severe droughts, which can result in water scarcity and crop failure. In other regions, more intense rainfall events can lead to flooding and soil erosion, which can degrade soil quality and reduce agricultural productivity.
- Additionally, rising temperatures can exacerbate water stress by increasing evaporation rates and reducing soil moisture, making it harder for plants to grow and thrive. This can lead to increased competition for water resources and reduced crop yields.
- Furthermore, climate change is contributing to the melting of glaciers and snowpacks, which serve as important sources of freshwater for many regions. As these sources of water disappear, it will become increasingly challenging to meet the water needs of growing populations.
- In addition to the impacts on water resources, climate change is also threatening soil health. Soil degradation, caused by erosion, salinization, compaction, and nutrient depletion, can reduce soil fertility and degrade its ability to support plant growth. This can have significant impacts on agricultural productivity, food security, and ecosystem health.

In conclusion, climate change is exacerbating water scarcity and soil degradation, which are major threats to sustainable development and human well-being. Addressing these challenges will require a comprehensive approach that includes improving water management practices, promoting sustainable agriculture, and reducing greenhouse gas emissions.

Water crisis in relation to Soil health and climate extreme

The water crisis, soil health, and climate extremes are interconnected issues that are affecting many parts of the world. Let's break down each of these issues:

Water Crisis:

The water crisis is a situation where the demand for water exceeds the available supply. This can be due to various factors such as climate change, population growth, and mismanagement of water resources. As a result, many regions are facing severe water shortages, which is affecting agriculture, industry, and the overall quality of life.

Soil Health:

Soil health refers to the ability of soil to sustain plant and animal life. The soil is the foundation of agriculture, and healthy soil is essential for producing nutritious crops. However, soil health is declining in many parts of the world due to various factors such as erosion, pollution, and overuse. This has a significant impact on food security and the environment.

Climate Extremes:

Climate extremes refer to unusual and extreme weather events such as droughts, floods, and heat waves. These events are becoming more frequent and severe due to climate change, which is caused by human activities such as burning fossil fuels and deforestation. Climate extremes have a significant impact on water availability and soil health, as they can cause erosion, soil degradation, and water pollution.

These three issues are interconnected because they are all related to the availability and quality of water. The water crisis and climate extremes are causing soil degradation, which further exacerbates the water crisis. Similarly, soil health is essential for water retention and filtration, which helps mitigate the effects of climate extremes.

To address these issues, we need to adopt a holistic approach that takes into account the interconnection between water, soil, and climate. This includes better management of water resources, promoting sustainable agriculture practices, and reducing greenhouse gas emissions to mitigate the effects of climate change. By taking a comprehensive approach, we can ensure the long-term sustainability of our natural resources and the health of our planet.

Water Crisis solution

The water crisis is a complex issue that requires a multi-faceted approach to solve. The provision of clean drinking water may be ensured by cleaning these lakes and implementing waste management. Pakistan must also build more storage facilities. Little and huge dams should be constructed that do not spark political conflict. The nation must also teach its citizens the importance of conserving water. Recommendations for improving formal water management and governance include improving monitoring and evaluation, capacity building, installing water meters, proper waste and wastewater management and regulation including building treatment plants, and raising awareness of water conservation (SBP, 2017; IUCN, 2014). Here are some potential solutions:

- Water conservation: Encouraging people to conserve water is a critical step in addressing the water crisis. Simple measures such as fixing leaks, using low-flow showerheads and toilets, and reducing outdoor water use can make a significant impact.
- Water recycling and reuse: Reusing wastewater for non-potable purposes such as irrigation and industrial processes can reduce the demand for freshwater sources. Additionally, technologies such as reverse osmosis and desalination can convert saltwater into potable water.
- Rainwater harvesting: Collecting rainwater from rooftops and other surfaces can be an effective way to supplement water supplies, especially in areas with low rainfall. Rainwater can be used for various purposes, including drinking, cooking, and irrigation.
- Sustainable agriculture: Agriculture is one of the largest users of water worldwide. Promoting sustainable agriculture practices such as drip irrigation, crop rotation, and the use of drought-resistant crops can reduce water usage while maintaining food production.
- Infrastructure improvements: Upgrading aging water infrastructure such as pipes, treatment plants, and storage facilities can improve water efficiency and reduce water loss due to leaks.
- Education and awareness: Educating people about the importance of water conservation and the impact of their water usage can promote a culture of responsible water use and help to reduce demand for water.
- Government policies and regulations: Governments can implement policies and regulations to promote water conservation and ensure equitable distribution of water resources. This may include pricing

mechanisms to discourage wasteful water use and regulations to protect water sources from pollution and depletion.

Pakistan was alerted to its worrying status by development and research groups. How Pakistani authorities can stop the unthinkable threat worse than terrorism has been the subject of a lengthy debate on social media. It is anticipated that it will have a whole new kind of impact on Pakistan's economy. In terms of temperature rise, changes in precipitation patterns, greater glacier melt, higher evaporation, and increased irrigation water needs, these are the main implications of climate change. (Asif, 2013). Moreover, new dams like the Diamer Basha and Mohmand dams are currently being built. The Pakistani people were urged by the government to make donations for the building project. The Pakistani government has asked for assistance from both domestic and foreign Pakistanis in financing $14 billion to build dams (Meribole, 2020). Due to Pakistan's climatic conditions, its rivers are also drying up. Punjab, which has five major rivers that feed the nation's water supply, is the most severely affected. The Sutlej and Beas rivers have already dried up and transformed into sweltering deserts (Khalid, 2018). Parts of Pakistan's glaciers, most notably the Ultar Glacier, have suffered as a result of environmental pollution and global warming. The glacier has provided fresh water and helped the nearby valley residents cultivate food. Nonetheless, they are currently denied it in a similar way to rural and urban cultures (Hadid, 2019). Pakistan is fortunate to have an abundance of water resources. It has abundant natural water resources, such as the Hindu Kush Himalayan mountain range, which supplies the nation with about 8.6 million cubic metres of water per year (Asim et. al., 2012). In spite of Pakistan having the largest irrigation system in the world, which irrigates 16 million hectares of arable land, there is a severe water shortage that prevents the system from being fully functioning. It is necessary to develop and put into action an efficient plan to address the growing water problem. In several places of the country, severe floods are destroying numerous farms and homes due to glacier melting and excessive rains in hilly terrain. Moreover, it causes water to be wasted. It's important to note that the water crisis is a global issue that requires a collective effort from individuals, governments, and organizations to address.

Water Crisis solution empowering Soil Decontamination

Soil or land contamination can have serious implications for water quality, as pollutants can leach into groundwater and surface water sources. Here are some potential solutions to address soil or land contamination:

- Remediation technologies: Technologies such as bioremediation, phytoremediation, and chemical oxidation can be used to remove or neutralize contaminants in the soil. These techniques can be effective in treating contaminated sites and restoring them to their original condition.
- Land use planning: Careful land use planning can help to prevent contamination by ensuring that potentially hazardous activities, such as industrial processes, are located away from sensitive areas such as water sources. Zoning regulations can also be implemented to prevent certain land uses in areas with high risk of contamination.
- Sustainable agriculture practices: Sustainable agricultural practices, such as reducing the use of pesticides and fertilizers, can reduce the risk of soil and water contamination. Crop rotation, cover cropping, and other techniques can also help to maintain soil health and reduce erosion, which can prevent pollutants from entering water sources.
- Education and awareness: Educating people about the risks and consequences of soil and water contamination can promote responsible behavior and encourage individuals and organizations to take steps to prevent contamination.
- Government regulations and enforcement: Governments can implement regulations to prevent contamination and ensure that companies and individuals responsible for contamination are held accountable. Enforcement mechanisms can include fines, penalties, and other measures to discourage harmful practices.
- Monitoring and early warning systems: Monitoring of water quality and early warning systems can help to detect contamination before it becomes a major problem. This can allow for quick response and intervention to prevent further contamination.

It's important to note that addressing soil or land contamination requires a long-term, sustained effort. Collaboration between individuals, organizations, and governments is crucial to prevent and remediate contamination, and to protect water resources for future generations.

Water conservation is solution to Soil Nutrients Depletion

Soil nutrient depletion can have significant impacts on water quality, as it can lead to erosion, decreased soil fertility, and increased runoff of pollutants into water sources. Here are some potential solutions to address soil nutrient depletion:

- Sustainable agriculture practices: Sustainable agricultural practices such as crop rotation, cover cropping, and conservation tillage can help to build soil health and reduce nutrient depletion. These techniques can also help to reduce erosion, which can prevent pollutants from entering water sources.
- Nutrient management: Careful management of nutrients such as nitrogen and phosphorus can help to reduce their loss from agricultural fields and prevent them from entering water sources. Techniques such as precision fertilization and the use of slow-release fertilizers can help to reduce nutrient runoff.
- Composting: Composting organic waste can help to enrich soil with nutrients and improve soil structure. This can help to reduce soil erosion and runoff, and improve the quality of water sources.
- Agroforestry: Agroforestry practices such as planting trees and shrubs in agricultural fields can help to build soil health, prevent erosion, and reduce nutrient depletion. These practices can also improve water quality by reducing the amount of pollutants entering water sources.
- Soil conservation: Practices such as terracing, contour farming, and grassed waterways can help to reduce soil erosion and prevent the loss of nutrients from agricultural fields. These techniques can help to improve soil health and water quality.
- Education and awareness: Educating people about the importance of soil conservation and nutrient management can promote responsible behavior and encourage individuals and organizations to take steps to prevent soil nutrient depletion.
- Government policies and regulations: Governments can implement policies and regulations to promote sustainable agriculture practices and prevent soil nutrient depletion. This may include incentives for farmers to adopt sustainable practices, regulations to limit nutrient runoff, and measures to protect sensitive areas such as water sources.

It's important to note that addressing soil nutrient depletion requires a long-term, sustained effort. Collaboration between individuals, organizations, and governments is crucial to prevent and remediate soil nutrient depletion, and to protect water resources for future generations.

Water Crisis solution empowering Soil Organic matter Restoration

Soil organic matter depletion can have significant impacts on water quality, as it can lead to decreased soil fertility, erosion, and increased runoff of pollutants into water sources. Here are some potential solutions to address soil organic matter depletion:

- Cover cropping: Cover cropping can help to maintain soil health and prevent erosion by protecting soil from erosion and adding organic matter to the soil. This can improve water quality by reducing the amount of pollutants entering water sources.
- Conservation tillage: Conservation tillage practices can help to reduce soil erosion and maintain soil health by minimizing soil disturbance and leaving crop residues on the soil surface. This can help to reduce nutrient loss and improve water quality.
- Composting: Composting organic waste can help to enrich soil with organic matter and improve soil structure. This can help to reduce soil erosion and runoff, and improve water quality.
- Agroforestry: Agroforestry practices such as planting trees and shrubs in agricultural fields can help to build soil organic matter, prevent erosion, and improve water quality by reducing the amount of pollutants entering water sources.
- Soil conservation: Practices such as terracing, contour farming, and grassed waterways can help to reduce soil erosion and prevent the loss of organic matter from agricultural fields. These techniques can help to improve soil health and water quality.
- Education and awareness: Educating people about the importance of soil conservation and organic matter management can promote responsible behavior and encourage individuals and organizations to take steps to prevent soil organic matter depletion.
- Government policies and regulations: Governments can implement policies and regulations to promote sustainable agriculture practices and prevent soil organic matter depletion. This may include incentives for farmers to adopt sustainable practices, regulations to limit erosion and nutrient runoff, and measures to protect sensitive areas such as water sources.

It's important to note that addressing soil organic matter depletion requires a long-term, sustained effort. Collaboration between individuals, organizations, and governments is crucial to prevent and remediate soil organic matter depletion, and to protect water resources for future generations.

Water Crisis solution empowering Soil Microbiota population

Soil microbiota plays a crucial role in maintaining healthy soil, which in turn is essential for ensuring a sustainable supply of clean water. Soil microbiota helps to improve soil structure and fertility, promote plant growth, and reduce erosion and nutrient loss. In addition, healthy soil also plays a vital role in water regulation and purification. Soil microbiota depletion can lead to a decline in soil health, which can exacerbate water scarcity and quality issues. To address this, here are some solutions that can help:

- Soil conservation practices: Implementing soil conservation practices such as reduced tillage, cover cropping, crop rotation, and the use of organic fertilizers can help improve soil health and increase soil microbiota populations.
- Soil restoration techniques: Techniques such as soil inoculation with beneficial microorganisms, composting, and biochar application can help restore soil microbiota populations.
- Water conservation practices: Implementing water conservation practices such as rainwater harvesting, drip irrigation, and water-efficient crops can help reduce water use and increase water availability.
- Land-use management: Proper land-use management practices can help reduce soil erosion and nutrient loss, which can improve soil health and water quality.
- Education and outreach: Educating farmers and the public on the importance of soil health and soil microbiota can help increase awareness and encourage the adoption of practices that promote soil health and water conservation.

Overall, addressing soil microbiota depletion is an important step towards ensuring a sustainable supply of clean water. By implementing these solutions, we can promote healthy soils, increase water availability, and mitigate the impacts of water scarcity and quality issues.

Solution of Soil threats to combat climate change

The water crisis and soil threats due to climate change are interconnected and complex problems that require multifaceted solutions. Here are some potential solutions to address these issues: Implement Sustainable Land Management Practices: Sustainable land management practices can help reduce soil erosion, increase water retention in soils, and improve soil fertility. These practices include agroforestry, conservation agriculture, and sustainable land use planning.

- Promote Efficient Irrigation Systems: Efficient irrigation systems can help conserve water by reducing waste and improving the distribution of water to crops. Drip irrigation, for example, can reduce water use by up to 60% compared to traditional flood irrigation methods.
- Increase Water Use Efficiency: Improving water use efficiency can help reduce the amount of water needed to produce crops. This can be achieved by selecting drought-tolerant crops, using crop rotation and intercropping, and implementing precision irrigation technologies.
- Develop Climate-Resilient Crop Varieties: Developing climate-resilient crop varieties can help farmers adapt to changing climate conditions. These varieties can be designed to tolerate drought, flood, or extreme temperatures, ensuring that farmers have a reliable source of food and income.
- Promote Rainwater Harvesting: Rainwater harvesting can help supplement irrigation water and reduce the pressure on groundwater sources. This can be achieved by building small-scale water catchment systems or implementing large-scale rainwater harvesting programs.
- Support Integrated Water Resource Management: Integrated water resource management involves coordinating the management of water resources across different sectors, such as agriculture, industry, and domestic use. This approach can help ensure that water resources are used sustainably and equitably.

The current predicament has also been impacted by climate change. The water situation has been exacerbated by the extreme heat, melting glaciers, and gas emissions. Due to the harsh weather, the rivers and seas are getting wetter every year. Just in Karachi in May 2018 did heatstroke cause the deaths of 65 people. Yet, in 2015, around 1200 people lost their lives as a result of the intense heat (Baloch, 2018). In recent years, the monsoon season has been unpredictable. The timing of the monsoon is impossible to forecast, according to weather specialists. The winter and monsoon seasons appear to have gotten shorter as a result of climate change. Year after year, their duration has been reduced. The length of winters has also looked to shorten recently. In most regions of the country, the winter season now only lasts for around two months instead of the nearly four months it used to last throughout the year. Social change also has an impact on greenhouse gas emissions and exceptionally hot weather (Meribole, 2020).

Overall, addressing the water crisis and soil threats due to climate change will require a combination of sustainable land management practices, efficient irrigation systems, climate-resilient crops, and integrated water resource management. By working together to implement these solutions, we can help ensure a more sustainable future for our planet.

Water Crisis solution empowering Soil health under climate extreme

Soil health and climate extremes are both important factors that can contribute to water scarcity and quality issues. Soil health plays a crucial role in water regulation, while climate extremes such as droughts and floods can exacerbate water scarcity and quality issues. Here are some solutions that can help address the water crisis caused by soil health and climate extremes:

- Sustainable soil management: Implementing sustainable soil management practices such as reducing tillage, planting cover crops, and using organic fertilizers can help improve soil health and increase water retention.
- Agroforestry: Integrating trees and other perennial plants into agricultural systems can help improve soil health and increase water availability.
- Soil erosion control: Implementing measures such as contour farming, terracing, and conservation tillage can help reduce soil erosion and improve soil health, which in turn can improve water quality.
- Water harvesting: Capturing and storing rainwater using techniques such as rainwater harvesting, runoff harvesting, and floodwater harvesting can help increase water availability during times of drought.
- Drought-resistant crops: Planting drought-resistant crops can help mitigate the impact of droughts on water availability.
- Flood management: Implementing flood management techniques such as constructing retention ponds and levees can help reduce the impact of floods on water quality.
- Climate adaptation measures: Implementing climate adaptation measures such as early warning systems, disaster preparedness, and water storage infrastructure can help mitigate the impact of climate extremes on water availability and quality.

Overall, addressing the water crisis caused by soil health and climate extremes requires a multifaceted approach that involves sustainable soil management, agroforestry, soil erosion control, water harvesting, drought-resistant crops, flood management, and climate adaptation measures. By implementing these solutions, we can promote healthy soils, increase water availability, and mitigate the impact of climate extremes on water scarcity and quality.

3. Environmental Crisis

The environmental crisis refers to a broad range of issues and challenges facing the natural environment due to human activities. These include climate change, biodiversity loss, deforestation, air and water pollution, soil degradation, and waste management. The environmental crisis has significant impacts on human health, well-being, and the economy, and it threatens the sustainability of the planet. Rising temperatures are fuelling environmental degradation, natural disasters, weather extremes, food and water insecurity, economic disruption, conflict, and terrorism. Sea levels are rising, the Arctic is melting, coral reefs are dying, oceans are acidifying, and forests are burning.

Here are some solutions that can help address the environmental crisis:

- **Reducing carbon emissions**: Reducing carbon emissions by transitioning to renewable energy sources, promoting energy efficiency, and reducing deforestation can help mitigate the impact of climate change.
- **Conservation of biodiversity**: Conserving biodiversity by protecting natural habitats, reducing over fishing and hunting, and promoting sustainable agriculture can help preserve ecosystems and the services they provide.
- **Sustainable resource use**: Promoting sustainable resource use by reducing waste, promoting recycling, and reducing consumption can help reduce the strain on natural resources.
- **Pollution control**: Reducing pollution through regulation and technology such as emissions controls on industrial facilities and promoting public transportation can help reduce the negative impact on air and water quality.
- **Education and awareness**: Increasing education and awareness about the environmental crisis can help foster behavior change and promote sustainable living practices.
- **International cooperation**: Promoting international cooperation and agreements can help address environmental challenges that cross

borders, such as climate change, biodiversity loss, and air and water pollution.

Overall, addressing the environmental crisis requires a concerted effort from individuals, governments, businesses, and civil society. By implementing these solutions, we can promote a sustainable and resilient environment for current and future generations.

Environmental crisis causes soil or land Contamination

Soil or land contamination is a significant environmental issue that can have severe impacts on human health, the environment, and the economy. Soil contamination can be caused by various human activities, such as industrial activities, agricultural practices, and improper waste disposal. Here are some solutions that can help address the environmental crisis caused by soil or land contamination:

- Soil remediation: Soil remediation involves the use of techniques such as bioremediation, phytoremediation, and soil vapour extraction to remove or neutralize contaminants in the soil.
- Soil conservation: Implementing soil conservation practices such as reduced tillage, cover cropping, and crop rotation can help prevent soil erosion and reduce the risk of contamination.
- Sustainable land use: Promoting sustainable land use practices such as organic farming, integrated pest management, and sustainable forestry can help reduce the use of harmful chemicals and prevent contamination.
- Site assessment and monitoring: Conducting site assessments and monitoring can help identify contaminated sites and track the progress of remediation efforts.
- Land-use management: Proper land-use management practices such as zoning regulations and environmental impact assessments can help prevent contamination and ensure that contaminated sites are properly remediated.
- Public education and awareness: Increasing public education and awareness about the risks of soil contamination and the importance of proper waste disposal can help prevent future contamination.

Overall, addressing the environmental crisis caused by soil or land contamination requires a combination of prevention, remediation, and proper management practices. By implementing these solutions, we can protect human health, preserve the environment, and promote sustainable development.

Environmental crisis causes nutrients Depletion

Nutrient depletion is an environmental crisis that refers to the loss of nutrients in soil, water, and other natural resources that are essential for plant growth and other ecological processes. Nutrient depletion can be caused by various abiotic factors like intensive rainfall, hail storms, fooloding, erosion, runoff, geological features of the soil, topography and other human activities, such as overuse of fertilizers, unsustainable agricultural practices, and deforestation. Here are some solutions that can help address the environmental crisis caused by nutrient depletion:

- **Sustainable agriculture**: Implementing sustainable agricultural practices such as crop rotation, intercropping, and conservation tillage can help reduce nutrient depletion and promote soil health.
- **Organic farming**: Promoting organic farming practices can help reduce the use of synthetic fertilizers and pesticides, which can contribute to nutrient depletion.
- **Composting**: Composting organic waste can help replenish nutrients in soil and reduce the need for synthetic fertilizers.
- **Soil conservation**: Implementing soil conservation practices such as reduced tillage, cover cropping, and agroforestry can help prevent soil erosion and nutrient depletion.
- **Nutrient management**: Implementing nutrient management practices such as precision farming, nutrient cycling, and integrated nutrient management can help optimize nutrient use and reduce the risk of nutrient depletion.
- **Sustainable forestry**: Implementing sustainable forestry practices such as selective harvesting and forest thinning can help maintain nutrient levels in soil and reduce the risk of nutrient depletion.
- **Public education and awareness**: Increasing public education and awareness about the importance of nutrient management and sustainable agricultural practices can help prevent future nutrient depletion.

Overall, addressing the environmental crisis caused by nutrient depletion requires a combination of prevention, management, and sustainable practices. By implementing these solutions, we can promote soil health, preserve natural resources, and ensure sustainable development for current and future generations.

Environmental crisis causes organic matter Depletion

Organic matter depletion is related to environmental crisis that refers to the loss of organic matter in soil, and other natural resources that are essential for plant growth, microbial population, soil aggregation and other ecological and biological processes. Organic matter depletion can be caused by continuous ploughing which loosen the soil and by rainfall, windstorms, runoff or any other ways lost carbon and organic matter. Various human activities, such as overuse of chemical fertilizers, lack of proper crop rotation, no use or less use of organic amendments and unsustainable agricultural practices, and deforestation may accelerate organic matter depletion in soil. Here are some solutions that can help address the environmental crisis caused by organic matter depletion:

- Sustainable agriculture: Implementing sustainable agricultural practices such as crop rotation, intercropping, and conservation tillage can help reduce organic matter depletion and promote soil health.
- Organic farming: Promoting organic farming practices can help reduce the use of synthetic fertilizers and pesticides, which can contribute to organic matter depletion.
- Composting: Composting organic waste can help replenish organic matter in soil and reduce the need for synthetic fertilizers.
- Soil conservation: Implementing soil conservation practices such as reduced tillage, cover cropping, and agroforestry can help prevent soil erosion and organic matter depletion.
- Carbon sequestration: Implementing carbon sequestration practices such as afforestation and reforestation can help increase the amount of organic matter in soil and reduce atmospheric carbon dioxide levels.
- Public education and awareness: Increasing public education and awareness about the importance of organic matter management and sustainable agricultural practices can help prevent future organic matter depletion.

Overall, addressing the environmental crisis caused by organic matter depletion requires a combination of prevention, management, and sustainable practices. By implementing these solutions, we can promote soil health, preserve natural resources, and ensure sustainable development for current and future generations.

Environmental crisis empowering soil Microbiota Depletion

Soil microbiota depletion is an environmental crisis that refers to the loss of beneficial microorganisms in soil, which play a crucial role in maintaining soil fertility, nutrient cycling, and plant growth. Soil microbiota depletion can be caused by various human activities, such as the use of synthetic fertilizers, pesticides, and herbicides, overuse of antibiotics, and pollution. Here are some solutions that can help address the environmental crisis caused by soil microbiota depletion:

- Organic farming: Promoting organic farming practices can help reduce the use of synthetic fertilizers, pesticides, and herbicides, which can contribute to soil microbiota depletion.
- Composting: Composting organic waste can help replenish beneficial microorganisms in soil and reduce the need for synthetic fertilizers.
- Soil conservation: Implementing soil conservation practices such as reduced tillage, cover cropping, and agroforestry can help maintain soil microbiota and promote soil health.
- Biofertilizers: Using biofertilizers such as mycorrhizal fungi, nitrogen-fixing bacteria, and plant growth-promoting rhizobacteria can help replenish soil microbiota and improve soil fertility.
- Integrated pest management: Implementing integrated pest management practices such as biological control and crop rotation can help reduce the use of pesticides, which can harm beneficial microorganisms in soil.
- Public education and awareness: Increasing public education and awareness about the importance of soil microbiota and the harmful effects of synthetic fertilizers and pesticides can help prevent future soil microbiota depletion.

Overall, addressing the environmental crisis caused by soil microbiota depletion requires a combination of prevention, management, and sustainable practices. By implementing these solutions, we can promote soil health, preserve natural resources, and ensure sustainable development for current and future generations.

Environmental crisis soil threats due to climate change

Climate change is a significant threat to soil health and can lead to various environmental crises. Here are some of the soil threats due to climate change:

- Soil erosion: Climate change can increase the intensity and frequency of rainfall events, which can lead to increased soil erosion. Erosion can cause soil to lose its fertility and reduce crop yields.
- Soil compaction: Droughts and extreme weather events can lead to soil compaction, making it difficult for plants to grow and for water to penetrate the soil.
- Soil salinization: Rising sea levels and saltwater intrusion can lead to soil salinization, making it difficult for plants to grow and reducing soil productivity.
- Soil nutrient depletion: Climate change can alter the nutrient cycle and reduce the availability of nutrients in soil, leading to nutrient depletion and reduced crop yields.
- Soil carbon loss: Climate change can accelerate the loss of soil organic carbon, leading to reduced soil fertility and increased greenhouse gas emissions.
- Soil biodiversity loss: Climate change can alter the distribution and abundance of soil microorganisms, leading to loss of biodiversity and reduced soil health.

To address these soil threats due to climate change, we need to implement measures such as sustainable agriculture practices, soil conservation, and carbon sequestration. These solutions can help maintain soil health, increase soil resilience, and mitigate the impacts of climate change on soil and the environment. Additionally, promoting public awareness and education on soil health and climate change can help in implementing these solutions and fostering a sustainable future.

Environmental crises impact on Soil health and climate extreme

Soil health and climate extremes are closely interconnected, and climate extremes can have significant impacts on soil health. Here are some ways that climate extremes can impact soil health:

- Drought: Droughts can cause soil to become dry and compacted, making it difficult for plants to grow and for water to penetrate the soil. This can lead to soil erosion, nutrient depletion, and reduced crop yields.
- Floods: Heavy rainfall and floods can cause soil erosion, nutrient leaching, and loss of soil organic matter. Floods can also lead to soil compaction and reduced soil aeration, which can negatively impact soil health.

- Heat waves: Heat waves can cause soil to dry out and increase the rate of soil organic matter decomposition, which can reduce soil fertility and carbon sequestration potential.
- Cold temperatures: Cold temperatures can cause soil to freeze and expand, leading to soil compaction and reduced soil structure. This can negatively impact soil water holding capacity and nutrient availability.

To maintain soil health in the face of climate extremes, we need to implement measures such as sustainable agriculture practices, soil conservation, and carbon sequestration. These solutions can help maintain soil resilience, increase soil water holding capacity, and mitigate the impacts of climate extremes on soil and the environment. Additionally, promoting public awareness and education on soil health and climate change can help in implementing these solutions and fostering a sustainable future.

3.1. Environmental Crisis solution

The environmental crisis is a complex issue that requires a multifaceted approach to address. Here are some solutions that can help address the environmental crisis:

- Reduce greenhouse gas emissions: To mitigate the impacts of climate change, we need to reduce our greenhouse gas emissions. This can be achieved through measures such as transitioning to renewable energy sources, improving energy efficiency, and reducing consumption.
- Promote sustainable practices: We need to promote sustainable practices across all sectors, including agriculture, forestry, fisheries, and industry. This includes measures such as sustainable resource management, waste reduction, and sustainable production and consumption.
- Protect and restore ecosystems: Protecting and restoring ecosystems such as forests, wetlands, and oceans can help maintain biodiversity, sequester carbon, and provide essential ecosystem services.
- Invest in research and innovation: Investing in research and innovation can help develop new technologies and solutions that can address environmental challenges, such as clean energy, sustainable agriculture, and waste management.
- Education and awareness: Education and awareness campaigns can help promote public understanding of the environmental crisis and

encourage individuals to take action to reduce their environmental impact.

- Policy and governance: Strong policies and governance frameworks can help regulate industries, protect natural resources, and promote sustainable practices.

Overall, addressing the environmental crisis requires a concerted effort from individuals, businesses, governments, and civil society. By implementing these solutions and working together, we can mitigate the impacts of the environmental crisis and ensure a sustainable future for all.

Environmental crisis solution and prevention of Soil or land Contamination

Soil or land contamination is a serious environmental crisis that requires careful management and remediation. Here are some solutions that can help address soil or land contamination:

- Site assessment and characterization: Conducting a site assessment and characterization is essential to identify the extent and severity of soil contamination. This information can help inform the selection of appropriate remediation methods.
- Soil remediation: Various soil remediation techniques can be used to remove or reduce the concentration of contaminants in the soil, including physical, chemical, and biological treatments. The choice of remediation method depends on the type and extent of contamination, as well as the desired outcome.
- Soil management: Managing contaminated soil can help minimize the spread of contaminants and reduce exposure to humans and wildlife. This includes measures such as covering the soil with a clean layer, restricting access to the contaminated area, and implementing erosion control measures.
- Sustainable land use: Encouraging sustainable land use practices, such as organic farming and urban agriculture, can help reduce the use of harmful chemicals and promote healthy soil management practices.
- Monitoring and verification: Monitoring and verification of soil remediation activities is essential to ensure the effectiveness of the remediation and to detect any potential re-contamination.
- Education and awareness: Education and awareness campaigns can help promote public understanding of soil contamination and encourage individuals to take action to prevent contamination and promote healthy soil management practices.

Overall, addressing soil or land contamination requires a coordinated effort from individuals, businesses, governments, and civil society. By implementing these solutions and working together, we can mitigate the impacts of soil contamination and ensure healthy and productive soil for future generations.

Environmental crisis solution prevent Soil Nutrients Depletion

Soil nutrient depletion is a serious environmental crisis that can have significant impacts on agricultural productivity and ecosystem health. Here are some solutions that can help address soil nutrient depletion:

- Soil testing and nutrient management planning: Conducting regular soil testing and developing nutrient management plans can help optimize soil fertility and minimize nutrient depletion. This includes measures such as soil amendments, crop rotations, and cover cropping.
- Sustainable agriculture practices: Promoting sustainable agriculture practices, such as reduced tillage, agroforestry, and integrated pest management, can help improve soil health, reduce nutrient depletion, and increase crop yields.
- Nutrient recycling and waste management: Recycling organic waste, such as crop residues and animal manure, can help replenish soil nutrients and reduce the need for synthetic fertilizers.
- Agro ecology: Implementing agro ecological practices, such as intercropping and agroforestry, can promote nutrient cycling, reduce soil erosion, and improve soil health.
- Soil conservation: Implementing soil conservation measures, such as contour ploughing and terracing, can help prevent soil erosion and maintain soil fertility.
- Research and innovation: Investing in research and innovation can help develop new technologies and practices that can address nutrient depletion and promote sustainable soil management.

Overall, addressing soil nutrient depletion requires a coordinated effort from individuals, businesses, governments, and civil society. By implementing these solutions and working together, we can mitigate the impacts of soil nutrient depletion and promote healthy and productive soil for future generations.

Environmental crisis solution of Soil Organic matter Depletion

Soil organic matter (SOM) depletion is a significant environmental crisis that can have negative impacts on soil health, productivity, and ecosystem services. Here are some solutions that can help address soil organic matter depletion:

- Conservation agriculture: Implementing conservation agriculture practices, such as reduced tillage, crop rotations, and cover cropping, can help increase SOM levels, reduce soil erosion, and promote soil health.
- Organic farming: Encouraging and supporting organic farming practices can help increase SOM levels, reduce the use of synthetic fertilizers and pesticides, and promote sustainable soil management practices.
- Composting and green manure: Recycling organic waste, such as food scraps, yard trimmings, and cover crops, through composting and green manure can help increase SOM levels and improve soil health.
- Agroforestry: Integrating trees and shrubs into agricultural landscapes through agroforestry can help promote SOM accumulation, reduce soil erosion, and provide additional ecosystem services, such as carbon sequestration and wildlife habitat.
- Soil restoration and reclamation: Restoring degraded and contaminated soils through reclamation and restoration practices can help increase SOM levels and improve soil health.
- Education and awareness: Education and awareness campaigns can help promote public understanding of the importance of SOM and encourage individuals to take action to prevent SOM depletion and promote healthy soil management practices.

Overall, addressing SOM depletion requires a coordinated effort from individuals, businesses, governments, and civil society. By implementing these solutions and working together, we can mitigate the impacts of SOM depletion and ensure healthy and productive soil for future generations.

Environmental crisis solution of Soil Microbiota Depletion

Soil microbiota depletion is a serious environmental crisis that can have negative impacts on soil health, productivity, and ecosystem services. Here are some solutions that can help address soil microbiota depletion:

Reduce tillage: Reducing or eliminating tillage can help maintain soil structure and increase microbial diversity in the soil.

- Organic farming practices: Encouraging and supporting organic farming practices, such as crop rotations, intercropping, and cover cropping, can help promote soil microbial diversity and health.
- Integrated Pest Management (IPM): Implementing IPM practices can help reduce the use of pesticides and herbicides that can harm soil microbes.
- Composting and organic matter amendments: Adding compost and other organic matter amendments to the soil can help promote microbial activity and diversity.
- Biochar application: Applying biochar, a form of charcoal made from organic matter, can help increase soil microbial diversity and activity.
- Restoring soil health: Restoring degraded and contaminated soils through soil restoration practices can help promote microbial diversity and activity.
- Education and awareness: Education and awareness campaigns can help promote public understanding of the importance of soil microbiota and encourage individuals to take action to prevent soil microbiota depletion and promote healthy soil management practices.

Overall, addressing soil microbiota depletion requires a coordinated effort from individuals, businesses, governments, and civil society. By implementing these solutions and working together, we can mitigate the impacts of soil microbiota depletion and ensure healthy and productive soil for future generations.

3.1.1. Environmental crisis solution of Soil threats due to climate change

Soil threats due to climate change, such as erosion, desertification, and soil degradation, are a significant environmental crisis that can have negative impacts on soil health, productivity, and ecosystem services. Here are some solutions that can help address soil threats due to climate change:

- Conservation agriculture: Implementing conservation agriculture practices, such as reduced tillage, crop rotations, and cover cropping, can help reduce erosion, promote soil health, and increase soil carbon sequestration.

- Agroforestry: Integrating trees and shrubs into agricultural landscapes through agroforestry can help reduce soil erosion, improve soil moisture retention, and provide additional ecosystem services, such as carbon sequestration and wildlife habitat.
- Sustainable land management: Adopting sustainable land management practices, such as avoiding overgrazing, reducing land clearing, and planting windbreaks, can help reduce soil degradation and desertification.
- Soil restoration and reclamation: Restoring degraded and contaminated soils through reclamation and restoration practices can help mitigate the impacts of climate change on soil health.
- Climate-smart agriculture: Implementing climate-smart agricultural practices, such as drought-resistant crops, water-efficient irrigation, and agroforestry, can help increase soil resilience and adaptability to changing climatic conditions.
- Education and awareness: Education and awareness campaigns can help promote public understanding of the impacts of climate change on soil health and encourage individuals to take action to mitigate these impacts.

Overall, addressing soil threats due to climate change requires a coordinated effort from individuals, businesses, governments, and civil society. By implementing these solutions and working together, we can mitigate the impacts of climate change on soil health and ensure healthy and productive soil for future generations.

3.1.2. Environmental crisis solution of Soil health and climate extreme

Soil health and climate extremes are a significant environmental crisis that can have negative impacts on soil health, productivity, and ecosystem services. Here are some solutions that can help address soil health and climate extremes:

- Increasing soil organic matter: Increasing soil organic matter through the addition of compost, cover cropping, and reduced tillage can help improve soil structure and increase water-holding capacity, which can help mitigate the impacts of drought and flooding.
- Soil conservation: Implementing soil conservation practices, such as contour plowing, terracing, and mulching, can help reduce soil erosion and improve soil health, which can increase resilience to climate extremes.

- Drought-resistant crops: Planting drought-resistant crops can help increase soil resilience to drought and reduce water demand.
- Water management: Implementing efficient water management practices, such as drip irrigation and rainwater harvesting, can help reduce the impacts of drought and increase soil moisture.
- Soil restoration and reclamation: Restoring degraded and contaminated soils through reclamation and restoration practices can help increase soil health and resilience to climate extremes.
- Sustainable land management: Adopting sustainable land management practices, such as avoiding overgrazing and reducing land clearing, can help reduce soil degradation and increase soil health and resilience.

Climate-smart agriculture: Implementing climate-smart agricultural practices, such as intercropping, agroforestry, and integrated pest management, can help increase soil resilience and adaptability to changing climatic conditions.

Overall, addressing soil health and climate extremes requires a coordinated effort from individuals, businesses, governments, and civil society. By implementing these solutions and working together, we can mitigate the impacts of climate extremes on soil health and ensure healthy and productive soil for future generations.

This chapter will provide an overview of quantities, qualities and productivity of land and water, and agriculture based on them in Central Asia. We have analysed available sources on the status of water and land, trends in their use and developments. The information and data came from the scientific literature, recent research reports, statistical databases and commission reports of United Nations Organizations such as the FAO and UNECE, international research and development projects, and the results of our own field work.

Conclusion

Despite significant progress in using South Asia's water and land resources more sustainably, many challenges and knowledge gaps stay. We conclude that a shift in scientific methodologies, including measurement methods, evaluation methods, models, and technological solutions, is needed. Central Asian scientists should have access to cutting-edge technologies for resource monitoring and management. They, as well as their work, must be integrated into the international scientific community. The examples supplied demonstrate how scientific collaboration and the inclusion of expertise from foreign partners can aid in the identification and resolution of problems. There is a list of added measurement and evaluation methods available, which will be presented in the following chapters. Some will require investment and site-specific studies before they can be implemented in South Asia. Others could be implemented right away and at no cost. This would need increased awareness and understanding among decision-makers and researchers, as well as the implementation of some initiatives. Most Pakistan's people are rural residents who rely on the country's natural resources for their livelihood. Most of the population may face catastrophic water shortages if water is consistently mishandled. Agriculture is the foundation of our economy, and a lack of water has an impact on both. Dams must solve this problem. For this national purpose, all the provinces must come together. There must be an end to provincial mistrust. Water cannot be manipulated for political purposes. The national water crisis must be resolved without involving politicians. Government must take note of it so that we can preserve this resource for future generations.

References

1. Risk Assessment Guidance for Superfund, Human Health Evaluation Manual, Office of Emergency and Remedial Response, U.S. Environmental Protection Agency, Washington D.C. 20450
2. George, Rebecca; Joy, Varsha; S, Aiswarya; Jacob, Priya A. "Treatment Methods for Contaminated Soils – Translating Science into Practice" (PDF). International Journal of Education and Applied Research. Retrieved February 19, 2016.
3. Zhang Z, Qu Y, Li S, Feng K, Wang S, Cai W, Liang Y, Li H, Xu M, Yin H, Deng Y. 2017. Soil bacterial quantification approaches coupling with relative abundances reflecting the changes of taxa. Sci Rep 7:4837.
4. World Urbanization Prospects, The 2018 Revision, ST/ESA/SER.A/ 420 (UN Department of Economic and Social Affairs, 2019).
5. Snapp, S. S., Nyiraneza, J., Otto, M., & Kirk, W. (2003). Managing Manure in Potato and Vegetable Systems (E2893). East Lansing: Michigan State University, MSU Extension. Retrieved from http://web2.msue.msu.edu/bulletins/ Bulletin/PDF/E2893.pdf
6. Overview of EPA's Brownfields Program. USEPA https://www.epa.gov/brownfields/overview-epas-brownfields-program (2022).

7. Imran, Amanullah & Abdel Rahman Al Tawaha (2022d): Indigenous organic resources utilization, application methods and sowing time replenish soil nitrogen and increase maize yield and total dry biomass, Journal of Plant Nutrition, DOI: 10.1080/01904167.2022.2067055.
8. Imran, Amanullah and Ibrahim Ortash. 2022b. Agronomic Practices Improved Cucumber Productivity, Nutrients Uptake and Quality. Gesunde Pflanzen. 10.1007/s10343-022-00634-1.
9. Imran, Amanullah, and Abdel Rahman Al Tawaha. 2022c. Regenerating potential of dual purpose rapeseed (Brassica napus L.) as influenced by decapitation stress and variable rates of phosphorous. Communication in soil science and plant analysis. Doi.org/10.1080/00103624.2022.2118297
10. Imran and Amanullah. (2022b). Integration of peach (Prunus persica L) remnants in combination with beneficial microbes and phosphorus differ phosphorus use efficiency, agronomic efficiency and partial factor productivity in soybean Vs maize crops. Journal of Plant Nutrition. doi.org/10.1080/01904167.2022.2099890
11. Imran, & Abdel Rahman Al Tawaha (2022): Humic Acid and Sulfur Integration Enhances Growth and Yield Related Traits of Brassica Napus L. Journal of Plant Nutrition. doi.org/10.1080/01904167.2022.212882
12. Imran. 2022a. Growing of Off-Season Tomato in High Tunnel and Its Nutritional Value Augmentation with Integrated Nutrients Management. Journal of Plant Nutrition. Doi: 10.1080/01904167.2022.2046062
13. Imran and Amanullah. 2021a. Assessment of chemical and manual weed control approaches for effective weed suppression and maize productivity enhancement under maize-wheat cropping system. Gesunde Pflanzen, DOI: 10.1007/s10343-021-00599-7.
14. Imran, Amanullah and Abdel Rehman Altawaha. 2022c. Carbon Assimilation and Dry Matter Partitioning in Soybean Ameliorates with the Integration of Nano-Black Carbon, Along with Beneficial Microbes and Phosphorus Fertilization. Journal of Plant Nutrition, DOI:10.1080/01904167.2022.2035753
15. Imran. 2022b. (Review) Phosphorus availability enhanced with combine application of organic amendments and beneficial microbes under soybean-wheat cropping system. Communication in soil science and plant analysis. doi.org/10.1080/00103624.2022.2034848
16. Imran.2022c. Management of Abiotic Stresses with Nano-black carbon is a tool for Crop Production. Journal of Plant Nutrition. doi.org/10.1080/01904167.2022.2046085

17. Imran and Amanullah. 2021b. Phosphorus biofortification and uptake in maize enhanced with integrated phosphorus management. Phosphorus, Sulfur, and Silicon and the Related Elements. DOI: 10.1080/10426507.2021.2022677.
18. Imran. 2021. The bioavailability of Phosphorus in composite vs. Hybrid maize differ with phosphorus and boron fertilization. Phosphorus, Sulfur, and Silicon and the Related Elements. https://doi.org/10.1080/10426507.2021.1920588.
19. Imran, Amanullah, Asad Ali khan, Tariq Mahmood, Abdel Rahman Al Tawaha and Samia Khanum. 2021a. Adequate Fertilization, Application Method and Sowing Techniques Improve Maize yield and related traits. Communication in soil science and plant analysis. DOI:10.1080/00103624.2021.1925688
20. Imran, Amanullah and Abdel Rahman M. Al Tawaha. 2021b. Management of Nano-black carbon, Phosphorous and Bio Fertilizer improve Soil organic carbon and ensilage biomass of soybean and maize. Communication in Soil Science and Plant Analysis. DOI.10.1080/00103624.2021.1966439
21. Imran and Amanullah. 2021c. Phosphorus and Boron Application Optimizing Biofortification Of phosphorus and Productivity of French Bean (Phaseolus Vulgaris L.). Communication In Soil Science and Plant Analysis. DOI:10.1080/00103624.2021.1971689
22. Imran, Amanullah and Abdel Rahman Al-Tawaha. 2021c. Carbon Sources Application Increase Wheat Yield And Soil Fertility. Communication in soil science and plant analysis DOI: 10.1080/00103624.2020.1865397.
23. Imran, Amanullah and Abdel Rehman Al-Tawaha.2020a. The Productivity of subsequent wheat enhanced with residual carbon sources and phosphorus under improved irrigation system. Communication in soil science and plant analysis, DOI:10.1080/00103624.2020.1763387
24. Imran, Amanullah, Muhammad Arif, Zahir Shah and Abdul Bari. 2020b. Soil application of Trichoderma and Peach (Prunus persica L.) residues possesses biocontrol potential for weeds and enhances growth and profitability of soybean (Glycine max). Sarhad Journal of Agriculture 36 (1) 10-20. DOI | http://dx.doi.org/10.17582/journal.sja/2020/36.1.10.20
25. Franzluebbers, A. J. (2010). Achieving soil organic carbon sequestration with conservation agricultural systems in the southeastern United States. Soil Science Society of America Journal. 74(2):347–357. Retrieved from

www.soils.org/ publications/sssaj/articles/74/2/347
26. Baker K, Cook RJ. Biological control of plant pathogens. San Francisco: Freeman WH Publisher; 1974.
27. Morra L, Pagano L, Iovieno P, Baldantoni D, Alfani A. Soil and vegetable crop response to addition of different levels of municipal waste compost under Mediterranean greenhouse conditions. Agron Sustain Dev. 2010;30:701–9.
28. O'Donnell AG, Colvan SR, Malosso E. Twenty years of molecular analysis of bacterial communities in soils and what have we learned about function? In: Bardgett RD, Usher MB, Hopkins DW, editors. Biological diversity and function in soils. New York: Cambridge University Press; 2005. p. 44–73.
29. Kibblewhite MG, Ritz K, Swift MJ. Soil health in agricultural systems. Trans R Soc Biol Sci. 2008;363:685–701.
30. Haggag WM, Saber MSM. Evaluation of three compost as multi-strain carriers for biofertilizer and biocontrol agents against Fusarium wilt disease of some legume plants. Arab J Biotechnol. 2000;3:133–44.
31. Alam, K., 2015. Farmers' adaptation to water scarcity in drought-prone environments: A case study of Rajshahi District, Bangladesh. Agricultural Water Management, 148, pp.196-206.
32. Yamada KD, Tomii K, Katoh K. Application of the MAFFT sequence alignment program to large data-reexamination of the usefulness of chained guide trees. Bioinformatics. 2016;32:3246–51.
33. Asif, M. (2013). Climatic change, irrigation water crisis and food security in Pakistan, Master Thesis, Uppsala University.
34. Hadid, Diaa. (2019). Maybe It will destroy everything': Pakistan's Melting Glacier Causes Alarm. NPR, Nov. 21,
35. https://www.npr.org/2019/11/21/779412377/maybe-it-willdestroy-everything-pakistan-s-melting-glaciers-cause-alarm, accessed on Aug. 12, 2020.
36. Khalid, Haroon (2018). With Pakistan's rivers dying, are its ancient cities running out of time?, Feb. 12, Dawn.
37. Asim, M., Vains, A. H., Yousaf, H., & Ramzan, M. A. (2012). Socio Economic Impact of Water Crisis on Agrarian Community in District Faisalabad, Pakistan. Mediterranean Journal of Social Sciences, 3(11), 235-235
38. Baloch, Shah Meer. (2018). Water crisis: How is Pakistan running dry.? June, 07, DW, available at https://www.dw.com/en/water-crisis-why-is-pakistan-running-dry/a44110280#:~:text=%22Pakistan%20receives%20around%2014

5%20million,because%20 we%20have%20few%20dams, accessed on August 10, 2020

39. Meribole, Joshua (2020). The Water Crisis In Pakistan. Borgen Magazine, April, 29. https://www.borgenmagazine.com/water-crisis-in-pakistan/, accessed on Aug. 12, 2020.
40. Xu LH, Ravnskov S, Larsen J, Nilsson RH, Nicolaisen M. Soil fungal community structure along a soil health gradient in pea fields examined using deep amplicon sequencing. Soil Biol Biochem. 2012;46:26–32.
41. Thelen, K. D., Fronning, B. E., Kravchenko, S. A., Min, D.-H., & Robertson, G. P. (2010). Integrating livestock manure with a corn-soybean bioenergy cropping system improves short-term carbon sequestration rates and net global warming potential. Biomass and Bioenergy, 34(7), 960–966. doi:10.1016/j.biombioe.2010.02.004
42. Adams, D., De Sousa, C. & Tiesdell, S. Brownfield development: a comparison of North American and British approaches. Urban. Stud. 47, 75–104 (2010).
43. Libera, A. et al. Climate change impact on residual contaminants under sustainable remediation. J. Contam. Hydrol. 226, 103518 (2019).
44. W.R. Teague, S. Apfelbaum, R. Lal, U.P. Kreuter, J. Rowntree, C.A. Davies, R. Conser, M. Rasmussen, J. Hatfield, T. Wang, F. Wang and P. Byck (2016) "The role of ruminants in reducing agriculture's carbon footprint in North America" Journal of Soil and Water Conservation March 2016, 71 (2) 156-164; DOI: https://doi.org/10.2489/jswc.71.2.156
45. R L Mulvaney, S A Khan, and T R Ellsworth (2009). "Synthetic nitrogen fertilizers deplete soil nitrogen: a global dilemma for sustainable cereal production" J Environ Qual. 2009 Oct 29;38(6):2295-314. doi: 10.2134/jeq2008.0527.
46. Jerry L. Hatfield and Charles L. Walthall (2015). "Soil Biological Fertility: Foundation for the Next Revolution in Agriculture" Communications in Soil Science and Plant Analysis Volume 46, 2015 - Issue 6
47. A Mottet, F Teillard, P Boettcher, G De' Besi, B Besbes (2018) "Review: Domestic herbivores and food security: current contribution, trends and challenges for a sustainable development" Animal. 2018 Dec;12(s2):s188-s198. doi: 10.1017/S1751731118002215. Epub 2018 Sep 14
48. Rijk, Ingrid J. C.; Ekblad, Alf (April 2020). Farquhar, James; Bao, Huiming; Thiemens, Mark (2000-08-04). "Atmospheric Influence of Earth's Earliest Sulfur Cycle". Science. 289 (5480): 756–758.

Bibcode:2000Sci...289..756F. doi:10.1126/science.289.5480.756. ISSN 0036-8075. PMID 10926533.

49. Canfield, Donald (2014). Oxygen. Princeton University Press. ISBN 9781400849888.
50. Falkowski, Paul (2015). Life's Engines. Princeton University Press. ISBN 9781400865727.
51. Kibblewhite, M. G., Ritz, K. and Swift, M. J. (2008). Soil health in agricultural systems. Philosophical Transactions of the Royal Society B: Biological Sciences 363(1492):685–701.
52. Jelen, Benjamin I.; Giovannelli, Donato; Falkowski, Paul G. (2016). "The Role of Microbial Electron Transfer in the Coevolution of the Biosphere and Geosphere". Annual Review of Microbiology. 70 (1): 45–62. doi:10.1146/annurev-micro-102215-095521. PMID 27297124.
53. McCarthy, N., Lipper, L. and Branca, G. (2011). Climate-smart agriculture: Smallholder adoption and implications for climate change adaptation and mitigation. Mitigation of Climate Change in Agriculture Working Paper 3:1–37
54. Agroconection (2021). Soil Erosion – Causes and Effects. How to prevent soil erosion - 4 Ways | Facts About Soil. http://www.agroconection.com/soil/soil-erosion-causes-and-effects/
55. Sinha, R. (2021). A complete info guide on how to prevent soil erosion. KuKun Home Intelligence. https://blog.mykukun.com/how-to-prevent-soil-erosion
56. Byjus (2021). What is Soil Erosion? – Definition, Causes, Effects, Prevention. Byjus Learning Program. https://byjus.com/biology/soil-erosion/ 5. Geology (2021). Reading: Causes of Soil Erosion. Module 4: Weathering and Soil Formation. Lumen – Geology. https://courses.lumenlearning.com/geo/chapter/reading-causes-of-soilerosion/
57. Sulaeman, D. & Westhoff, T. (2020). The Causes and Effects of Soil Erosion, and How to Prevent It. World Resources Institute. https://www.wri.org/insights/causes-and-effects-soilerosion-and-how-prevent-it
58. Singh, M., Sarkar, B., Biswas, B., Bolan, N. and Churchman, G. J. (2017). Relationship between soil clay mineralogy and carbon protection capacity as influenced by temperature and moisture. Soil Biology and Biochemistry 109:95–106.
59. Rao, Subba. Soil Microbiology. Fourth ed. Enfield: Science Publishers, 1999.

60. Lal, R. (2011). Soil health and climate change: An overview. In Soil Health and Climate Change (Eds., Singh, B. P., Cowie, A. L. and Chan, K. Y.), Soil Biology Series (29):3–24. Springer-Verlag, Berlin
61. Islam, Waqar; Saqib, Hafiz Sohaib Ahmad; Adnan, Muhammad; Wang, Zhenyu; Tayyab, Muhammad; Huang, Zhiqun; Chen, Han Y.H. (2021-08-11). "Differential response of soil microbial and animal communities along the chronosequence of Cunninghamia lanceolata at different soil depth levels in subtropical forest ecosystem". Journal of Advanced Research. 38: 41–54. doi:10.1016/j.jare.2021.08.005. ISSN 2090-1232. PMC 9091736. PMID 35572399.
62. Vrieze, Jop de (2015). "The littlest farmhands". Science. 349 (6249): 680–683. Bibcode:2015Sci...349..680D. doi:10.1126/science.349.6249.680. ISSN 0036-8075. PMID 26273035 Wood, Martin. Soil Biology. New York: Chapman and Hall, 1989.
63. "Carbon and nitrogen cycling in a lead polluted grassland evaluated using stable isotopes (δ13C and δ15N) and microbial, plant and soil parameters". Plant and Soil. 449 (1–2): 249–266. doi:10.1007/s11104-020-04467-7. ISSN 0032-079X. S2CID 212689936.
64. Heavy metals in soils. B. J. Alloway (3rd ed.). Dordrecht: Springer. 2012. ISBN 978-94-007-4470-7. OCLC 801654870
65. Roesch LF, Fulthorpe RR, Riva A, Casella G, Hadwin AK, Kent AD, Daroub SH, Camargo FA, Farmerie WG, Triplett EW. 2007. Pyrosequencing enumerates and contrasts soil microbial diversity. ISME J 1:283–290.
66. Lebeis SL, Rott M, Dangl JL, Schulze-Lefert P. 2012. Culturing a plant microbiome community at the cross-Rhodes. New Phytol 196:341–344.
67. Tan, Z. X.; Lal, R.; Wiebe, K. D. (2005-06-14). "Global Soil Nutrient Depletion and Yield Reduction". Journal of Sustainable Agriculture. 26 (1): 123–146. doi:10.1300/J064v26n01_10. ISSN 1044-0046. S2CID 26750583.
68. Dalal, R. C., & Probert, M. E. (1997). Soil Nutrient Depletion. Retrieved February 10, 2021, from https://www.researchgate.net/profile/Ram_Dalal/publication/291868075_Soil_nutrient_depletion/links/5727d8b808aee491cb414dd4.pdf
69. Lal, R. (2009). "Soil degradation as a reason for inadequate human nutrition". Food Security. 1 (1): 45–57. doi:10.1007/s12571-009-0009-z. ISSN 1876-4517. S2CID 32798094.

70. Whitcomb, Sarah J.; Heyneke, Elmien; Aarabi, Fayezeh; Watanabe, Mutsumi; Hoefgen, Rainer (2014), Hawkesford, Malcolm J.; Kopriva, Stanislav; De Kok, Luit J. (eds.), "Mineral Nutrient Depletion Affects Plant Development and Crop Yield", Nutrient Use Efficiency in Plants, Cham: Springer International Publishing, vol. 10, pp. 205–228, doi:10.1007/978-3-319-10635-9_8, ISBN 978-3-319-10634-2, retrieved 2022-09-25
71. Birnie-Gauvin, Kim; Peiman, Kathryn S.; Raubenheimer, David; Cooke, Steven J. (2017). "Nutritional physiology and ecology of wildlife in a changing world". Conservation Physiology. Oxford University Press (OUP). 5 (1): cox030. doi:10.1093/conphys/cox030. ISSN 2051-1434. PMC 5516125. PMID 28740638.
72. Both refer to water bodies instead of soil, because an overabundance of nutrients will usually be washed out from the soil "Biodiversity". WHO. December 3, 2012. Retrieved February 19, 2020.
73. Sherrard, Mark E.; Elgersma, Kenneth J.; Koos, Jordan M. A.; Kokemuller, Catherine M.; Dietz, Hannah E.; Glidden, Alec J.; Carr, Christina M.; Cambardella, Cynthia A. (2019). "Species composition influences soil nutrient depletion and plant physiology in prairie agroenergy feedstocks". *Ecosphere. Wiley. 10 (7).* doi:10.1002/ecs2.2805. ISSN 2150-8925.
74. Andrews-Speed P, Bleischwitz R, Boersma T, Johnson C, Kemp G, Van Deveer SD (2012) The global resource nexus. The struggles for land, energy, food, water, and minerals. Transatlantic Academy Washington, DC May 2012, p 100. http://www.gmfus.org/wp-content/blogs.dir/1/ files_mf/ta2012report_apr12_web.pdf. Accessed on 6 Dec 2012
75. Gupta R, Kienzler K, Martius C, Mirzabaev A, Oweis T, de Pauw E, Qadir M, Shideed K, Sommer R, Thomas R, Sayre K, Carli C, Saparov A, Bekenov M, Sanginov S, Nepesov M, Ikramov R, (2009) Research Prospectus: a vision for sustainable land management research in Central Asia. ICARDA Central Asia and Caucasus Program. Sustainable Agriculture in Central Asia and the Caucasus Series No.1. CGIAR-PFU, Tashkent, Uzbekistan. 84 pp
76. Mueller L, Schindler U, Mirschel W, Shepherd TG, Ball BC, Helming K, Rogasik J, Eulenstein F, Wiggering H (2010) Assessing the productivity function of soils. A Rev Agron Sustain Dev 30(3):601–614

77. Pauw E de (2007) Principal biomes of Central Asia. pp 3–24 In Climate change and terrestrial carbon sequestration in Central Asia. Lal R, Suleimenov M, Stewart BA, Hansen DO, Doraiswamy P (eds) Taylor and Francis Group, New York, NY, USA
78. Saparov A, Mirzakeev EK, Sharypova TM, Saparov GA, Abuduwaili J (2013) Irrigation erosion of irrigated soils in the foothills of southern Kazakhstan. J Arid Land 5(2):166–171 doi: 10.1007/s40333-013-0157-z
79. Smalley IJ, Mavlyanova NG, Rakhmatullaev KL, Shermatov MS, Machalett B, O'Hara Dhand K, Jefferson IF (2006) The formation of loess deposits in the Tashkent region and parts of Central Asia; and problems with irrigation, hydrocollapse and soil erosion. Quatern Int 2006:152–153
80. Suleimenov M, Saparov A, Akshalov K, Kaskarbayev Z (2012) Land degradation issues in Kazakhstan and measures to address them: research and adoption. Pedologist 2012:373–381
81. Vazhenin IG (1987) Methodological guidelines for soil surface mapping by levels of pollution by industrial emissions (Menolbxecrbe peroveylawbb go rapnoupaabpaodayb. goxdeyyouo gorpoda go ypodyzv paupzpyeyyocnb gpovsikeyysvb ds,pocavb). Moscow, p 18
82. M. Deng, D. Qin, and H. Zhang, "Public perceptions of climate and cryosphere change in typical arid inland river areas of China: facts, impacts and selections of adaptation measures," Quaternary International, vol. 282, pp. 48–57, 2012.
A. J. McMichael, D. H. Campbell-Lendrum, C. F. Corvalán, and K. L. Ebi, "Andrew githeko, joel D. Scheraga, and alistair woodward," Climate Change and Human Health: Risks and Responses, World Health Organization, Geneva, Switzerland, pp. 1–340, 2003.
83. R. E. Woodruff, T. McMichael, C. Butler, and S Hales, "Action on climate change: the health risks of procrastinating," Australian & New Zealand Journal of Public Health, vol. 30, no. 6, pp. 567–571, 2006.
84. M. S. Bacha, M. Muhammad, Z. Kılıç, M. Nafees, and M. Nafees, "The dynamics of public perceptions and climate change in Swat valley, khyber Pakhtunkhwa, Pakistan," Sustainability, vol. 13, no. 8, pp. 1–22, 2021.
85. A. Guterres, "Secretary-General's video message on flash appeal in support of Pakistan flood response plan for Pakistan," 2022, https://www.un.org/sg/en/content/sg/statement/2022-08-30/secretary-generals-video-message-flash-appeal-support-of-pakistan-flood-response-plan-for-pakistan.

86. Su-L. Tan, “Pakistan is bearing the brunt of the climate crisis despite ‘small carbon footprint, ’ minister says, “ CNBC,” vol. 6, 2022.
87. W. Ishaque, S. Ahmed Mahesar, and I. Hussain Sahito, “Influence of climate change on Pakistan’s national security,” Grassroots, vol. 49, no. 2, pp. 66–72, 2015.

CHAPTER THREE

SOLUTION TO OPPOSING COMPONENTS OF SOIL CRISES AND FOOD SECURITY

IMRAN[1,2,3]

[1]DEPARTMENT OF AGRONOMY, THE UNIVERSITY OF AGRICULTURE, PESHAWAR, PAKISTAN,
[2]RHIZOSPHERE LABORATORY, UNIVERSITY OF CUKUROVA, FACULTY OF AGRICULTURE, DEPARTMENT OF SOIL SCIENCE AND PLANT NUTRITION, ADANA, TURKEY,
[3]MINISTRY OF AGRICULTURE, ENTENSION WING, KHYBER PAKHTUNKHWA, PAKISTAN

Abstract

As a result of any global change, some of the larger and wider changes in the soil are progressive, especially the gradual decrease in soil fertility and physical properties subsequent to increased environmental temperature. Low levels crops of crop production, low water-use efficiency, and generally similar or somewhat high rainfall can lead to soil degradation and nutrient deprivation. Larger but less extensive changes in the soil, including greater biological activity and intermittent reduction in soils, are expected. Other changes due to climate change (temperature and rainfall) are expected to contribute relatively well to mineral composition, organic matter content or structural stability of many soils. In some fragile soils, the nature of the dominant soil formation may be inferior, with increasing, decreasing or more severe seasonal rainfall. In most cases, changes in soil through direct human action, on-site or off-site, far outweigh the effects of climate change. Soil management measures designed to improve the soil's productivity, and may lead the soil capacity to be sufficient to prevent any reduction of agricultural land through climate change. The land of natural areas, or other land that has a low intensity of management such as semi-natural forests used for digging wood and other products, is less easily protected from the effects of climate change. Such soils are also less vulnerable to climate change than they are to human activities, such as pollution from acid accumulation, or excessive nutrient depletion on site, such as very little input agriculture. To disarm the world's soil against any negative impact of climate change, or external nutrient deficiencies or over-pollution, or drought or high intensity rains, the best that can happen to them is: manage their soil by maintaining maximum physical flexibility by maximizing the off-ground cover through a stable, proportional soaking system. Using an integrated plant nutrient management system to balance nutrient input and uptake during crops or over the years, while maintaining soil nutrient levels substantially and sometimes to minimize losses. Occasionally high demands should be high enough to eliminate them.

Keywords: soil fertility, climate change, soil degradation, erosion, loss of nutrients

1 Introduction

Soils are the support of primary production in terrestrial ecosystems (Imran et al., 2021a) and also provide ecosystem service. At the ecological and human time scales, soil resources are non-renewable. Agricultural production of food and fiber for the human population is strongly dependent on soil resources (Akhtar et al. 2015a; Imran 2018; Khan et al. 2015). Therefore, soil conservation is a critical factor to ensure increasing food demands. Soils can be a carbon (C) source and sink (Lal, 2019). Therefore, soils contribute to the regulation of the carbon cycle and its consequent effects on climate change (Amonette and Joseph 2009; Armor et al. 2005; Arshad et al. 2008; Asad et al. 2015; Ayaz, et al. 2000; Babar et al. 2016). Changes in land use constitute the driving force that determines the soil's role as a source or as a sink of C. Tilling of land has led to a loss of organic carbon (OC) and an immediate increase in atmospheric carbon emissions, whereas the reforestation of croplands and covering surface with plant species causes an increase in carbon sequestration (Babar et al. 2016; Baber et al. 2015; Badridze, et al., 2009; Bates et al., 1973(Ortas and Yucel, 2020)). However, sequestration of C under forestry or other non-agricultural land uses often only slowly replaces OC lost through cropping, and such differences in time scale between loss or change of soil properties due to man and their reversal are commonly several orders of magnitude different (Imran et al., 2021b). Furthermore, the properties of soils are sensitive to climate change (Bohnert and Jensen. 1996; Bos, et al., 2005; Bruinsma,

1963; Brussaard et al., 2007; Capell, and Doerffling. 1993; Chimenti, et al., 2006(Ortaş et al., 2016)). In mineral soils, the approximate relationship between organic matter and soil organic C is 1.724 x % OC = % organic matter (OM) which is important to estimate soil biological fertility. Temperature increase would cause a higher decomposition rate of the OM. Degradation of OM significantly caused soil fertility degradation and finally increases atmospheric carbon dioxide (CO2) emission. Increased drought would have the opposite effect (Conde, 2011; Cramer, et al., 2011; Imran 2018; Imran and Khan, 2017; Imran and Khan, 2015; Duan, et al., 2007; Farahbakhsh, 2012).

The possible increase in plant productivity due to the fertilizing effect of an increase in atmospheric CO_2 would lead to an increase of OC inputs to the soil, especially with intensive cropping (assuming no hydrological or nutritional limitations). In natural ecosystems where productivity is limited manly by N, P and other mineral nutrients (Ortas and Bykova, 2020)shortage, increased atmospheric N deposition would increase OC inputs to the soil (Imran. 2021). On the contrary, decreased productivity due to intensification of water stress leads to losses of soil OC. The foreseeable increase in the occurrence of forest fires would cause losses of OC and would increase the risk of erosion (Flowers, 2004; Gasco, et al., 2012; Goncalves et al., 2001; Gonzalez et al., 2018). All over the world, in recent years, significant decreases have been achieved in the OC amounts of the soils with the large forests caused by the increasing drought due to climate changes. Increased soil erosion causes the loss of upper soil horizons richer in OC. In areas in which forest fires are a recurring phenomenon, like in the Mediterranean basin, the production of highly stable forms of OC during combustion of biomass (charcoal) can contribute to the stabilization of C in the medium term. All these processes do not exclude each other and some of the feedback may be positive, causing a multiplying effect (Heans, 1984; Imran 2018; Imran and Khan, 2015). OC is intimately linked with the natural fertility and productivity of soils) as a source of macronutrients, especially C, N and P) concerning the substrate of the soil's microbial activity) humified carbon contributes greatly to the capacity to retain nutrients and pollutants (capacity for cationic and anionic exchange capacity). Humic substances of lower molecular weight (fulvic acids) improve the solubility of some essential micronutrients, and toxic metals (Imran et al., 2016; Imran et al., 2015; Imran et al., 2017; Imran et al., 2018; ISTA. 2005). It is a critical factor in the development and maintenance of soil structure and the stability of soil aggregates and, consequently, of the physical properties that depend on these factors: water infiltration capacity, water holding capacity for plants, aeration, compaction, erodibility (Izhar et al., 2019; Imran and Khan, 2015; Imran et al., 2015; Imran et al., 2014(Ortaş et al., 2017a)). Crop and soil management practices have a significant impact on carbon sequestration and aggregate formation (Ortas et al., 2019).

Another process that would probably be affected by climate change is soil salinization. Projected increasing in evapotranspiration and drought would raise the water table, saline intrusion, and accumulation of salts in the rooting soil depth in arid and semiarid lands (Imran et al., 2021c). These processes can be reduced with the application of appropriate farming techniques, tilling, irrigation control, and management of organic amendments, along with the reforestation of barren lands. In short, those measures promoting soil fertility would set off a multiplying effect in the ecosystems' capacity to fix atmospheric carbon in the long-term (Imran and Khan, 2015; Imran et al., 2015; Akhtar et al., 2015b; Khan et al., 2016; Zhan et al., 2011). Of course, land use will almost certainly change as a consequence of climate change, opening opportunities to novel crops and varieties adapted to the new conditions, including the corresponding changes in cropping systems (Jackson, 1962; Jagatheeswari and Ranganathan. 2012; Joseph and Jini, 2011; Kamara, et al., 2014). In this context, the sustainability of nature and agriculture depends on choosing plant species that can adapt to climate changes.

The main potential changes in soil-forming factors (forcing variables) directly resulting from the global change would be in organic matter supply from biomass, soil temperature regime, and soil hydrology, the latter because of shifts in rainfall zones as well as changes in potential evapotranspiration. Soil changes because of a potential rise in sea level resulting from a net reduction in Antarctic ice cap volume and ocean warming are discussed in Brammer and Brinkman (1990) and are summarized at the end of this paper (Kammann, et al., 2011; Khan, et al., 2010; Imran et al., 2019; Imran, Amanullah. 2018; Imran et al., 201; Imran et al., 2015; Lashari et al., 2013; Lehmann and Joseph. 2009).

The biggest single change in soils expected as a result of these postulated forcing changes would be a gradual improvement in fertility and physical conditions of soils in humid and sub-humid climates (Imran and Amanullah· 2021; Imran et al., 2021c). Another major change would be the poleward retreat of the permafrost boundary, discussed by Goryachkin and Targulian (1990). Other widespread changes would be in degree rather than in kind of Certain tropical soils with low physico-chemical activity, such as in the Amazon region, may undergo a radical change from one major soil-forming process to another (Sombroek, 1990), as discussed below under processes in soils.

The temperature changes but particularly in rainfall to be expected as a result of global warming are subject to major uncertainties for several reasons. Different global circulation models do not lead to mutually consistent results (an example for Europe is given by Santer, 1985), and they are not yet adequately verified. Also, the interaction with changes in location and intensity of major ocean currents and resultant possible modifications in sea surface temperatures is still most uncertain, as well as the interaction with possible major changes in cloudiness and land cover by cower crop or reforestation and the resulting changes in albedo and actual evapotranspiration (Lehmann, et al., 2011; Lei, et al., 2011; Lu. And Vonshak, 2002; Maestre, et al., 2007; Martinez and Manzur 2005; Imran et al., 2017; Imran et al., 2015; Imran, 2017; Imran, 2018).

Indirect effects of climate change on soils may be greater than direct effects of higher temperatures or greater rainfall variability and larger or smaller rainfall totals on soils. For example, indirect effects of climate change on soils through CO2-induced increases in growth rates or water-use efficiencies, sea-level rise, climate-induced decrease or increase in vegetative cover, or a change in human influence on soils due to changes in options for the farmer (Mazhar, et al., 2016; MINFAL 2007; Miri and Mirjalili, 2013; Mittler, 2006).

2 Possible changes in imposing variables

A gradual, continuing rise in atmospheric CO_2 concentration entailing increased photosynthetic rates and water-use efficiencies of vegetation and crops, hence increases in organic matter supplies to soils (Fig. 2 & Fig. 3). Minor increases in soil temperatures in the tropics and subtropics; moderate increases and extended periods in which soils are warm enough for microbial activity (warmer than about 5°C) in temperate and cold climates, parallel to the changes in air temperatures and vegetation zones as summarized by Emanuel *et al.* (1985).

Minor increases in evapotranspiration in the tropics to major increases in high latitudes caused both by temperature increase and by extension of the growing period (Palm, and Volkenburgh, 2012; Palta et al., 2006; Pant et al., 2015). Increases in amount and variability of rainfall in the tropics; possible decrease in rainfall in a band in the subtropics poleward of the present deserts; minor increases in amount and variability in temperate and cold regions. Peak rainfall intensities could increase in several regions (Islam et al., 2015; Munns et al., 2006; Munns, 2002; Munns, and Tester, 2008; Nazarli et al., 2011; Neto et al., 2004). Recently mainly in semi-arid climate conditions, there is an increase in the number and amount of sudden precipitation per unit area. The intensive falling precipitation which caused the destruction of plants, the deterioration of the soil and erosion, and decreased in soil fertility. After rainfall revers are charring tons of organic matter, soil and other organic material to sees. A gradual sea-level rise causing deeper and longer inundation in river and estuary basins and on levee back slope, and brackish-water inundation leading to encroachment of vegetation that accumulates pyrite in soils near the coast (Paz et al., 2014; Peng, et al., 2008; Ping et al., 2007; Prado et al., 2000; Raghothama, 1999).

Figure 1. Picture, after heavy rain tons of organic material, timbers, soils are carried to see (August, 2021, Black see part of Turkey. Anonymous, 2021).

3 Processes and mineralogical change in soils

The most rapid processes of physical and manly chemical or mineralogical change under changing external conditions would be loss of salts and nutrient cations where leaching increases, and salinization where net upward water movement occurs because of increased evapotranspiration or decreased rainfall or irrigation water supply (Paz et al., 2014; Peng, et al., 2008; Ping et al., 2007; Prado et al., 2000; Raghothama, 1999). The clay mineral composition and the mineralogy of the coarser fractions would generally change little, even over centuries. Exceptions would be the transformation of X-ray amorphous material into the clay mineral halloysite when a volcanic soil previously under perennially moist conditions becomes subject to periodic drying, or the gradual dehydration of goethite to hematite in soils subject to higher temperatures or severe drying, or both. Changes in the surface properties of the clay fraction, while generally slower than salt movement, can take place much faster than changes in bulk composition or crystal structure (Imran et al., 2021d). Such surface changes have a dominant influence on soil physical and chemical properties (Brinkman, 1985, 1990). Clay minerals and fraction (especially high mineral surface area values of 300 and 700 m^2 per g) are very important to keep organic compound to increase soil organic material. Minerals significantly influence the accumulation and stabilization of organic carbon (OC). Also clay minerals significantly influence the accumulation and stabilization of soil organic carbon (Singh et al., 2016). Soil organic carbon/material is also very important to mitigate atmospheric CO2 concentration. Results of Churchman et al. (2020) shown that addition of clay to soils led to increased OC content. Soil OC is the major sources of soil fertility and productivity.

Changes in the clay mineral surfaces or the bulk composition of the clay fraction of soils are brought about by a small number of transformation processes, listed below (Brinkman, 1982). Each of these processes can be accelerated or inhibited by changes in external conditions due to global change (Raza 2012; Rengasamy, 2002; Roelofs et al., 2008).

- Hydrolysis by water containing carbon dioxide, which removes silica and basic cations;
- Chelation, which dissolves and removes especially aluminum and iron by chelating organic acids;
- Ferrolysis, a cyclic process of clay transformation and dissolution mediated by alternating iron reduction and oxidation, which decreases the cation exchange capacity by aluminum interlaying in swelling clay minerals;
- Dissolution of clay minerals by strong mineral acids, producing acid aluminum salts and amorphous silica;
- Reverse weathering, i.e., clay formation and transformation under neutral to strongly alkaline conditions, which may create, e.g., montmorillonite, palygorskite, or analcime.

Hydrolysis and chelation may be accelerated by increased leaching rates. Ferrolysis may occur where soils are subject to reduction and leaching in alternation with oxidation: in a warmer world, this may happen over larger areas than at present, especially in high latitudes and in monsoon climates. Dissolution by strong acids would occur, e.g., where sulphidic materials in coastal plains are oxidized with an improvement of drainage; however, a rise in sea level would reduce the likelihood of this occurring naturally. Reverse weathering could begin in areas drying out during global warming and would continue in most presently arid areas (Rolf et al., 2004; Rubio et al., 2001; Savvides et al., 2015; Shahida et al., 2016; Shamsi and Kobraee. 2013; Shamsuddin et al., 2015; Shazma et al., 2016).

These processes would influence the surface properties of the clay fraction only over centuries, even with the changes envisaged as a consequence of global warming. By contrast, direct human action can vastly accelerate some of these processes as is evident, for example, from the severe effects of acid rain on sandy soils in parts of Europe (Van Breemen, 1990) or the extremely rapid ferrolysis in soils seasonally inundated by water level fluctuations in the Volta lake in Ghana (Amatekpor, 1989).

Not only can the speed of soil formation be accelerated by human action, but also, albeit much more locally, it's very nature or direction. In most places, the natural soil-forming processes are not fundamentally changed, but there are certain threshold situations, generally with fragile soils, where even a small change in external conditions may cause a major, and adverse, change from one dominant soil-forming process to another. The four examples summarized below (from Sombroek, 1990) illustrate a change from hydrolysis to chelation (Ferralsols to Podzols); irreversible hardening of the subsoil; clay illuviation forming a dense subsoil in originally homogeneous, porous Ferralsols; and salinization (Sidari, et al., 2008; Karim and Imran. 2019; Tank and Saraf. 2010; Tiwari et al., 2011; Tombesi, et al., 2018; Ulfat et al., 2016).

The yellowish sandy Ferralsols and Ferralic Arenosols of Eastern Amazonia, Kalimantan, and the Zaire basin may rapidly change into Podzols or Albic Arenosols (giant Podzols) with even small increases in total rainfall or stronger seasonality, or increased input of acidic organic matter. An increase in effective rainfall due to climate change may cause a major increase in the extent of Podzols formed from present-day yellowish sandy Ferralsols where, presently, Podzols occur in patches within the Ferralsols area (Lucas *et al.,* 1987; Dubroeucq and Volkoff, 1988). The imperfectly drained loamy Plinthosols on the flat interfluves of Western

Amazonia would change into shallow, droughty soils with an irreversibly hardened subsoil if subject to drying out with climate change. Recently very often it has been seen that increasing temperature due to climatic changes and consequently decreasing soil moisture which causes the soil structure degradation.

Figure 2. Tengible view of field and fertility status as influenced by environemtal factors (precipitation, temperature and Light)

The deep reddish, porous loamy to clayey Ferralsols of the transition zones between forest and savanna in Eastern Africa, stable under the present vegetation, maybe leached so far that a denser subsoil with washed-in clay is formed below unstable topsoil with little organic matter, as already observed where the land was cleared several decades ago; the same may happen over more extensive areas under sparser vegetation brought about by a somewhat drier climate (Sidari, et al., 2008; Karim and Imran. 2019; Tank and Saraf. 2010; Tiwari et al., 2011; Tombesi, et al., 2018; Ulfat et al., 2016).

The silty Fluvisols in the broad river valleys of the Sudano-Sahelian zone of West Africa, such as the interior delta of the Niger river, may become saline or sodic upon even minimal change in precipitation and flooding regimes - as exemplified by current human actions with the same soil-hydrological implications (Sombroek and Zonneveld, 1971).

4 Effects of higher CO_2 on soil fertility, physical conditions, and productivity

Higher atmospheric CO_2 concentration increases growth rates and water-use efficiency of crops (Fig. 3) and natural vegetation in so far as other factors do not become limiting (Van et al., 2001; Wang et al., 2007; Wang and Xu. 2013; Wang and Jin. 2007; Wolfe 2007). The higher temperature optima of some plants under increased CO_2 would tend to counteract adverse effects of temperature rise, such as increased night time respiration. The shortened growth cycle of a given species because of higher CO_2 and temperature would be compensated for in natural vegetation by adjustments in species composition or dominance (Imran et al., 2020b). In agroecosystems, the choice of longer-duration cultivars or changes in cropping patterns could eliminate unproductive periods that might arise because of the shorter growth cycle of the main crop (Wang and Xu. 2013; Wang and Jin. 2007; Wolfe 2007).

There will be adequate time to adjust to the changes since these are expected to occur over decades, rather than years or days as in all present experimental situations. The increased productivity is generally accompanied by more litter or crop residues (Imran et al., 2020a; Ilyas et al., 2020), a greater total root mass and root exudation, increased mycorrhizal colonization, and activity of other rhizosphere (Ortaş et al., 2019) or soil micro-organisms, including symbiotic and root-zone N, fixers (Fig. 3). The latter would have a positive effect on N supply to crops or vegetation (Van et al., 2001; Wang et al., 2007; Wang and Xu. 2013; Wang and Jin. 2007; Wolfe 2007). The increased microbial and root activity in the soil would entail higher CO_2 partial pressure in soil air and CO_2 activity in soil water, hence increased rates of plant nutrient release (e.g., K, Mg, micronutrients) from weathering of soil minerals. Similarly, the mycorrhizal activity would lead to better phosphate, zinc, copper and nitrogen uptake (Ortas, 2020). These effects would be in synergy with better nutrient uptake by the more intensive root system due to higher atmospheric CO_2 concentration. There is no *a priori* reason why the degree of synchrony between nutrient release and demand by crops or natural vegetation would be subject to major changes under high CO_2 conditions (Xu et al., 2008; Yeo et al., 1990; Younis et al., 2015; Zakaria et al., 2012; Zargar et al., 2017).

The greater microbial activity tends to increase the number of plant nutrients cycling through soil organisms. The increased production of root material (at similar temperatures) tends to raise soil organic matter content, which also entails the temporary immobilization and cycling of greater quantities of plant nutrients in the soil. Higher C/N ratios in the litter, reported by some workers under high CO_2 conditions, would entail slower decomposition and slower remobilization of the plant nutrients from the litter and uptake by the root mat and would provide more time for incorporation into the soil by earthworms, termites, etc. Higher soil temperatures would counteract increases in 'stable' soil organic matter content but would further stimulate microbial activity (Xu et al., 2008; Yeo et al., 1990; Younis et al., 2015; Zakaria et al., 2012; Zargar et al., 2017).

In all experimental situations, whether chamber-type or free-air enrichment, CO_2 increases are rapid or sudden, often to double ambient concentration, sometimes higher. The consequently rapid increases in soil organic matter dynamics and soil micro-organisms may cause temporary competition for plant nutrients. These temporary effects have on occasion been reported as negative factors affecting plant response to elevated CO_2. However, increased organic matter dynamics and microbial activity in soils are positive for the soil-plant system when CO_2 concentrations rise gradually over decades, as currently and in the recent past (Tank and Saraf. 2010; Tiwari et al., 2011; Tombesi, et al., 2018; Ulfat et al., 2016; Van et al., 2001; Wang et al., 2007; Wang and Xu. 2013;(Wang et al., 2020)). Future experiments could be set up to compensate for the temporary effects caused by the suddenness of the CO_2 increase, for example by artificially higher soil organic matter contents estimated to be near equilibrium with each stepwise higher CO_2 concentration, in a range between 350 and 600 ppm.

Increased microbial activity due to higher CO_2 concentration and temperature produces greater amounts of polysaccharides and other soil stabilizers (Figure. 2). Increases in the litter or crop residues, root mass, and organic matter content tend to stimulate the activity of soil macro-fauna, including earthworms, with consequently improved infiltration rate and bypass flow by the greater number of stable bio pores (Tank and Saraf. 2010; Tiwari et al., 2011; Tombesi, et al., 2018; Ulfat et al., 2016; Van et al., 2001; Wang et al., 2007; Wang and Xu. 2013). The greater stability and the faster infiltration increase the resilience of the soil against water erosion and consequent loss of soil fertility. The increased proportion of bypass flow also decreases the nutrient loss by leaching during periods with excess rainfall. This refers to the available nutrients in the soil, including well-incorporated fertilizers or manure, but not to fertilizers broadcast on the soil surface. These are subject to lose by runoff or leaching.

- Gradually rising CO_2 as in this century and *transient* global change scenarios.
- Soils with some weatherable minerals at least in the subsoil or substratum within rooting depth.
- Extreme weather events may disrupt some relationships in the figure, so any major increase in their frequency or intensity may counteract the positive effects shown.
- Species composition adjusts, or choices indicated are made to adjust, to the newly attainable biomass or crop production under increased atmospheric CO_2, compensating for shortened growth cycles of existing species or crops. The figure does not include the positive effects of higher temperatures on the length of growing periods in temperate or boreal climates.

These changes increase the resilience of the soil against physical degradation and nutrient loss by increased intensity, seasonality, or variability of rainfall, as well as against some of the unfavorable changes in rate or direction of soil-forming processes discussed in the next sections (Ping et al., 2007; Prado et al., 2000; Raghothama, 1999; Rajalakshmi et al., 2015; Rashid et al., 2003).

If the partial pressure of CO_2 in the soil air would rise, and that of O_2 decrease to levels impairing root function, part of the benefits indicated would not materialize. The improved gas exchange with the atmosphere through increased numbers of stable bio pores would tend to keep CO_2 and O_2 in the soil at 'safe' levels, at least in naturally or artificially well-drained soils. Wetland crops such as rice or jute have their gas exchange mechanisms and would not be affected; neither would natural wetland vegetation (Munns, 2002; Munns, and Tester, 2008; Nazarli et al., 2011; Neto et al., 2004; Palm, and Volkenburgh, 2012; Palta et al., 2006).

The positive effect on weathering rate and plant nutrient availability would occur in soils with significant amounts of weatherable minerals, not in very deeply and strongly weathered or otherwise very poor soils.

5 Effects of rainfall and temperature changes in different climates

In the humid tropics and monsoon climates, increased intensities of rainfall events and increased rainfall totals would increase leaching rates in well-drained soils with high infiltration rates and would cause temporary flooding or water-saturation, hence reduced organic matter decomposition, in many soils in level or depressional sites (Munns, 2002; Munns, and Tester, 2008; Nazarli et al., 2011; Neto et al., 2004; Palm, and Volkenburgh, 2012; Palta et al., 2006). This may affect a significant proportion of especially the better soils in Sub-Saharan Africa, for example. They would also give rise to greater amounts and frequency of runoff on soils in sloping terrain, with sedimentation downslope and, worse, downstream. Locally, there would be increased chances of the mass movement in the form of landslides or mudflows in certain soft sedimentary materials (Such as in picture... in Turkey as well), discussed below. Soils most resilient against such changes would have adequate cation exchange capacity and anion sorption to minimize nutrient loss during leaching flow and have high structural stability and a strongly heterogeneous system of continuous macropores to maximize infiltration and rapid bypass flow through the soil during high-intensity rainfall (Imran et al., 2019; Imran and Amanullah. 2018; Imran et al., 201; Imran et al., 2015; Lashari et al., 2013).

In subtropical and other subhumid or semi-arid areas, the increased productivity and water-use efficiency due to higher CO_2 would tend to increase ground cover, counteracting the effects of higher temperatures. In recent years, in may part of the water deficiency areas withdrawals of groundwater are higher than the rainwater recharge to aquifer due to continuous expansion of irrigated agriculture in the areas. In Central Anatolia, Konya-Turkey, in regions where the annual precipitation is 210 mm as a result of too much ground water withdrawal potholes are formed, and as a result, plant production cannot be done. For future sustainable water and food security. If there would be locally much less rainfall and increasing intra- and inter-annual variability, these could lead to less dry-matter production and hence, in due course, lower soil organic matter contents (Imran et al., 2019; Imran and Amanullah. 2018; Imran et al., 201; Imran et al., 2015; Lashari et al., 2013). Periodic leaching during high-intensity rainfall with less standing vegetation could desalinize some soils in well-drained sites, cause increased runoff in others, and lead to soil salinization in depressional sites or where the groundwater table is high. Soils most resilient against the effects of such increasing aridity and rainfall variability would have high structural stability and a strongly heterogeneous system of continuous macropores (the same as in the tropics); hence a rapid infiltration rate, as well as a large available water capacity and a deep groundwater table (Imran 2018; Imran and Khan, 2017; Imran and Khan, 2015; Duan, et al., 2007; Farahbakhsh, 2012).

Higher temperatures, particularly in arid conditions, entail a higher evaporative demand. Where there is sufficient soil moisture, for example in irrigated areas, this could lead to soil salinization if land or farm water management, or irrigation scheduling or drainage are inadequate. On the other hand, recent experiments by the Salinity Laboratory, Riverside, California, point to increased salt tolerance of crops under high atmospheric CO_2 conditions (E.V. Maas, pers. comm.; Bowman and Strain, 1987). Since under abiotic soil conditions atmospheric CO_2 concentration is increase which may have negative effects on soil quality and plant production. Application of biological processes such as mycorrhizal fungi and other microbial strains as inoculants provide a cost-effective long-term solution for coping with saline and drought conditions (Ortas et al., 2021).

In temperate climates, minor increases in rainfall totals would be expected to be largely taken up by increased evapotranspiration of vegetation or crops at the expected higher temperatures, so that net hydrologic or chemical effects on the soils might be small. The negative effect on soil organic matter contents of a temperature rise might be more than compensated by the greater organic matter supply from vegetation or crops growing more vigorously because of the higher photosynthesis, the greater potential evapotranspiration, and the higher water-use efficiency in a high-CO_2 atmosphere (Imran 2018; Imran and Khan, 2017; Imran and Khan, 2015; Duan, et al., 2007; Farahbakhsh, 2012). The temperate zone would thus be likely to have the smallest changes in soils, even in poorly buffered ones, directly caused by the effects of global change. A minor and probably slow, but very visible, change could be a reddening of presently brown soils where increased periods with high summer temperatures would coincide with dry conditions so that the iron oxide haematite would be stable over the presently dominant goethite. This mineralogical change might decrease the intensity and amount of phosphate fixation. An overview of such changes, with emphasis on temperate climate zones, is given by Buol *et al.* (1990).

In boreal climates, the gradual disappearance of large extents of permafrost and the reduction of frost periods in extensive belts adjoining former permafrost is expected to improve the internal drainage of soils in vast areas, with probable increases in leaching rates. The appreciable increase in the period when the soil temperature is high enough for microbial activity would lead to lower organic matter contents, probably not fully compensated by increased primary production through somewhat higher net photosynthesis and the longer growing period (Imran 2018; Imran and Khan, 2017; Imran and Khan, 2015; Duan, et al., 2007; Farahbakhsh, 2012). Depletion of the OM content is one of the major causes underlying the declining agronomic productivity in soils of the semi-arid region (Khresat et al., 2008; Ryan et al., 2008). Especially heavy tillage and breakdown of soil aggregates exacerbate mineralization of SOM and emission of CO_2 (Lal, 2010; Larionova et al., 2010). Ander long term organic fertilizer conditions application of organic fertilizers increased soil aggregation and enhanced C sequestration, and the rate and total amount of soil organic carbon (Ortas et al., 2013).

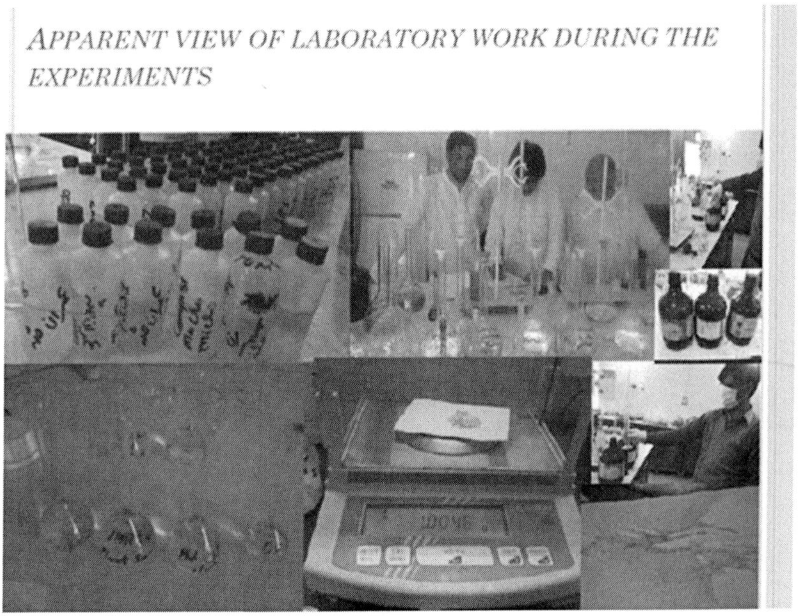

Figure 3. Tengible view of soil and water testing laboratory, analysing soil fertility and nutrients status as influenced by environemtal factors (precipitation, temperature and Light)

Table 1. Initial soil inorganic, organic carbon concentration (in 0-15 cm) and effects of organic and inorganic fertilizers application on total soil organic carbon pool (Ortas et al., 2013).

Treatments	**Total soil carbon (%)**		**Inorganic carbon (%)**		**SOC (%)**		**SOC pool (Mg ha^{-1})**	
Control	4.61	±0.01b	3.80	±0.01a	0.81	±0.00b	17.69	±0.27b
Min. Ferti.	4.83	±0.10ab	3.86	±0.07a	0.97	±0.05b	20.99	±1.83ab
Animal Manure	5.04	±0.07a	3.74	±0.07a	**1.30**	**±0.11a**	**23.08**	**±2.49a**
Compost	5.02	±0.38a	3.76	±0.20a	1.26	±0.20a	21.40	±3.87ab
Com + Myc.	4.86	±0.05ab	3.75	±0.05a	1.11	±0.10a	21.42	±1.01ab

Mean of three replicates ± SD. Means in the same column followed by the same letter represent significant differences ($p \leq 0.05$) among treatments at the same depth.

Paradoxically, the extent of soils subject to periodic reduction could well increase in level areas, despite the greater leaching, because of increased periods when the soils are water-saturated but also sufficiently warm for microbial activity. Soils most resilient against such effects, including the leaching of nutrients and periodic soil reduction, would have similar characteristics as the most resilient ones in other climates: adequate cation exchange capacity and anion sorption to minimize nutrient loss during leaching flow, high structural stability, and a strongly heterogeneous system of continuous macropores to maximize rapid bypass flow during periods with excess meltwater (Armor et al., 2005; Arshad et al., 2008; Asad et al., 2015; Ayaz, et al., 2000; Babar et al. 2016; Babar et al. 2016; Baber et al. 2015; Badridze, et al., 2009).

6 Some properties of clay surfaces

Soils with naturally high structural stability, for example, Ferralsols and Nitisols, occupy sizeable areas in the tropics. The former is widespread, among others, in northern South America, the latter in East Africa (Armor et al., 2005; Arshad et al., 2008; Asad et al., 2015; Ayaz, et al., 2000; Babar et al. 2016; Babar et al. 2016; Baber et al. 2015; Badridze, et al., 2009). The clay fraction in such soils generally has oxidic surfaces: mainly iron (III) and aluminium oxides or hydroxides, while the bulk of the clay fraction may have different compositions. The oxidic surfaces could form from parent materials with moderate or high iron contents under long-continued hydrolysis by water (containing carbon dioxide).

At the other extreme soil with a very low structural stability, or with a severe hazard of failure under load or shock (so-called quick clays). The surfaces of the clay minerals in these soils are generally covered by amorphous, gel-like material with a high silica content (McKyes et al., 1974). Such material may have originated in earlier periods when the soils were strongly saline and, presumably, subject to processes of reverse weathering (Fig. 1). Examples are the quick clays of the Champlain Sea sediments in Ontario and Quebec and parts of Scandinavia. Such soils are most likely to generate high proportions of runoff and suspended sediment, but also most liable to mudflows once sloping sites are water-saturated to appreciable depth. Some Andosols are thixotropic and have similarly low stability because of their similar composition, derived from volcanic materials (tuff).

Most soils fall somewhere between these extremes. Vertisols, for example, have moderate or low structural stability, and clay surfaces that are mainly silica, but with generally small amounts of the amorphous coating. In Planosols, if formed by ferrolysis, the clay fraction in the upper, eluvial horizons has been partly decomposed with a residue of amorphous silica, but the remaining smectite or illite has been interlayered with aluminum hydroxide polymers, which has decreased the swell-shrink potential and the cation exchange capacity of the clay fraction. Concurrently, parts of the free iron oxides have been reduced and leached out. The net effect of these changes generally is a decrease in structural stability.

7 Resilience against physical and chemical soil degradation

As discussed, most soils do not have a high intrinsic resilience against physical soil degradation by, for example, high-intensity rainfall. In natural conditions in humid climates, it is the complete soil cover near ground level combined with the perforating activity of the soil fauna that makes the soil-vegetation system resilient against physical degradation (Armor et al., 2005; Arshad et al., 2008; Asad et al., 2015; Ayaz, et al., 2000; Babar et al. 2016; Babar et al. 2016; Baber et al. 2015; Badridze, et al., 2009).

In the Rhine river plain in the Netherlands, for example, most of the originally calcareous alluvial soils have been decalcified within a millennium or so. Only in small areas on the highest levels of that age that have continually remained under forest, soils are still calcareous and even have lime pseudomycelia (filaments) indicative of less humid soil conditions, and abundant vertical macropores produced by earthworms. In these soils, faunal activity is high because of the adequate litter supply: the resulting macropores remain open, protected against rain impact by litter and undergrowth: and excess water from heavy rain passes to the substratum through the macropores without leaching lime from most of the soil mass (Akhtar et al., 2015a; Imran 2018; Imran and Khan, 2015; Imran et al., 2015; Akhtar et al., 2015b; Khan et al., 2016).

Research to increase the resilience of soils against any adverse effects of climate change could be done in conjunction with that on soil resilience against direct adverse human impacts (Fig 4). Until site-specific management procedures have been elaborated, soil and crop (including trees and pasture) management should aim to maintain soil cover and organic matter supply to soil biota, while minimizing mechanical disturbance by heavy traffic, cultivation, or excessive grazing intensity. Such kind of management may also help to conserve plant nutrients (in soils not flooded for wetland cultivation) since the stable, heterogeneous system of bio pores produced by the soil fauna would favor bypass flow of any excess moisture and thus decrease leaching through the soil mass (Akhtar et al., 2015a; Imran 2018; Imran and Khan, 2015; Imran et al., 2015; Akhtar et al., 2015b; Khan et al., 2016). A single management recipe would not be generally applicable in different conditions.

Soil and crop management may also affect soil properties, such as enhancing the SOC pool, the development of the soil structure as well. Ortas et al. (2019) reported that manage SOC and soil microorganisms such as mycorrhizal fungi which are powerful tool to fix more CO_2 from atmosphere to soil through plant leaves and roots to make the soil more qualitative. Plant roots and mycorrhizal hyphae can physically protect the soil carbon from microbial decomposition through aggregation (Golchin et al., 1994; Jastrow et al., 1998). Root exudates organic compounds such as polysaccharides, carbohydrates, lignin and lipids have significant effects on aggregation. Aggregation have a significant effect on soil fertility and keeping carbon fixation.

Minimizing damage by certain termite species harming crop performance may necessitate a period without residues on the soil, for example; or crop residues may be needed for feed or fuel. Management methods for wetland need to be developed that make optimum use of any increased potential productivity while minimizing secondary effects such as increased CH_4 or N_2O emission from the reduced soil. Such factors, and others, should be taken into account in designing an optimum management strategy for any specific natural and cultural environment.

8 Resilience against soil reduction (anoxic conditions)

Soil reduction, which would limit land suitability for dryland crops, or strong reduction, which would be liable to produce toxins even for wetland crops, may take place once the soil is water-saturated long enough for microbial action to exhaust the oxygen remaining in the soil when water-saturation started. Another necessary condition is the presence of sufficient readily decomposable organic matter as an energy source for microbial activity (Akhtar et al., 2015a; Imran 2018; Imran and Khan, 2015; Imran et al., 2015). In most soils, during reduction the redox status is stabilized at an Eh about 100-200 mV near neutrality by the Fe^{2+} - $Fe(OH)_3$ equilibrium, except where the content of readily decomposable organic matter is very high or the content of free iron (III) oxides very low. In such cases, negative

Eh values may occur, and toxic hydrogen sulphide or low-molecular organic compounds - including methane - may be formed.

Figure 4. Soil fertlity with increased Co_2 concentration and organic amendments influenced crop productivity.

Resilience against soil reduction in practice depends on the drainage conditions since most soils have sufficient organic matter for reduction to start within about a week after water-saturation. Soils most resilient against a reduction in conditions of increased rainfall variability and incidence of high-intensity rainfall have similar properties as those resilient against the negative effects of other perturbations: high infiltration rate, high structural stability and a permanent heterogeneous system of tubular macropores, good external drainage (Baber et al. 2015; Badridze, et al., 2009; Bates et al., 1973; Bohnert and Jensen. 1996; Bos, et al., 2005).

9 Soil reaction (pH)

Most soils would not be subject to rapid pH changes resulting from climate change (Fig. 1 & Fig. 2). Exceptions might be found in potential acid sulphate soils, extensive in some coastal plains and estuaries if they become subject to increasingly long dry seasons. Even though most of such soils are clays with moderate or high cation exchange capacity, the amounts of the acid liberated in such soils upon oxidation generally exceed this rapid buffering capacity (Baber et al. 2015; Badridze, et al., 2009; Bates et al., 1973; Bohnert and Jensen. 1996; Bos, et al., 2005). Therefore, pH values may temporarily reach 2.5 to 3.5 and a small part of the clay fraction may be decomposed as indicated under *Processes in soils,* above. This then buffers the pH generally between 3.5 and 4 in the long run. Depending on the efficiency with which the excess acid formed can be leached out, the period of extreme acidity and aluminum toxicity may last between less than a year and several decades.

In calcareous soils, soil reaction may range between about 8.5 and 7 depending on the partial pressure of CO_2 in the soil; this range is maintained against leaching of basic cations by the different soil processes as long as a few percent of finely distributed lime remain. Buffering in non-calcareous soils is less strong but depends on the cation exchange capacity at soil pH. In soils with variable-charge surfaces of the clay fraction, this decreases with acidification (Arshad et al., 2008; Asad et al., 2015; Ayaz, et al., 2000; Babar et al. 2016).

It should be noted that the simple modelling of accelerated $CaCO_3$ leaching under a doubled atmospheric CO_2 concentration generally does not hold. In most soils, the ongoing decomposition of organic matter maintains CO_2 concentrations in the soil air far above atmospheric concentration even now, and $CaCO_3$ solubility is determined by the partial pressure of CO_2 in soil air and its activity in soil water, rather than in the atmosphere. Similarly, in rhizosphere high CO2 is produced by soil organism's activity which is change the rhizosphere pH as well. Leaching of lime is thus positively related to the rate of organic matter decomposition, negatively to gas diffusion rate, and positively to the amount of water percolating through the soil (Arshad et al., 2008; Asad et al., 2015; Ayaz, et al., 2000; Babar et al. 2016).

In conditions where leaching is accelerated by climate change, it would be possible to find relatively rapid soil acidification after a long period with little apparent change, as has been the case - but after a shorter latent period - in some soils in Europe that have been subject to acid rain for several decades. The soil might be steadily depleted of basic cations, but a pH change may start or may become more rapid, once certain buffering pools are nearly exhausted. Such non-linear and time-delayed effects have been discussed in the context of soil and water pollution by Stigliani (1988); they are also expected to occur in various ways at different times after increased temperatures and changed rainfall patterns will have been operative.

10 Effects of a rising sea level on soils in coastal areas

The probable effects on soil characteristics of a gradual eustatic rise in sea-level will vary from place to place depending on several local and external factors, and interactions between them (Brammer and Brinkman, 1990). In principle, a rising sea level would tend to erode and move back existing coastlines. However, the extent to which this happens will depend on the elevation, the resistance of local coastal materials, the degree to which they are defended by sediments provided by river flow or longshore drift, the strength of longshore currents and storm waves, and on human interventions which might prevent or accelerate erosion (Kammann, et al., 2011; Khan, et al., 2010; Imran et al., 2019; Imran, Amanullah. 2018; Imran et al., 201; Imran et al., 2015; Lashari et al., 2013; Lehmann and Joseph. 2009; Lehmann, et al., 2011; Lei, et al., 2011).

In major deltas, such as those of the Ganges-Brahmaputra and the major Chinese rivers, sediment supplies delivered to the estuary will generally be sufficient to offset the effects of a rising sea level. Such deltaic aggradation could decrease, however, under three circumstances:

- where human interventions inland, such as large dams or successful soil conservation programs, drastically reduce sediment supply to the delta: e.g., the construction of the Aswan high dam in 1964 has led to coastal erosion and increased flooding of lagoon margins in the Nile delta (Stanley, 1988). Similarly, about ten dams built on the Tigris and Euphrates rivers in the Southeast of Turkey play a serious role in preventing erosion as well as water reservoirs;
- where construction of embankments within the delta interrupts sediment supply to adjoining back swamps, exposing them to submergence by a rise in sea level: e.g., embankments along the lower Mississippi River have cut off sediment supplies to adjoining wetlands which formerly offset land subsidence occurring due to compaction of underlying sediments (Day and Templet, 1989);
- where land subsidence occurs due to the abstraction of water, natural gas, or oil: e.g., as is presently happening in Bangkok and the northern part of the Netherlands.

In coastal lowlands that are insufficiently defended by sediment supply or embankments, tidal flooding by saline water will tend to penetrate further inland than at present, extending the area of perennially or seasonally saline soils. Where *Rhizophora* mangrove or *Phragmites* vegetation invades the area that would over several decades lead to the formation of potential acid sulphate soils. The impedance of drainage from the land by a higher sea level and by the correspondingly higher levels of adjoining estuarine rivers and their levees will also extend the area of perennially or seasonally reduced soils and increase normal inundation depths and durations in the river and estuary basins and on levee backslope (Imran et al., 2015; Imran, 2017; Imran, 2018; Mazhar, et al., 2016; MINFAL 2007; Miri and Mirjalili, 2013; Mittler, 2006; Kammann, et al., 2011; Khan, et al., 2010; Imran et al., 2019; Imran, Amanullah. 2018; Imran et al., 201; Imran et al., 2015; Lashari et al., 2013; Lehmann and Joseph. 2009; Lehmann, et al., 2011; Lei, et al., 2011). In sites that become perennially wet, soil organic matter contents will tend to increase, resulting eventually in peat formation. On the other hand, where coastal erosion removes an existing barrier of mineral soils or mangrove forest, higher storm surges associated with a rising sea level could allow sea-water to destroy existing coastal eustatic peat swamps, which might eventually be replaced by freshwater or saltwater lagoons (Kammann, et al., 2011; Khan, et al., 2010; Imran et al., 2019; Imran, Amanullah. 2018; Imran et al., 201; Imran et al., 2015; Lashari et al., 2013; Lehmann and Joseph. 2009; Lehmann, et al., 2011; Lei, et al., 2011).

The probable response of low-lying coastal areas to a rise in sea level can be estimated in more detail based on the geological and historical evidence of changes that occurred during past periods when sea level was rising eustatically or in response to tectonic or isostatic movements: e.g., around the southern North Sea (Jelgersma, 1988); in the Nile delta (Stanley, 1988); on the coastal plain of the Guyanas (Brinkman and Pons, 1968); in the Musi delta of Sumatera (Brinkman, 1987). Contemporary evidence is available in areas where land levels have subsided as a result of recent abstraction of water, natural gas, or oil from sediments underlying coastal lowlands. Further studies of such contemporary and palaeoenvironments are needed together with location-specific studies to better understand the change processes, identify appropriate responses and assess their technical, ecological, and socio-economic implications (e.g., Warrick and Farmer, 1990).

11 Conclusions

Some major and widespread soil changes expected as a result of any global change are positive, especially the gradual increases in soil fertility and physical qualities consequent on increased atmospheric CO_2. The increased productivity and water-use efficiency of crops and vegetation, and the generally similar or somewhat higher rainfall indicated by several global circulation models, not fully counteracted by higher evapotranspiration, would be expected to lead to widespread increases in ground cover, and consequently better protection against runoff and erosion. Major but less widespread soil changes, including greater biological activity and increased extent of periodic reduction in soils, would be expected where permafrost would disappear. In unprotected low-lying coastal areas, gradual encroachment of *Rhizophora* mangroves or *Phragmites* following more extensive brackish-water inundation may give rise to the formation of potential acid sulphate soil layers after several decades. Deeper and longer-duration flooding of basins and levee back slopes in the adjacent river and estuary plains could lead to more extensive reducing conditions and increased organic matter contents, and locally to peat formation. Other changes due to climate change (temperature and precipitation) are expected to be relatively well buffered by the mineral composition especially high surface clay minerals, the organic matter content, or the structural stability of many soils. However, decreases in cover by vegetation or annual or perennial crops, caused by any locally major declines in rainfall not compensated by CO_2 effects, could lead to soil structure degradation and decreased porosity, as well as increased runoff and erosion on sloping sites and by the concomitant more extensive and rapid sedimentation. Changes in options available to land users because of climate change may have similar effects. In certain fragile soils, the nature of the dominant soil-forming process may change for the worse with increased, decreased, or more strongly seasonal rainfall. In most cases, changes in soils by direct human action, on-site or off-site (whether intentional or unintended), are far greater than the direct climate-induced effects. Soil and crop management measures designed to optimize the soil's sustained productive capacity would therefore be generally adequate to counteract any degradation of agricultural land by climate change. Soils of nature areas, or another land with a low intensity of management such as semi-natural forests used for extraction of wood and other products, are less readily protected against the effects of climate change but such soils, too, are threatened less by climate change than by human actions - off-site, such as pollution by acid deposition, or on-site, such as excessive nutrient extraction under very low-input agriculture. Under several climate conditions, low-input ecological soil and crop management using organic matter, biochar, cover crops and microorganism's uses are strongly increase soil organic carbon and mitigates atmospheric CO_2 concentration.

To armor the world's soils against any negative effect of climate change, or other extremes in external circumstances such as nutrient depletion or excess (pollution), or drought or high-intensity rains, the best that land users could do, would be:

- to manage their soils to give them maximum physical resilience through a stable, heterogeneous pore system by maintaining a closed ground cover as much as possible;
- to use an integrated plant nutrient management system to balance the input and offtake of nutrients over a cropping cycle or the years, while maintaining soil nutrient levels low enough to minimize losses and high enough to buffer occasional high demands.

Human action and management have been emphasized in these conclusions because most of the world's land is used and, to different degrees, managed rather than under natural conditions.

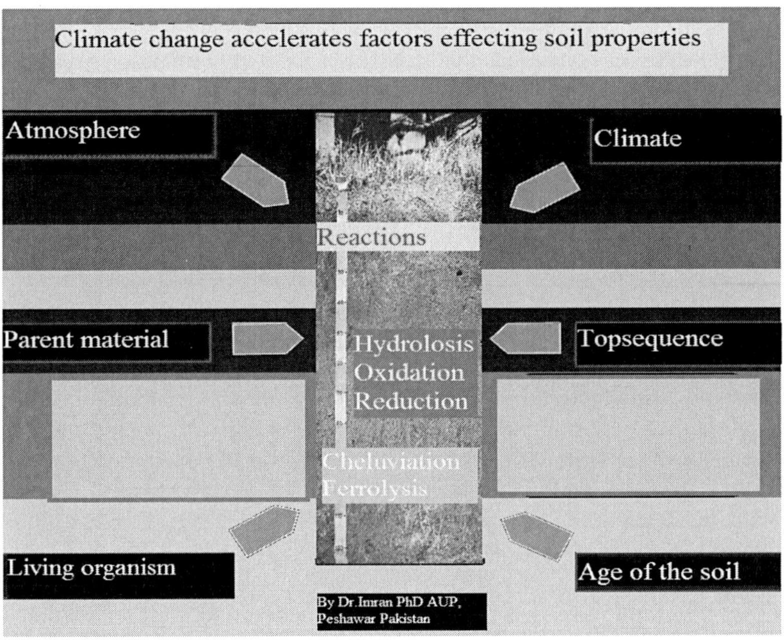

Figure 5. Soil properties response to various climatic factors

12 Summary

As a result of any global change, some of the larger and wider changes in the soils are progressive, especially the gradual decrease in soil fertility and physical properties after increased environmental temperature. Low productivity of crops and water use efficiency generally similar or somewhat high rainfall, can lead to soil degradation and nutrient deprivation. Expect larger but less extensive changes in the soil, including greater biological activity and intermittent reduction in soils. Other changes due to climate change (temperature and rainfall) are expected to contribute relatively well to mineral nutrient composition, organic matter content, or structural stability of many soils. In some fragile soils, the nature of the dominant soil formation may be inferior, with increasing, decreasing, or more severe seasonal rainfall. In most cases, changes in the soil through direct human action, on-site or off-site, far outweigh the effects of climate change. Soil and crop management measures are designed to improve the soil's productivity and may lead the soil capacity to be sufficient to prevent any reduction of agricultural land through climate change. The land of natural areas, or other lands that have a low intensity of management such as semi-natural forests used for digging wood and other plant products, is less easily protected from the effects of climate change. Such soils are also less vulnerable to climate change than they are to human activities, such as pollution from acid accumulation, or excessive nutrient depletion on site, such as very little input agriculture. To disarm the world's soil against any negative impact of climate change, or external nutrient deficiencies or over-pollution, or drought or high-intensity rains. As a results soils degradation and soil structure is broken-down and soil organic matter is mineralized and more carbon is flux to atmosphere which is caused climate change. The best that can happen to them is: manage their soil by maintaining maximum physical flexibility by maximizing the off-ground cover through a stable, proportional soaking system. Using an integrated plant nutrient management system to balance nutrient input and offtake during crops or over the years, while maintaining soil nutrient levels substantially and sometimes to minimize losses. Occasionally high demands should be high enough to eliminate them.

References

1. A. Khan, M. N. Khan, I. uallah, S. Shah, I. R. Arshad, I. M. A. Zeb, and Imran. 2015. Effect of potash application on growth, yield and

yield components of spring maize hybrids. Pure and Applied Biology Vol. 4, Issue 2, 195-203.

2. Abdel Rahman M. Al-Tawaha, Nusrat Jahan…Imran, et al 2020. Growth, Yield and Biochemical Responses in Barley To DAP And Chitosan Application Under Water Stress. *Journal of Ecological Engineering*. 21(6). 86-93. Doi. 10.12911/22998993/123251.
3. Akhtar, S. S., M. N. Andersen, and F. Liu. 2015a. Biochar mitigates salinity stress in potato. *Journal of Agronomy and Crop Sciences* 201:321–400.
4. Akhtar, S. S., M. N. Andersen, and F. Liu. 2015b. Residual effects of biochar on improving growth, physiology and yield of wheat under salt stress. *Agricultural Water Management* 158:61–8.
5. Imran, Amanullah and Abdel Rahman M. Al Tawaha. 2021. Management of Nano-black carbon, Phosphorous and Bio Fertilizer improve Soil organic carbon and ensilage biomass of soybean and maize. *Communication in Soil Science and Plant Analysis.* DOI.10.1080/00103624.2021.1966439
6. Imran, Amanullah and Abdel Rehman Altawaha. 2021. Carbon Assimilation and Dry Matter Partitioning in Soybean Ameliorates with the Integration of Nano-Black Carbon, Along with Beneficial Microbes and Phosphorus Fertilization. *Journal of Plant Nutrition*
7. Imran. 2021. The bioavailability of Phosphorus in composite vs. Hybrid maize differ with phosphorus and boron fertilization. *Phosphorus, Sulfur, and Silicon and the Related Elements. https://doi.org/10.1080/10426507.2021.1920588.*
8. Imran, Amanullah, Asad Ali khan, Tariq Mahmood, Abdel Rahman Al Tawaha and Samia Khanum[4]. 2021. Adequate Fertilization, Application Method and Sowing Techniques Improve Maize yield and related traits. *Communication in soil science and plant analysis*. DOI:10.1080/00103624.2021.1925688
9. Imran and Amanullah. 2021. Phosphorus and Boron Application Optimizing Biofortification Of phosphorus and Productivity of French Bean (*Phaseolus Vulgaris* L.). *Communication In Soil Science and Plant Analysis*.
10. Imran, Amanullah and Abdel Rahman Al-Tawaha. 2021. Carbon Sources Application Increase Wheat Yield and Soil Fertility. *Communication in soil science and plant analysis DOI: 10.1080/00103624.2020.1865397.*
11. Imran, Amanullah and Abdel Rehman Al-Tawaha.2020. The Productivity of subsequent wheat enhanced with residual carbon

sources and phosphorus under improved irrigation system. *Communication in soil science and plant analysis*, DOI:10.1080/00103624.2020.1763387

12. Imran, Amanullah, Muhammad Arif, Zahir Shah and Abdul Bari. 2020. Soil application of *Trichoderma* and Peach (*Prunus persica* L.) residues possesses biocontrol potential for weeds and enhances growth and profitability of soybean (Glycine max). *Sarhad Journal of Agriculture* 36 (1) 10-20. DOI | http://dx.doi.org/10.17582/journal.sja/2020/36.1.10.20
13. Mohammad Ilyas, Gohar Ayub, Imran*, Azmat Ali Awan and Manzoor Ahmad.2020. Calcium and Boron effect on Production and Quality of Autumn Potato under Chilling Temperature. *Communication in soil science and plant analysis*. DOI: 10.1080/00103624.2020.1854286
14. Akhtar, S. S., M. N. Andersen, M. Naveed, Z. A. Zahir, and F. Liu. 2015. Interactive effect of biochar and plant growth-promoting bacterial endophytes on ameliorating salinity stress in maize. *Functional Plant Biology* 42:770–81.
15. Amatekpor, J.K. 1989. The effect of seasonal flooding on the clay mineralogy of a soil series in the Volta lake drawdown area, Ghana. *Land Degradation and Rehabilitation* **1**: 89-100.
16. Amir Zaman Khan, Imran, Asim Muhammad, Aiman Khalil, Hasina Gul, Habib Akbar and Said Wahab.2016. Impact of fertilizer priming on seed germination behavior and vigor of maize. Pure and Applied Biology. Vol. 5, Issue 4, pp744-751. http://dx.doi.org/10.19045/bspab.2016.50093
17. Amonette, J. E., and S. Joseph. 2009. Characteristics of biochar: Microchemical properties. In *Biochar for environmental management: Science and technology*, eds. J. Lehmann and S. Joseph, 33–52. Earthscan, London: Sterling.
18. Armor, N. B., K. B. Hamed, A. Debez, and C. Grignonand Crabbedly. 2005. Physiological and antioxidant responses of perennial halophyte *Crithmum maritimum* to salinity. *Plant Science*4:889–99.
19. Arshad, M., B. Shaharoona, and T. Mahmood. 2008. Inoculation with *Pseudomonas* spp. containing ACC-deaminase partially eliminates the effects of drought stress on growth, yield, and ripening of pea (*Pisum sativum* L.). *Pedosphere* 18:611–20.
20. Asad Ali Khan, Imran* Fawad Ali Shah Inamullah, Laiq Zada Muhammad Naeem Muhammad Nouman Khan.2015. Phenological Traits of Rice as Influenced by Seedling Age and Number of Seedling

per Hill under Temperate Region. J. Bio, Agric and H. care ISSN 2224-3208 (Paper) ISSN 2225-093X (Online).

21. Ayaz, F. A., A. Kadioglu, and T. R. Turgut. 2000. Water stress effects on the content of low molecular weight carbohydrates and phenolic acids in *Cienanthe setosa*. *Candian Journal of Plant Sciences* 80:373–8.
22. Babar Iqbal, Bashir Ahmad, Inam Ullah, Imran, Asad Ali Khan, Shazma Anwar, Asad Ali, Khurrum Shahzadand Shehryar Khan. 2016. Effect of phosphorus, sulphur and different irrigation levels on phenological traits of Triticale. Pure and Applied Biology. Vol. 5, Issue 2, pp 303-310. http://dx.doi.org/10.19045/bspab.2016.50039
23. Babar Iqbal, Mohammad Tariq Jan, Zar Muhammad, Asad Ali Khan, Shazma Anwar, Imran and Khurrum Shahzad. 2016. Phenological traits of Maize influenced by integrated management of compost and fertilizer Nitrogen. J. Pure & Applied biology. 5(1): 58-63.
24. Babar Iqbal[1], Muhammad Tariq Jan[1], Inamullah[1], Imran*[1], Asad Ali Khan*[1], Zar Muhammad[2], Shazma Anwar[1], Shehryar Khan[1], Shamsuddin Baqa[1], Amjad Usman[3] .2015. Integrated Management of Compost Type and Fertilizer-N In Maize. J. Pure & Applied biology Vol. 4, Issue 4.
25. Baber Iqbal, M.T. Jan, Imran, Zar M, A.A. Khan, S. Anwar and K. Shahzad 2015. Growth and phenology of maize as affected by integrated management of compost and fertilizers Nitrogen. J. Pure & Applied biology Vol. 4, Issue 4.
26. Badridze, G., A. Weidner, F. Asch, and A. Borner. 2009. Variation in salt tolerance within Georgian wheat germplasm collection. *Genetic Resources and Crop Evolution* 56:1125–30.
27. Bates, L. S., R. P. Waldren, and I. D. Teare. 1973. Rapid determination of free proline for water stress studies. *Plant and Soil* 39:205–7.
28. Bohnert, H. J., and R. G. Jensen. 1996. Strategies for engineering water stress tolerance in plants. *Trends in Biotechnology* 14:89–97.
29. Bos, C., B. Juillet, H. Fouillet, L. Turlan, S. Dare, C. Luengo, R. Nto unda, R. Benamouzig, N. Gausseres, V. Tome, and C. Gaudichon. 2005. Postprandial metabolic utilization of wheat protein in humans. *The American Journal of Clinical Nutrition* 81:87–94.
30. Bowman, W.D. and Strain, B.R. 1987. Interaction between CO_2 enrichment and salinity stress in the C4 non-halophyte *Andropogon glomeratus* (Walter) BSP. *Plant Cell Environ.* **10**: 267-270.
31. Brammer, H. and Brinkman, R. 1990. Changes in soil resources in response to a gradually rising sea-level. Chapter 12. In: Scharpenseel *et al.* (eds.). 1990. pp. 145-156.

32. Brinkman, R. 1982. Clay transformations: aspects of equilibrium and kinetics. In: *Soil Chemistry. B. Physicochemical Models. Developments in Soil Science 5B.* G.H. Bolt (ed.). 2nd ed. Elsevier, Amsterdam. pp. 433-458.
33. Brinkman, R. 1985. Mineralogy and surface properties of the clay fraction affecting soil behavior and management. In: *Soil Physics and Rice.* T. Woodhead (ed.). International Rice Research Institute, Los Baños, Philippines. pp. 161-182.
34. Brinkman, R. 1987. Sediments and soils in the Karang Agung area. In: *Some Aspects of Tidal Swamp Development with Special Reference to the Karang Agung Area, South Sumatra Province.* R. Best, R. Brinkman and J.J. van Roon. Mimeo. World Bank, Jakarta. pp. 12-22.
35. Brinkman, R. 1990. Resilience against climate change? Soil minerals, transformations and surface properties, Eh, pH. In: Scharpenseel *et al.* (eds.). 1990. pp. 51-60.
36. Brinkman, R. and Pons, L.J. 1968. A pedo-geomorphological classification and map of the Holocene sediments in the coastal plain of the three Guyanas. *Soil SurveyPaper No. 4.* Soil Survey Institute (Staring Centre), Wageningen. 40 p, separate map.
37. Bruinsma, J. 1963. The quantitative analysis of chlorophyll a and b in plant extract. *Photochemistry and Photobiology* 2:241–9.
38. *Brussaard, Lijbert; de Ruiter, Peter C.; Brown, George G. (2007). "Soil biodiversity for agricultural sustainability". Agriculture, Ecosystems and Environment. 121 (3): 233–244.* doi:10.1016/j.agee.2006.12.013.
39. Buol, S.W., Sanchez, P.A., Kimble, J.M. and Weed, S.B. 1990. Predicted impact of climatic warming on soil properties and use. *American Soc. Agron. Special Publ.* **53**: 71-82.
40. Capell, B., and K. Doerffling. 1993. Genotype specific difference in chilling tolerance of maize in relation to chilling induce changes in water status and abscisic acid accumulation. *Physiologia Plantarum* 88:638–46.
41. Chimenti, C. A., M. Marcantonio, and J. A. Hall. 2006. Divergent selection for osmotic adjustment results in improved drought tolerance in maize (*Zea mays* L.) in both early growth and flowering phases. *Field Crops Research* 95:305–15.
42. *Conde, Artur (2011).* "Membrane Transport, Sensing and Signaling in Plant Adaptation to Environmental Stress" (PDF). *Plant & Cell Physiology. 52 (9): 1583–1602.*
43. doi:10.1093/pcp/pcr107. PMID 21828102 – *via Google Scholar.*

44. *Cramer, Grant R; Urano, Kaoru; Delrot, Serge; Pezzotti, Mario; Shinozaki, Kazuo (2011-11-17).* "Effects of abiotic stress on plants: a systems biology perspective". *BMC Plant Biology. 11:163.* doi:10.1186/1471-2229-11-163. ISSN 1471-2229. PMC 3252258. PMID 22094046.
45. Day, J.W. and Templet, P.H. 1989. Consequences of sea level rise: implications from the Mississippi delta. *Coastal Management* 17: 241-257.
46. Duan, D. Y., W. Q. Li, X. J. Liu, H. P. Ouyang, and P. An. 2007. Seed germination and seedling growth of *Suaeda salsa* under salt stress. *Annales Botanici Fennici* 44:161–9.
47. Dubois, M., K. Gilles, J. K. Hammiltron, P. A. Robers, and F. Smith. 1951. A colorimetric method for the determination of sugars. *Nature* 168:167–8.
48. Dubroeucq, D. and Volkoff, B. 1988. Evolution des couvertures pédologiques sableuses à podzols géants d'Amazonie (Bassin du haut Rio Negro). *Cahiers ORSTOM, Série Pédologie* **24** (3): 191-214.
49. Emanuel, W.R., Shugart, H.H. and Stevenson, M.P. 1985. Climatic change and the broad-scale distribution of terrestrial ecosystem complexes. *Climatic Change* **7**: 29-43.
50. Farahbakhsh, H. 2012. Germination and seedling growth in un-primed and primed seeds of fennel as affected by reduced water potential induced by NaCl. *International Research Journal of Applied and Basic Sciences* 3:737–44.
51. Flowers, T. J. 2004.Improving crop salt tolerance. *Journal of Experimental Botany* 55:307–19.
52. Gasco, G., Paz-Ferreiro, and A. Mendez. 2012. Thermal analysis of soil amended with sewage sludge and biochar from sewage sludge pyrolysis. *Journal of Thermal Analysis and Calorimetry*108:769–75.
53. Giannopolitis, C. N., and S. K. Ries. 1977. Superoxide dismutase occurance in higher plants. *Plant Physiology* 59:304–14.
54. *Goncalves-Alvim, Silmary J.; Fernandez, G. Wilson (2001). "Biodiversity of galling insects: historical, community and habitat effects in four neotropical savannas". Biodiversity and Conservation. 10: 79–98.* doi:10.1023/a:1016602213305
55. *González-Villagra, Jorge; Rodrigues-Salvador, Acácio; Nunes-Nesi, Adriano; Cohen, Jerry D.; Reyes-Díaz, Marjorie M. (2018). "Age-related mechanism and its relationship with secondary metabolism and abscisic acid in Aristotelia chilensis plants subjected to drought stress". Plant Physiology and Biochemistry. 124: 136–145.*

56. doi:10.1016/j.plaphy.2018.01.010. ISSN 0981-9428. PMID 29360623.
57. Goryachkin, S.V. and Targulian, V.O. 1990. Climate-induced changes of the boreal and subpolar soils. In: Scharpenseel *et al.* (eds.). pp. 191-209.
58. Heans, D. L. 1984. Determination of total organic carbon in soils by an improved chromic acid digestion and spectrophotometric procedure. *Communications in Soil Science and Plant Analysis* 15:1191–213.
59. Imran 2018. Climate Change Is Threat toward Agronomy (Base of Food, Fiber System), and Food Security. Food Nutr J: FDNJ-160. DOI: 10.29011/2575- 7091. 100060
60. Imran 2018. Phosphorous fertilization influenced weeds attributes and phenological characteristics of mungbean cultivars (vigna radiata l.) Russian Agricultural Sciences, 2018, Vol. 44, No. 3, pp. 229–234.
61. Imran 2018. Physiological and Morphological Traits of Agronomic Crops Influenced By Climate Change. Mod Concep Dev Agrono.1(4). MCDA.000524.
62. Imran and Asad Ali Khan . 2015. Grain Yield and Phenology of Maize Cultivars Influenced by Various Phosphorus Sources Food Sci and Quality Management. ISSN 2224-6088 (Paper) ISSN 2225-0557
63. Imran and ASAD ALI KHAN. 2015. Biochar Application and Shoot Cutting Duration (Days) Influenced Growth, Yield and Yield Contributing Parameters of Brassica Napus L. J. Bio, Agric and H. care ISSN 2224-3208 (Paper) ISSN 2225-093X (Online).
64. Imran and Asad Ali Khan. 2015. Influence of Compost Application and Seed Rates on Production Potential of Late Sown Maize on High Elevation in Swat -Pakistan. J. Environ and E. Science ISSN 2224-3216 (Paper) ISSN 2225-0948.
65. Imran and ASAD ALI KHAN. 2015. Phenological Charateristics of Brassica Napus L. as Influenced by Biochar Application and Shoot Cutting Duration (Days). Civi and Environ Res. ISSN 2224-5790 (Paper) ISSN 2225-0514 (Online
66. Imran and Khan, A. A. (2017). Canola Yield And Quality Enhanced With Sulphur Fertilization Russ. Agricult. Sci. 43: 113 10.3103/S1068367417020100
67. Imran, Asad Ali Khan, Fayaz Ahmad and Irfanullah. 2015. Nitrogen Levels, Tillage Practices and Irrigation Timing Influenced Yeild, Yeild Components and Oil Contents Of Canola. CivI and Environ Res. ISSN 2224-5790 (Paper) ISSN 2225-0514 (Online

68. Imran, A. A. Khan, I. Ullah, H. zada, F. Ahmad, S.T. Shah, A. Usman and I. Ullah. 2015. Yield and yield attributes of rapeseed cultivars as influence by sulfur level under swat valley conditions. J. Pure & Applied biology Vol. 4, Issue 3.
69. Imran, A.A. Khan, I.U. Khan and S. Naveed. 2016. Weeds density and late sown maize productivity influenced by compost application and seed rates under temperate environment. Pak. J. Weed Sci. Res. 22(1): 169-181.
70. Imran, A.A. Khan, Inamullah and Luqman. 2015. Weeding stages and their effect on yield and yield components of rice in upper Swat, Pakistan. Pak. J. Weed Sci. Res. 21(4): 555-563.
71. Imran, Abdul Bari, Roshan Ali, Naeem Ahmad, Zulfiqar Ahmad, Muhammad Idrees Khattak, Akhtar Ali, Fayaz Ahmad, Inayatullah Khan and Shahida Naveed. 2017. Traditional Rice Farming Accelerate CH4 & N2O Emissions Functioning As A Stronger Contributors of Climate Change. Int J Environ Sci Nat Res. 2017; DOI: 10.19080/ARTOAJ.2017.09.555765
72. Imran, Amanullah and Abdel Rehman Al-Tawaha.2020. The Productivity of subsequent wheat enhanced with residual carbon sources and phosphorus under improved irrigation system. *Communication in soil science and plant analysis*, DOI:10.1080/00103624.2020.1763387
73. Imran, Amanullah, A. Bari and R. Ali. 2018. Peach sources, phosphorous and beneficial microbes enhance productivity of soybean. Soy. Res. 16(2): 39-48.
74. Imran, Amanullah, A.A Khan, Abdul bari…. 2019. Production statistics and modern technology of maize cultivation in Khyber Pakhtunkhwa Pakistan. Plant sci. Arc.2 (2) 1-12.
75. Imran, Amanullah, Asad Ali khan, Tariq Mahmood, Abdel Rahman Al Tawaha and Samia Khanum[4]. 2020. Adequate Fertilization, Application Method and Sowing Techniques Improve Maize yield and related traits. *Communication in soil science and plant analysis*.
76. Imran, Amanullah, Muhammad Arif, Zahir Shah and Abdul Bari. 2020. Integration of Peach (*Prunus persica* L.) residues, beneficial microbes and phosphorus enhance phenology, growth and yield of soybean. ISSN 1068-3674, *Russian Agricultural Sciences*, 2020, Vol. 46, No. 3, pp. 223–230. © Allerton Press, Inc., 2020.
77. Imran, Amanullah, Muhammad Arif, Zahir Shah and Abdul Bari. 2020. Soil application of *Trichoderma* and Peach (*Prunus persica* L.) residues possesses biocontrol potential for weeds and enhances growth

and profitability of soybean (Glycine max). *Sarhad Journal of Agriculture* 36 (1) 10-20. DOI | http://dx.doi.org/10.17582/journal.sja/2020/36.1.10.20

78. Imran, Amanullah. 2018. Review. Global impact of climate change on water, soil resources and threat towards food security: evidence from Pakistan. Adv Plants Agric Res. 2018;8(5):350–355. DOI: 10.15406/apar.2018.08.00349
79. Imran, ASAD ALI KHAN .2015. Effect of transplanting dates on yield and yield components of various rice genotypes in hilly area cold climatic region of khyber pakhtunlhwa-Pakistan. j. Bio, agric and h. Care issn 2224-3208 (paper) issn 2225-093x (online).
80. Imran, Asad Ali Khan Fayaz Ahmad .2015. Phenology of Various Rice Genotypes as Affected by Different Transplanting Dates under Cold Climatic Region of Khyber Pakhtunkhwa-Pakistan. J. Environ and E. Science ISSN 2224-3216 (Paper) ISSN 2225-0948.
81. Imran, Asad Ali Khan, Fayaz Ahmad and Irfanullah. 2015. Influence of Hydrated Calcium Sulphate (CaSO4.2H2O) and Nitrogen Levels on Water Infiltration Rate and Maize Varieties Productivity in Rainfed Area of Swat, Pakistan. Chem and Envrn Res.ISSN 2224- 3224 (Print) ISSN 2225- 0956 (Online)
82. Imran, Asad Ali Khan, Hayat Zada, Fayaz Ahmad and Irfanullah.2015. Graine Yield and Yield Components of Wheat Cultivar "Siran 2010" As Affected by Phosphorous Levels under Rain fed Condition. J. N. Sci. Res. ISSN 2224-3186 (Paper) ISSN 2225-0921 (Online)
83. Imran, Asad Ali Khan, Inamullah Inam & Fayaz Ahmad. (2016). Yield and yield attributes of Mungbean (*Vigna radiata* L.) cultivars as affected by phosphorous levels under different tillage systems, *Cogent Food & Agriculture* (2016), 2: 1151982.
84. Imran, Asad Ali Khan, Irfanullah and Fayaz Ahmad.2014. Production potential of rapeseed (*brassica napus* l.) As influenced by different nitrogen levels and decapitation stress under the rainfed agro-climatic condition of swat Pakistan, J. Glob. Innov. Agric. Soc. Sci. ISSN (Online): 2311-3839; ISSN (Print): 2312-5225
85. Imran, Asad Ali Khan, Kashif Akhtar, Sajjad Zaheer, Shah Faisal, Shahzad Ali. 2015. Rice Seedling Characteristics of Various Genotypes Influenced by Different Sowing Dates in Swat-Pakistan. J. Environ and E. Science ISSN 2224-3216 (Paper) ISSN 2225-0948.
86. Imran, Fazal Maula, Muhammad Uzair and Hayat Zada.2015. Farmers Income Enhancement through Off-Season Vegetables Production

under Natural Environment in Swat-Pakistan J. Environ and E. Science ISSN 2224-3216 (Paper) ISSN 2225-0948.

87. Imran, I. Hussain, I. khattak, A. Rehman[2], F. Ahmad and S.T. shah .2015. *Roots nodulation, yield and yield contributing parameters of mungbean cultivars as influenced by different phosphorous level in swat-Pakistan.* J. Pure & Applied biology Vol. 4, Issue 4.
88. Imran, I. Khattak, I. Hussain[1], A. Rehman, S. Anwar[1], F. Ahmad[3], A. A. Khan[1] and H. Zada. 2015. Growth and Yield of Maize Hybrids as effected by different Sowing Dates in Swat Pakistan. J. Pure & Applied biology Vol. 4, Issue 4.
89. Imran, Muhammad Uzair, Fazal Maula, Marco Vacirca, Salvatore Farfaglia and Muhammad Nauman Khan. 2015. Introduction and promotion of off-season vegetables production under natural environment in hilly area of upper swat-Pakistan. J. Bio, Agric and H. care.
90. Imran, Nasar Jamal, Ashfaq Alam and Asad Ali Khan. 2017. Grain Yield, Yield Attributes of Wheat and Soil Physio-Chemical Charicteristics Influenced by Biochar, Compost and Inorganic Fertilizer Application. Agri Res & Tech: Open Access J. DOI: 10.19080/ARTOAJ.2017.10.555795
91. Imran, Shahida Naveed, Asad Ali Khan and Inayat Khattak. 2015. Impact Of Phosphorus Levels And Seed Rates On Growth And Yield Of Late Sown Maize On High Elevation In Swat, Pakistan. Pakistan J. Agric. Res. Vol. 28 No. 4:406-413
92. Imran, Zada H, Naveed S, Khattak I And Ahmad S. 2016. Variable Rates Of Phosphorous Application Influenced Phenological Traits Of Green Gram (*Vignaradiata* L.) J Agri Res 2016, 1(3).
93. Imran. 2017. Climate Change Is a Real Fact confronting to Agricultural Productivity. Int J Environ Sci Nat Res. 2017; 3(3): 555613. DOI: 10.19080/IJESNR.2017.03.555613.
94. Imran. 2018. Ecological Environmental Variability Influence Growth and Yield Potential of Rice under Northern Climatic Scenario. Russian Agricultural Sciences, 2018, Vol. 44, No. 1, pp. 18–24.
95. Imran. 2018. Organic Matter Amendments Improve Soil Health, Productivity and Profitability of Maize and Soybean. Ann Rev Resear. 1(3): 555564.
96. Imran.2015. Effect of Germination on Proximate Composition of Two Maize Cultivars. J. Bio, Agric and H. care ISSN 2224-3208 (Paper) ISSN 2225-093X (Online).
97. Imran.2015. Influence of Nitrogen Levels and Decapitation Stress on Biological Potential of Rapeseed (Brassica Napus L) under Water

Difficit Condition of Swat-Pakistan. J. N. Sci. Res. ISSN 2224-3186 (Paper) ISSN 2225-0921 (Online)

98. Iqbal A, Amanullah, Ali A, Iqbal M, Ikramullah and Imran. 2017. Integrated use of phosphorus and organic matter improve fodder yield of moth bean (Vigna aconitifolia (Jacq.) under irrigated and dryland conditions of Pakistan. Journal of AgriSearch 4(1): 10-15
99. ISTA. 2005. *International rules for seed testing*. Zurich, Switzerland: International Seed Testing Association.
100. Izhar Ali, Asad Ali Khan, Imran, inamullah, Aman Khan .... 2019. Humic acids and Nitrogen levels optimizing productivity of green gram (vigna radiate L.). Russian Agriculture Sciences, 45(1), 43-47.
101. Jackson, M. 1962. *Soil chemical analysis*, 496. London: Constable and Company Limited.
102. Jagatheeswari, D., and P. Ranganathan. 2012. Influence of Mercuric Chloride on seeds germination, seedling growth and biochemical analysis of green gram (*Vigna radiata* (L.) Wilczek. Var. Vamban-3). *International Journal of Pharmaceutical and Biological Archives* 3:291–5.
103. Jelgersma, S. 1988. A future sea-level rise: its impacts on coastal lowlands. In: *Geology and Urban Development. Atlas of Urban Geology.* Vol. 1. UN-ESCAP. pp. 61-81.
104. Joseph B., and D. Jini. 2011. Development of salt stress-tolerant plants by gene manipulation of antioxidant enzymes. *Asian Journal of Agricultural Research* 5:17–27.
105. Kamara, A., A. Kamara, M. M. Mansaray, and P. A. Sawyerr. 2014. Effects of biochar derived from maize stover and rice straw on the germination of their seeds. *American Journal of Agriculture and Forestry* 2:246–9.
106. Kammann, C. I., S. Linsel, J. W. Gobling, and H. Koyro. 2011. Influence of biochar ondrought tolerance of *Chenopodium quinoa* willd and on soil-plant relation. *Plant and Soil* 345:195–210.
107. Khan, M. A., E. Islam M. U. Shirazi, S. Mumtaz, S. M. Mujtaba, M. A. Khan, A. Shereen, M. Y. Ashrafand, and G. M. Kaleri. 2010. Physiological responses of various wheat genotypes to salinity. *Pakistan Journal of Botany* 42:3497–505.
108. Lashari M.S., Y. Liu, W. Li, J. Pan, G. Pan, J. Zheng, G. Zheng, X. Z hang, and X. Yu. 2013. Effect of amendment of biochar manure compost in conjunction with pyroligneous solution on soil quality and wheat yield of a salt stressed cropeland from central China Great Plain. *Food Crop Research* 144:113–8.

109. Lehmann J., and S. Joseph. 2009. Biochar for environmental management: An introduction. In *Biochar for environmental management: Science and technology*, eds. J. Lehmann and S Joseph, 1–12. London: Earthscan.
110. Lehmann, J., M. C. Rillig, J. Thies, C. A. Masiello, W. C. Hockaday, and D. Crowley. 2011. Biochar effects on soil biota. *Soil Biology and Biochemistry* 43:1812–36.
111. *Lei, Gang; Shen, Ming; Li, Zhi-Gang; Zhang, Bo; Duan, Kai-Xuan; Wang, Ning; Cao, Yang-Rong; Zhang, Wan-Ke; Ma, Biao (2011). "EIN2 regulates salt stress response and interacts with a MA3 domain-containing protein ECIP1 in Arabidopsis". Plant, Cell & Environment.34 (10):1678–1692.* doi:10.1111/j.13653040.2011.02363.x. ISSN 1365-3040.
112. Lu. Congming, A. Vonshak. (2002). Effects of salinity stress on photosystem II function in cyanobacterial Spirulina platensis cells. Physiol. Plant 114 405-413.
113. Lucas, Y., Boulet, R., Chauvel, A. and Veillon, L. 1987. Systèmes sols ferrallitiques-podzols en région amazonienne. In: *Podzols et podzolisation.* D. Righi and A. Chauvel (eds.). AFES-INRA, Paris. pp. 53-68.
114. *Maestre, Fernando T.; Cortina, Jordi; Bautista, Susana (2007). "Mechanisms underlying the interaction between Pinus halepensis and the native late-successional shrub Pistacia lentiscus in a semi-arid plantation". Ecography. 27 (6):776–786.* doi:10.1111/j.0906-7590. 2004.03990.x.
115. Martinez-Beltran J, Manzur CL. (2005). Overview of salinity problems in the world and FAO strategies to address the problem. Proceedings of the international salinity forum, Riverside, California, April 2005, 311–313.
116. Mazhar, R., N. Ilyas, M. Saeed, F. Bibi, and N. Batool. 2016. Biocontrol and salinity tolerance potential of *Azospirillum lipoferum* and its inoculation effect in wheat crop. *International Journal Of Agriculture and Biology* 3:494–500.
117. McKyes, E., Sethi, A. and Yong, R.N. 1974. Amorphous coatings on particles of sensitive clay soils. *Clays and Clay Minerals* **22**: 427-433.
118. Ministry of Food, Agriculture and Livestock, 2007. *Agricultural statistics of Pakistan*. Islamabad: Ministry of Food, Agriculture and Livestock, Government of Pakistan.
119. Miri, Y., and S. A. Mirjalili. 2013. Effect of salinity stress on seed germination and some physiological traits in primary stages of growth in purple coneflower (*Echinacea purpurea*). *International Journal of Agronomy and Plant Production* 4:142–6.

120. *Mittler, Ron (2006). "Abiotic stress, the field environment and stress combination". Trends in Plant Science. 11 (1): 15–19.* doi:10.1016/j.tplants.2005.11.002. PMID 16359910.
121. Mohammad Ilyas, Gohar Ayub, Imran, Azmat Ali Awan and Manzoor Ahmad.2020. Effect of Calcium and Boron on Production and Quality of Autumn Potato under Chilling Temperature. *Communication in soil science and plant analysis*. DOI: 10.1080/00103624.2020.1854286
122. Muhammad Islam[1], Shazma Anwar[1], Saqib Bashir[1], Wajid Ali Khattak[1], Imran, Murad Ali[2] and Nadar Khan .2015. Growth and Yield Components of Wheat Varieties As Affected By Dual Purpose Practices. J. Pure & Applied biology.
123. Munns, R. R., A. James, and A. Lauchli. 2006. Approaches to increasing the salt tolerance of wheat and other cereals. *Journal of Experimental Botany* 57:1025–43.
124. Munns, R. 2002. Comparative physiology of salt and water stress. *Plant Cell Environment*25:239–50.
125. Munns, R., and M. Tester. 2008. Mechanisms of salinity tolerance. *Annual Review of Plant Biology* 59:651–81.
126. Nazarli, H., F. Faraji, and M. R. Zardashti. 2011. Effect of drought stress and polymer on osmotic adjusment and photosynthetic pigments of sunflower. *Carcetriari Agronomica in Maldova* 1:45.
127. *Neto, Azevedo; De, André Dias; Prisco, José Tarquinio; Enéas-Filho, Joaquim; Lacerda, Claudivan Feitosa de; Silva, José Vieira; Costa, Paulo Henrique Alves da; Gomes-Filho, Enéas (2004-04-01). "Effects of salt stress on plant growth, stomatal response and solute accumulation of different maize genotypes". Brazilian Journal of Plant Physiology. 16(1): 31 38.* doi:10.1590/S1677-04202004000100005. ISSN 1677-0420.
128. Olsen, S. R., and L. E. Sommers. 1982. Phosphorous. In *Methods ofsoil analysis: Chemical and microbiological properties*, ed. D. Bunton, 403–27. 2nd ed. Madison, WI: Soil Science Society of America.
129. *Palm, Brady; Van Volkenburgh (2012). "Serpentine tolerance in Mimuslus guttatus does not rely on exclusion of magnesium". Functional Plant Biology.*
130. Palta, Jiwan P. and Farag, Karim. September 2006. "Methohasds for enhancing plant health, protecting plants from biotic and abiotic stress related injuries and enhancing the recovery of plants injured as a result of such stresses." United States Patent 7101828.
131. *Pant, Bikram Datt; Burgos, Asdrubal; Pant, Pooja; Cuadros-Inostroza, Alvaro; Willmitzer, Lothar; Scheible, Wolf-Rüdiger*

(2015). "The transcription factor PHR1 regulates lipid remodeling and triacylglycerol accumulation in Arabidopsis thaliana during phosphorus starvation". *Journal of Experimental Botany. 66 (7): 1907–1918.* doi:10.1093/jxb/eru535. ISSN 0022-0957. PMC 4378627. PMID 25680792.

132. *Pant, Bikram-Datt; Pant, Pooja; Erban, Alexander; Huhman, David; Kopka, Joachim; Scheible, Wolf-Rüdiger (2015). "Identification of primary and secondary metabolites with phosphorus status-dependent abundance in Arabidopsis, and of the transcription factor PHR1 as a major regulator of metabolic changes during phosphorus limitation". Plant, Cell & Environment. 38 (1): 172–187.* doi:10.1111/pce.12378. ISSN 1365-3040. PMID 24894834.
133. Paz-Ferreiro, J., S. Fu, A. Mendez, and G. Gasco. 2014. Interactive effects of biochar and the earhworm *Pontoscolex corethrurus* on plant productivity and soil enzymes activities, *Journal of Soils Sediments*, 14:483–94.
134. Peng, Y., Z. Gao, Y. Gao, G. Liu, L. Shengand D. Wang. 2008. Ecophysiological characteristics of alfalfa seedlings in response to various mixed salt-alkaline stresses. *Journal of Integrative Plant Biology* 50:29–39.
135. *Ping Ji, Geo K. Ron M. (2007). "Understanding Abiotic Stress Tolerance Mechanisms: Recent Studies on Stress Response in Rice". Journal of Integrative Plant Biology. 49 (6): 742–750.* doi:10.1111/j.1744-7909.2007.00495.x.
136. Prado, F. E., C. Boero, M. Gallarodo, and A. J. Gonzalez. 2000. Effect of NaCl on germination, growth and soluble sugar content in *Chenopodium quinoa* wil seeds. *Botanical Bulletin - Academia Sinica Taipei* 14:27–34.
137. *Raghothama, K. G. (1999). "Phosphate Acquisition". Annual Review of Plant Physiology and Plant Molecular Biology. 50 (1): 665–693.* doi:10.1146/annurev.arplant.50.1.665. PMID 15012223.
138. Rajalakshmi, A. G., S. K. Kumar, C. D. Bharathi, R. Karthika, K. D. Visalakshi, R. Meera, and S. Mohanapriya. 2015. Effect of biochar in seed germination: *In-vitro* study. *International Journal of Biosciences and Nanosciences* 2:132–6.
139. Rashid, A., Q. Saleem, A. Nazir, and H. S. Kazım. 2003. Yield potential and stability of nine wheat varieties under water stress conditions. *International Journal of Agriculture and Biology* 5:7–9.
140. Raza, A. 2012. Grain and field annual 2012 gain report. Global Agricultural Information Network. *USDA Foreign Agrilcultural services* 1–3.

141. Rengasamy, P. 2002. Transient salinity and subsoil constraints to dryland farming in Australian sodic soils: An overview. *Australian Journal of Experimental Agriculture* 42:351–61.
142. *Roelofs, D, Jiwan P and kwangy I.. (2008). "Functional ecological genomics to demonstrate general and specific responses to abiotic stress". Functional Ecology. 22: 8–18.* doi:10.1111/j.1365-2435.2007.01312.x.
143. *Rolf D, Vinebrooke K., Mittler J. (2004). "Impacts of multiple stressors on biodiversity and ecosystem functioning: the role of species co-tolerance". OIKOS. 104 (3): 451–457.* doi:10.1111/j.0030-1299.2004.13255.x.
144. *Rubio, Vicente; Linhares, Francisco; Solano, Roberto; Martín, Ana C.; Iglesias, Joaquín; Leyva, Antonio; Paz-Ares, Javier (2001).* "A conserved MYB transcription factor involved in phosphate starvation signaling both in vascular plants and in unicellular algae". *Genes & Development. 15 (16): 2122–2133.* doi:10.1101/gad.204401. ISSN 08909369. PMC 312755. PMID 11511543.
145. Santer, B. 1985. The use of general circulation models in climate impact analysis - a preliminary study of the impacts of a CO_2-induced climatic change on West European agriculture. *Climatic Change* **7**: 71-93.
146. *Savvides, Andreas (2015).* "Chemical Priming of Plants Against Multiple Abiotic Stresses: Mission Possible?". *Trends in Plant Science. 21 (4): 329 340.* doi:10.1016/j.tplants.2015.11.003. hdl:10754/596020.
147. Scharpenseel, H.W., Schomaker, M. and Ayoub, A. (eds.) 1990. Soils on a warmer earth. Effects of expected climatic change on soil processes, with emphasis on the tropics and subtropics. *Developments in Soil Science 20.* Elsevier, Amsterdam. xxii + 274 p.
148. Scholander, P. F., H. T. Hammel, E. D. Bradstreet, and E. A. Hemminogsen. 1965. Sap pressure in vascular plants: Negative hydrostatic pressure can be measured in plants. *Plant Sciences*148:339–46.
149. Shahida Naveed, Muhammad Ibrar, Inayat Khattak, Imtiaz Khan, Imran, Muhammad Isfhaq Khan and Haroon Khan.2016. Anthelmintic, Antilice, Insecticidal, Cytotoxic and Phytotoxic Potential of Ethanolic Extracts of Two Wild Medicinal Plants *Iphiona Grantioides* and *Plucheaarguta.* J. Woulfenia (23) 11: 13-25

150. Shamsi, K., and S. Kobraee. 2013. Biochemical and physiological responses of three wheat cultivars (*Triticum aestivum* L.) to salinity stress. *Annals of Biological Research* 4:180–5.
151. Shamsuddin Baqa[1], Amir Zaman Khan[1], Inamullah[1], Imran*[1], Asad Ali Khan*[1], Shazma Anwar[1], Babar Iqbal[1], Shehryar Khan[1], Amjad Usman[2]. 2015. Influence of Farm Yard Manure and Phosphorus Application on Yield and Yield Components of Wheat. J. Pure & Applied biology Vol. 4, Issue 4.
152. Imran, Amanullah and Abdel Rahman M. Al Tawaha. 2021a. Management of Nano-black carbon, Phosphorous and Bio Fertilizer improve Soil organic carbon and ensilage biomass of soybean and maize. *Communication in Soil Science and Plant Analysis.* DOI.10.1080/00103624.2021.1966439
153. Imran, Amanullah and Abdel Rehman Altawaha. 2021b. Carbon Assimilation and Dry Matter Partitioning in Soybean Ameliorates with the Integration of Nano-Black Carbon, Along with Beneficial Microbes and Phosphorus Fertilization. *Journal of Plant Nutrition*
154. Imran. 2021. The bioavailability of Phosphorus in composite vs. Hybrid maize differ with phosphorus and boron fertilization. *Phosphorus, Sulfur, and Silicon and the Related Elements. https://doi.org/10.1080/10426507.2021.1920588.*
155. Imran, Amanullah, Asad Ali khan, Tariq Mahmood, Abdel Rahman Al Tawaha and Samia Khanum[4]. 2021c. Adequate Fertilization, Application Method and Sowing Techniques Improve Maize yield and related traits. *Communication in soil science and plant analysis.* DOI:10.1080/00103624.2021.1925688
156. Imran and Amanullah· 2021. Phosphorus and Boron Application Optimizing Biofortification Of phosphorus and Productivity of French Bean (*Phaseolus Vulgaris* L.). *Communication In Soil Science and Plant Analysis.*
157. Imran, Amanullah and Abdel Rahman Al-Tawaha. 2021d. Carbon Sources Application Increase Wheat Yield And Soil Fertility. *Communication in soil science and plant analysis DOI: 10.1080/00103624.2020.1865397.*
158. Imran, Amanullah and Abdel Rehman Al-Tawaha.2020b. The Productivity of subsequent wheat enhanced with residual carbon sources and phosphorus under improved irrigation system. *Communication in soil science and plant analysis*, DOI:10.1080/00103624.2020.1763387
159. Imran, Amanullah, Muhammad Arif, Zahir Shah and Abdul Bari. 2020a. Soil application of *Trichoderma* and Peach (*Prunus persica* L.)

residues possesses biocontrol potential for weeds and enhances growth and profitability of soybean (Glycine max). *Sarhad Journal of Agriculture* 36 (1) 10-20. DOI | http://dx.doi.org/10.17582/journal.sja/2020/36.1.10.20

160. Mohammad Ilyas, Gohar Ayub, Imran*, Azmat Ali Awan and Manzoor Ahmad.2020. Calcium and Boron effect on Production and Quality of Autumn Potato under Chilling Temperature. *Communication in soil science and plant analysis*. DOI: 10.1080/00103624.2020.1854286
161. Shazma Anwar, Israeel, Babar Iqbal, Asad Ali Khan, Imran, Wajid Ali Shah, Mohammad Islam, Wajid Ali Khattak, Ikramullah, Wasim Akram and Wasim Abbas. 2016. Nitrogen and phosphorus fertilization of improved varieties for enhancing phenological traits of wheat. Pure and Applied Biology. Vol. 5, Issue 3, pp511-519. http://dx.doi.org/10.19045/bspab.2016.50065
162. Sidari, M., C. Mallamaci, and A. Muscolo. 2008. Drought, salinity and heat differently affect seed germination of Pinus pinea. *Journal of Forest Research* 13:326–330.
163. Sombroek, W.G. 1990. Soils on a warmer earth: the tropical regions. In: Scharpenseel *et al.* (eds.). 1990. pp. 157-174.
164. Sombroek, W.G. and Zonneveld, L.S. 1971. Ancient dune fields and fluviatile deposits in the Rima-Sokoto river basin (N.W. Nigeria). Stiboka (Staring Centre). *Soil Survey Paper No. 5.* Wageningen. 109 p.
165. Stanley, D.J. 1988. Subsidence in the Northeastern Nile delta: rapid rates, possible causes, and consequences. *Science* **240**: 497-500.
166. Stigliani, W.M. 1988. Changes in valued 'capacities' of soils and sediments as indicators of nonlinear and time-delayed environmental effects. *Environmental Monitoring and Assessment* **10**: 245-307.
167. Syed Arshad Karim and Imran. 2019. Reduction in substrate moisture content reduce final yield of wheat. Biomed J. Sci & Tec Res. 1-5.
168. Tank, N, and M. Saraf. 2010. Salinity-resistant plant growth promoting rhizobacteria ameliorates sodium chloride stress on tomato plants. *Journal of Plant Interaction* 5:51–8.
169. Tiwari S., P. Singh, R. Tiwari, K. K. Meena, M. Yandigeri, D. P. Singh, and D. K. Arora. 2011. Salt-tolerant rhizobacteria-mediated induced tolerance in wheat (*Triticum aestivum*) and chemical diversity in rhizosphere enhance plant growth. *Biological Fertile Soils* 47:907–16.
170. *Tombesi, Sergio; Frioni, Tommaso; Poni, Stefano; Palliotti, Alberto (2018). "Effect of water stress "memory" on plant behavior during*

subsequent drought stress". Environmental and Experimental Botany. 150: 106–114. doi:10.1016/j.envexpbot.2018.03.009. ISSN 0098-8472.

171. Ulfat Samreen, Muhammad Ibrar, Lalbadshah, Shahida Naveed, Imran and Inayat Khatak. 2016. Ethnobotanical study of subtropical hills of Darazinda, Takht-e-Suleman range F.R D.I. Khan, Pakistan. Pure and Applied Biology. Vol. 5, Issue 1, pp149-164. http://dx.doi.org/10.19045/bspab.2016.50020
172. Van Breemen, N. 1990. Impact of anthropogenic atmospheric pollution on soils. In: *Climate Change and Soil Processes.* UNEP, Nairobi. pp. 137-144.
173. Van Hoorn, J. W., K. Nader, A. Hamdy, and M. Mastrorilli. 2001. Effect of salinity on yield and nitrogen uptake of four grain legumes and on biological nitrogen contribution from the soil. *Agriculture Water Management* 51:87–98.
174. *Wang, W.; Vinocur, B.; Altman, A. (2007). "Plant responses to drought, salinity and extreme temperatures towards genetic engineering for stress tolerance". Planta. 218 (1):1–14.* doi:10.1007/s00425-003-1105-5. PMID 14513379.
175. Wang, G., and Z. Xu. 2013. The effects of biochar on germination and growth of wheat in different saline-alkali soil. *Asian Agricultural Research* 5:116–9.
176. Wang. H., and J. Jin. 2007. Effects of zinc deficiency and drought on plant growth and metabolism of reactive oxygen species in maize (*Zea mays* L.). *Agricultural Sciences in China*6:988–95.
177. Warrick, R. and Farmer, G. 1990. The greenhouse effect, climatic change and rising sea level: implications for development. *Trans. Inst. Br. Geogr. N.S.* **15**: 5-20.
178. Wolfe, A. 2007. "Patterns of biodiversity." Ohio State University, 10.1016/j.agee.2006.12.013.
179. Xu, D., J. Huang, S. Guo, X. Yang, Y. Bao, H. Tang, and H. Zhang. 2008. Over-expression of a TFIII-type zinc finger protein gene ZFP252 enhances drought and salt tolerance in rice (*Oryza sativa*L.). *Federation of European Biochemical Societies Letters* 582:1037–43.
180. Yeo, A. R., M. E. Yeo, S. A. Flowers, and T. J. Flowers. 1990. Screening of rice (*Oryza sativa* L.) genotypes for physiological characters contributing to salinity resistance, and their relationship to overall performance. *Theoretical Applied Genetics* 79:377–84.
181. Younis, U., M. Athar, S. A. Malik, M. H. RazaShah, S. Mahmood. 2015. Biochar impact on physiological and biochemical attributes of

spinach *Spinacia oleracea* (L.) in nickel contaminated soil. *Global Journal Environment Science Management* 1:245–54.
182. Zakaria, M., S. Daniel, V. Murphy, and L. K. Abbott. 2012. Biochars influence seed germination and early growth of seedlings. *Plant Soil* 353:273–87.
183. *Zargar, Sajad Majeed; Nagar, Preeti; Deshmukh, Rupesh; Nazir, Muslima; Wani, Aijaz Ahmad; Masoodi, Khalid Zaffar; Agrawal, Ganesh Kumar; Rakwal, Randeep (2017). "Aquaporins as potential drought tolerance inducing proteins: Towards instigating stress tolerance". Journal of Proteomics. 169: 233–238.* doi:10.1016/j.jprot.2017.04.010. ISSN 1874-3919.
184. Zhan, H. X., Z. J. Chang, A. L. Wei, X. J. Zhang, and X. Li. 2011. Impact of drought to wheat physiological index. *Journal of Shanxi Agriculture Sciences* 39:1049–51.
185. Churchman, G. J., Singh, M., Schapel, A., Sarkar, B., and Bolan, N. (2020). CLAY MINERALS AS THE KEY TO THE SEQUESTRATION OF CARBON IN SOILS. *Clays and Clay Minerals* **68**, 135-143.
186. Golchin, A., Oades, J. M., Skjemstad, J. O., and Clarke, P. (1994). Soil-structure and carbon cycling. *Australian Journal of Soil Research* **32**, 1043-1068.
187. Jastrow, J. D., Miller, R. M., and Lussenhop, J. (1998). Contributions of interacting biological mechanisms to soil aggregate stabilization in restored prairie. *Soil Biology & Biochemistry* **30**, 905-916.
188. Khresat, S., Al-Bakri, J., and Al-Tahhan, R. (2008). Impacts of land use/cover change on soil properties in the Mediterranean region of northwestern Jordan. *Land Degradation & Development* **19**, 397-407.
189. Lal, R. (2010). Managing Soils and Ecosystems for Mitigating Anthropogenic Carbon Emissions and Advancing Global Food Security. *Bioscience* **60**, 708-721.
190. Lal, R. (2019). Carbon Cycling in Global Drylands. *Current Climate Change Reports* **5**, 221-232.
191. Larionova, A. A., Kurganova, I. N., de Gerenyu, V. O. L., Zolotareva, B. N., Yevdokimov, I. V., and Kudeyarov, V. N. (2010). Carbon dioxide emissions from agrogray soils under climate changes. *Eurasian Soil Science* **43**, 168-176.
192. Ortas, I. (2020). "Mycorrhizas in fruit nutrition: Important breakthroughs."
193. Ortas, I., Akpinar, C., and Lal, R. (2013). Long-term impacts of organic and inorganic fertilizers on carbon sequestration in aggregates of an Entisol in Mediterranean Turkey. *Soil Science* **178**, 12-23.

194. Ortas, I., Bulutekin, B., and Öztürk, F. (2019). RHIZOSPHERE AND NON-RHIZOSPHERE SOIL STRUCTURE DEVELOPMENT UNDER PHOSPHORUS FERTILIZATION APPLICATION. *In* "Фундаментальные концепции физики почв: развитие, современные приложения и перспективы", pp. 369-379.
195. Ortas, I., and Bykova, A. (2020). Effects of long-term phosphorus fertilizer applications on soil carbon and CO(2)flux. *Communications in Soil Science and Plant Analysis* **51**, 2270-2279.
196. Ortas, I., Rafique, M., and Çekiç, F. (2021). Do Mycorrhizal Fungi Enable Plants to Cope with Abiotic Stresses by Overcoming the Detrimental Effects of Salinity and Improving Drought Tolerance? *In* "Symbiotic Soil Microorganisms", pp. 391-428. Springer.
197. Ortas, I., and Yucel, C. (2020). Do mycorrhizae influence cover crop biomass production? *Acta Agriculturae Scandinavica Section B-Soil and Plant Science* **70**, 657-666.
198. Ortaş, I., Lal, R., and Kapur, S. (2017). Carbon Sequestration and Mycorrhizae in Turkish Soils. *In* "Carbon Management, Technologies, and Trends in Mediterranean Ecosystems", pp. 139-149. Springer.
199. Ortaş, I., Rafique, M., and Iqbal, M. T. (2019). Mycorrhizae Resource Allocation in Root Development and Root Morphology. *In* "Plant Microbe Interface", pp. 1-26. Springer.
200. Ortaş, I., Razzaghi, S., and Rafique, M. (2016). Arbuscular mycorrhizae: Effect of rhizosphere and relation with carbon nutrition. *In* "Plant-Microbe Interaction: An Approach to Sustainable Agriculture" (D. Choudhary, A. Varma and N. Tuteja, eds.), pp. 125-152. Springer, Singapore.
201. Ryan, J., Masri, S., Ibrikci, H., Singh, M., Pala, M., and Harris, H. C. (2008). Implications of cereal-based crop rotations, nitrogen fertilization, and stubble grazing on soil organic matter in a Mediterranean-type environment. *Turkish Journal of Agriculture and Forestry* **32**, 289-297.
202. Singh, M., Sarkar, B., Biswas, B., Churchman, J., and Bolan, N. S. (2016). Adsorption-desorption behavior of dissolved organic carbon by soil clay fractions of varying mineralogy. *Geoderma* **280**, 47-56.
203. Wang, X. Y., He, T. H., Gen, S. Y., Zhang, X. Q., Wang, X., Jiang, D., Li, C. Y., Li, C. S., Wang, J. L., Zhang, W. Y., and Li, C. D. (2020). Soil properties and agricultural practices shape microbial communities in flooded and rainfed croplands. *Applied Soil Ecology* **147**.

CHAPTER FOUR

MYCORRHIZA: A SOLUTION TO THE CRISES OF SOIL HEALTH, PLANT NUTRITION AND FOOD SECURITY

IMRAN[1,2,3], ASAD ALI KHAN[1],
HAYAT ZADA[3] AND IBRAHIM ORTAS[2]
[1]DEPARTMENT OF AGRONOMY, THE UNIVERSITY OF AGRICULTURE, PESHAWAR, PAKISTAN
[2]RHIZOSPHERE LABORATORY, UNIVERSITY OF CUKUROVA, FACULTY OF AGRICULTURE, DEPARTMENT OF SOIL SCIENCE AND PLANT NUTRITION, ADANA, TURKEY,
[3]MINISTRY OF AGRICULTURE, ENTENSION WING, GOVT OF KHYBER PAKHTUNKHWA, PAKISTAN

Abstract

Management of soil and crop nutrients through integrated approaches is an important aspect in which mainly organic, inorganic, and chemical fertilizers are provided. However, the role of microorganisms, like nutrients solubilizer, nutrients mobilizer and other beneficial microbes has been neglected. Immobile elements like phosphorus, zinc, and copper are supported by the symbiotic relationship between plant roots and mycorrhizal fungi, but under certain conditions, more mobile ions like sulphur, calcium, sodium, potassium, iron, magnesium, manganese, chloride, and nitrogen are also taken up by the plant roots from the soil. Mycorrhizae promotes plant growth, development, stress tolerance, soil remediation, carbon sequestration, food safety, and agricultural sustainability. Our findings revealed that AMF significantly improved the growth mediated by drought and other environmental factors by enhancing photosynthesis, mineral nutrients, the glomalin content of the soil, the content of essential oils, and chemical composition, in addition to upregulating antioxidant systems and the enzymatic activity of phenylalanine ammonia-lyase and lipoxygenase. AMF enhanced soil quality and crop production, which was particularly evident in the increase of crop biomass and soil nutrients availability. The present outcomes suggested that the symbiotic association between AMF and crop plants may help to strengthen the fabrication of defensive compounds that ultimately help to relieve abiotic and biotic stresses and may also act as biological control of crop production constraints. Mycorrhizal inoculation greatly improves the root's ability to absorb nutrients and water under changing environments. Consequently, in this work, the role of mycorrhizal inoculation in nutrient absorption, as well as its important contribution to plant greater biomass and food security, are discussed in relation to mycorrhizal and plant symbiosis.

Keywords: Mycorrhiza, abiotic stresses, nutrients uptake, nutrients flow, nutrients flux, plant biomass and growth, yield, agriculture sustainability

Introduction

Arbuscular mycorrhiza (AM) association is the most widespread plant–fungal symbiosis on earth, in the majority of extant land plants, and can be traced back to the Early Devonian, more than 400 million years ago, which corresponds to the onset of the colonization of terrestrial habitats by plants (Sun et al., 2022; Zhang et al., 2022). Prior to the discovery of agriculturally beneficial microbes, particularly AMF, farmers transported fertile soil from one field to another, enhancing crop yield. The farmers were not aware of the indigenous or exogenous application of nutrients solubiliser and mobilizers by adding the soil of one field to another for fertility enhancement (Imran 2021). Then the farmers realized that clover planted soil significantly increase the yield and plant biomass as compared to other crops planted soil. The inclusion of some soil microorganisms in this phenomenon is not a novel, which began in early 1903. As a result, many P solubilizing microorganisms (PSMs) like bacteria, fungi, Trichoderma and mycorrhiza identified for the above said task. Later on it was reported that some microbes inhabit plant growth and development like beneficial bacteria and fungi. That promise opened in close association with the plants (Izhar et al., 2021). Some decades later, Russia and the Eastern European countries consisted of two primarily microbial products "Azotobacterin" and "Fasobacterin" containing Azotobacter crococum, nitro-fixer, (NF) and Bacillus megaterium (PSB). For 1958, it was reported that in Russia about

10 million hectares were treated with these microbial preparations and that there was some increase in production for different crops. The development of effective microbial products aiming to improve nutritional efficacy (Fig 1). Rhizobium-based bio-inoculant known as "nitrogen fixing bacteria" has a commercial history as microbial biofilizers. Since then, researchers from academic institutions and business companies have studied the potential of various microorganisms like fungus (AMF), bacteria (azotobacter), and then cyanobacteria to stimulate plant growth and nutritional. A few years later, a P-solubilizing bacteria-based microbial inoculant was developed (Ali et al., 2022).

Mycorrhizae improve the efficiency of mineral intake in soils where such minerals may be poor or otherwise less accessible (Imran & Amanullah, 2022a). It is widely accepted that the use of balanced fertilizers with integration of soil microbes is the key to higher yield, agriculture sustainability and highest net returns. Soil microbial population and roots colonization plays a vital role in the retention, breakdown, and incorporation of plant remnants, nutrient cycling, and their influence on soil structure, soil texture, soil porosity and ultimately on plant growth and higher biomass production (Imran 2021). Under adverse conditions (drought, salinity, nutrients depletion, pH etc), root growth and root function impairment reduce the root system's ability to obtain nutrients (Sun et al., 2022). Previous studies indicated that intense drought stress always has adverse effects on the synthesis of essential oil compounds and secondary metabolites in agronomic crops as well as in medicinal plants (Zhang et al., 2022; Wang et al., 2022). However, under the prevailing stress conditions, including drought, the input of high fertilizers does not constantly guarantee optimum plant growth, yield and biomass and eventually the defence system (uppermost level of secondary metabolites). To meet the prevailing climate threats, especially drought, most plants establish various defenses' techniques to survive and maintain their growth, maintenance, development and reproduction. Among defence techniques, the use of beneficial microbes, such as arbuscular mycorrhizal fungi (AMF), clearly proved to help the host plant through the symbiotic association of roots (Imran and Amanullah 2022b). AMF species belong to the subphylum Glomeromycota of the phylum Mucoromycota. AMF are well-known for improving the association with various plant species and consequently improve the contents of secondary metabolites and thus improve plant nutrition in order to enhance growth and development. This improvement changes the synthesis of different molecule families and terpenoids, glucosinolates, and essential oil and phenolic compounds and the phytoalexins. Further, the potential effects of AMF on the synthesis of secondary metabolites in different crops and also in medicinal plants have already been investigated in several studies. Moreover, AMF is supposed to be the real alternative to improve the growth and development of aromatic plants which are usually found in low fertility soil. AMF can provide better plant nutrition and essential oil quality (Yang et al., 2022; Imran et al., 2022a). In addition, plants of Piper nigrum L. cv. Bragantina revealed the differences in leaf essential oils chemical compositions by increasing elemol (11.4–23.2%) and α- bisabolol (32.3–48.5%) after AMF inoculation. Various studies have reported the links between AMF and tobacco, while others have been shown that the inoculation with AMF improves plant growth, augments photosynthesis, and increases the concentration of mineral nutrients and antioxidant functioning (Li et al., 2022; Imran et al., 2022b). However, the potential role of AMF inoculation on the synthesis of secondary metabolites in tobacco under drought stress further augmented the evidence of the defence system. Additionally, the plants' symbiotic association with root colonization through AMF may enhance the contents of phenolic compounds and also the plants' defense enzymes, e.g., phenylalanine ammonia-lyase (PAL) and lipoxygenases (LOXs). Lipoxygenases are abundant in various plant processes, e.g., plant development and growth, as well as reactions to pest and pathogen damage and response to several environmental factors (Pan et al., 2023).

Many studies documented that the activation of the lipoxygenases pathway enhances important functions in the primary and secondary metabolism in the plants (Imran et al., 2022c, Imran et al., 2022d). The cox enzymes (CaLOX1 and DkLOX3) are responsible for resistance against abiotic stresses, e.g., drought, higher salinity by altering the stress-responsive marker genes' expressions, abscisic acid, production of reactive oxygen species, and lipid peroxidation in Arabidopsis. Phenylalanine ammonia-lyase is an enzyme extensively found in higher plants and performs its role in accomplishing primary and secondary metabolism. During AMF symbiosis, these enzymes have been investigated by researchers due to their impact on the colonization process as they produce plant biomarkers in the physiological state (Liu et al., 2021). Moreover, AMF creates a specific compound called nitrogen-linked glycoprotein called glomalin. Economic crop, and drought stress has a significant effect on crop quality and yield. It is pertinent to learn the mechanism and production technologies of the tobacco crop (Yang et al., 2022). However, little attention has been paid to AMF and its consequences on tobacco plants' drought tolerance. Therefore, here, the present study hypothesized that AM inoculation would stimulate tobacco plant growth to enhance mineral nutrients' uptake, regulate enzymatic activities, and defend against drought stress. The potential of

AMF inoculation on crop growth, photosynthesis and chlorophyll fluorescence parameters, antioxidant enzyme activities, osmolyte metabolism, contents of glomalin, and amount of essential oil in various agronomic crops improved under drought stress (Li et al., 2022; Sun et al., 2022; Zhang et al., 2022).

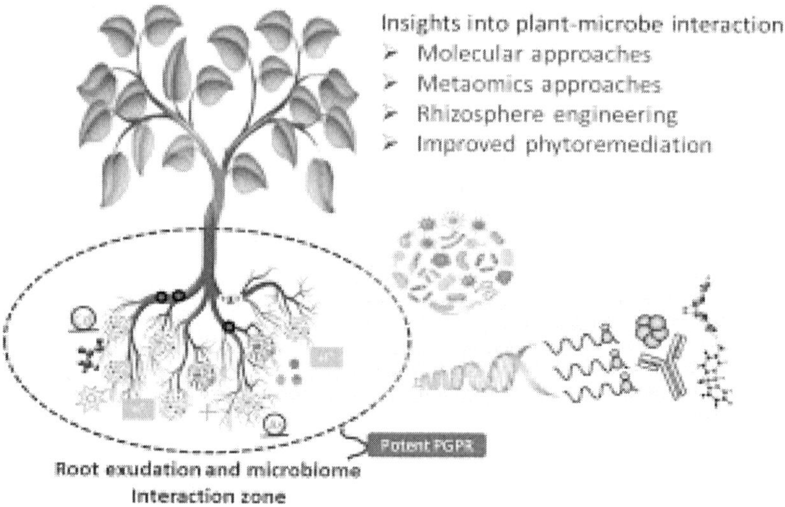

Figure 1. Plant microbe's interactions and mechanism of nutrients uptake from hyphosphere

The AMF is an important microbial component of farming systems, making it one of the most commonly used symbols on earth (more than 80% of the plants concerned, identified as the nutrient flow between the two living plants or microbes, the symbolic relationship involved in the bilateral biological exchange (Imran et al., 2022e). The mathematical model estimated the contribution of AMF hyphae to plants through plants has shown that plants rely solely on P nutrition. You may have to rely on the pathway of mycorrhizal. Morphologically, the AMF functions as an extension of the root system via its mycelium network and allows the plant to expand outside the boundaries of the rhizosphere, and this function is likely to prevail in the case of P deficiency (Yang et al., 2022; Imran. 2022b). In this respect, the AMF provides an additional p-optical pathway (AMF pathway.

Abiotic stresses, such as drought, are the key challenges for growth and production, leading to an annual of billions of dollars loss. Globally, in arid and semiarid zones, plants were affected by drought in various ways that include water shortage and transpiration reduction, which leads to oxidative stress (Li et al., 2022). Various cellular organelles generate toxic radicals that include hydroxyl ion (OH^-), superoxide (O_2^-), and hydrogen peroxide (H_2O_2) effect on mechanical and functional constancy to oxidize the related nucleic acids, proteins, and lipids. To combat abiotic stresses, plants adopt different physiological as well as molecular mechanisms that include the stimulation of antioxidant enzymes, i.e., catalase (CAT), peroxidase (POD), superoxide dismutase (SOD), ascorbate peroxidase (APX), etc (Lu et al., 2022).

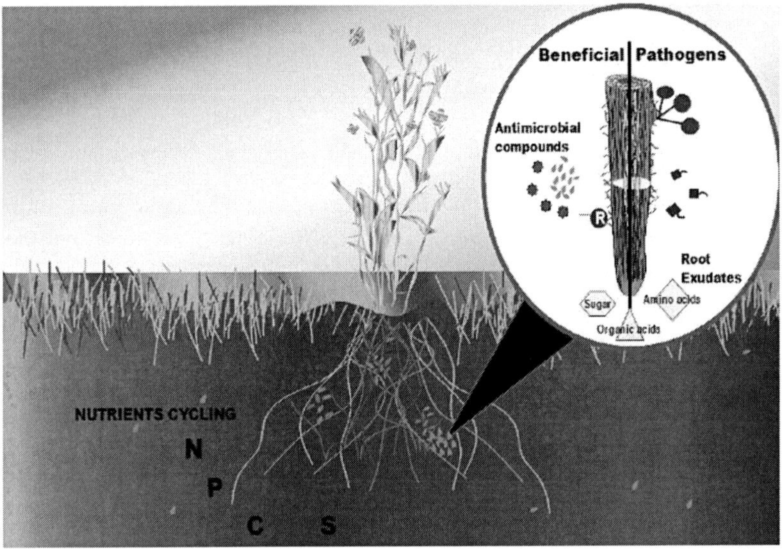

Figure. 2. Mycorrhiza interaction with crop plants in rhizospheric region and nutrients uptake

AM fungus consumes hexoses produced by the plant host's photosynthesis. Carbon can be transported from plant to fungus through arbascules or intraradical hyphae, respectively (Ding et al., 2022). AMF generates secondary hexose in intraradical mycelium. Trehalose and glycogen are produced from hexose inside the mycelium. Trehalose and glycogen are two types of quickly synthesizing and rapidly decaying carbon stores that can regulate intracellular sugar concentrations (Imran. 2022a; Imran. 2022b,). Through intraradical hexosis, the oxidative pentose phosphate pathway, which produces pentose for nucleic acids, is entered. About 25% of the carbon that travels from the plant to the fungus is stored in the extraradical hyphae. According to a study, the percentage of total fixed carbon that is allocated to mycorrhizal fungus is typically about 10%. However, up to 20% of the carbon in the host plant can also be transferred to the AM fungi (Imran.2022c). In addition, in return for nutrients from the plant, mycorrhizal fungus acquire photosynthesized carbon. Additionally, carbon encourages the transfer and absorption of nutrients. The host plant invested a significant amount of carbon in the mycorrhizal network, which increased the pool of subsurface organic carbon. Increased plant carbon availability to AMF boosts the fungi's capacity to absorb and transfer minerals and nutrients to the plant. Similarly, the fungus's ability to absorb and transmit phosphorus decreases as the amount of photosynthates available to it. Different AMF species have varying levels of ability to provide phosphorus to the plant (Izhar et al., 2022).

Mycorrhizas have been shown to promote plant absorption of nutrients, particularly phosphorus, which has been cited as its principal advantage. AMF are symbionts that provide little phosphorus while consuming a lot of carbon (Imran and Amanullah 2021a). Greater storage, increased nutrient flow into mycorrhizae, altered root environments, and increased surface area in contact with soil may all contribute to this. Lipid biosynthesis also includes intra-radical mycelium. After that, lipids can be exported or stored in extraradical hyphae where they can be used for metabolism or storage (Imran and Amanullah 2021b). In extra radical mycelium, a process termed as gluconeogenesis converts lipids into hexoses.

Understanding Mycorrhizal Fungi Functions

The finding that AM fungus showed strong affinity with inorganic phosphate (Pi) transporters was a significant leap in our understanding of fungal activity. Nitrogen is a key ingredient of AM fungus, and it has been found that several genes are involved in the transport of both ammonium and amino acids. (Imran et al., 2021b; Imran et al., 2021c). However, arginine is likely the favored chemical for long-distance transport to the host plant. A glutamine-synthesizing gene, linked to the breakdown of arginine is expressed more in intraradical mycelium, whereas Glomus intraradice is expressed preferentially in extraradical hyphae. Although the transfer of carbon from plants to AM fungi was demonstrated in the 1960s, its molecular mechanisms are still unclear. Surprisingly, with the exception of one gene from the non-AM *Geosiphon pyriforme* hosting intracellular cyanobacteria, so far no hexoid transporters responsible for C uptake from host cells have been identified in Glomeromycota. Availability of expressed sequence tag libraries from different phases of the life cycle of *G. intraradices* will provide a more complete picture of fungal activity. *G. intraradice* genes involved in transcription, protein synthesis and metabolism of quiescent and germinating spores, as well as in interactions with wild-type and AM-defective (Myc) Medicago truncatula mutants.

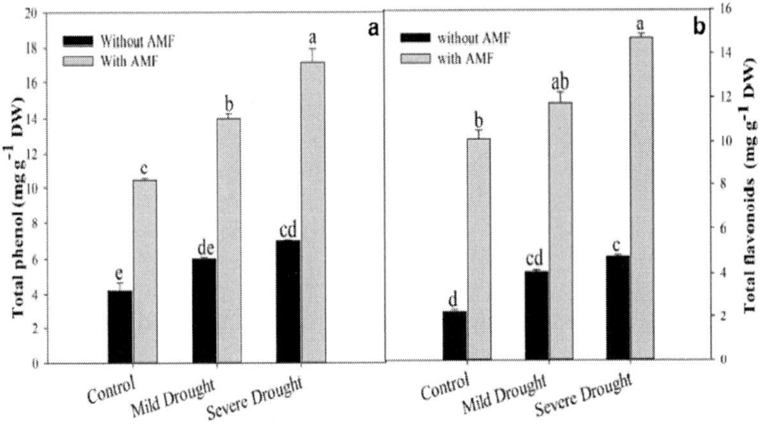

Figure 3. Effect of AMF inoculation on the content of a total phenols and b flavonoids under drought stress.

Although an increasing number of studies have been conducted on specific fungal genes, the genetic traits that discriminate AM fungus from an EM fungus or pathogen are currently unknown (Imran et al., 2020a). Interestingly, it's a *G. intraradice* gene like the STE12 complements the non-invasive mutant of the pathogenic Colletotrichum lindemuthianum, restoring its infectivity. Similarly, the Era-like GTPase encoding gene in the Magnaporthe oryzae rice pathogen and required for virulence found to be similar to the Gin-N protein. These findings suggest that some of the molecular mechanisms underlying the invasion of plant tissues are shared by pathogenic and symbiotic fungi, despite their distant phylogenetic relationships and diverse trophic habits (Liu et al., 2021; Yang et al., 2022). Plant roots release exudates containing sugars, organic acids, and amino acids that may attract microbes. In exchange, they protect the plant against pathogens releasing antimicrobial compounds; or increase nutrient uptake (Imran et al., 2020b). On the other hand, these carbon-containing compounds can also attract pathogens. They can compete for nutrients, infect the plant, and affect the rhizosphere microbial community (Fig 2).

The Special Biological Properties of Mycorrhizal Fungi

Mycorrhizas can increase significantly plant uptake of phosphorus and trace elements. They do this by extending the volume of soil explored by the plant. Mycorrhizal fungi are characterised by very thin hyphae, which are between 1 and 10 thousandths of a millimetre in width. Under harsh condition or phosphorus deficient condition, mycorrhiza improve plant defence system and may AMF inoculation under drought stress enhance content of a total phenols and flavonoids in various agronomic crops (Fig 3). There are a number of indicators that show how the AMF infection affects phosphorus intake and plant growth, including the shoot/root ratio, phosphorus depletion in the soil, and phosphorus inflow. The availability of phosphate in the rhizosphere may be quantitatively examined by looking at the depletion of phosphate while taking into account the P taken from the soil by plants. Phosphorus should be depleted at the soil-root contact as plants increase their intake of the mineral. While research into the relevance of fungal symbiosis is in in its infancy, the difficulty of both types to grow together, their vertical transmission, and their small genome sizes suggest that this ancient AM partner and fungal speciation may have had evolutionary significance. All of these unorthodox characteristics have so far restricted the possibility of applying traditional methods to AM fungi such as genetic transformation, mutant generation and classification, as well as large-scale transcriptomic and genomic studies. Since the sequencing methods are rapidly improving, the AM fungus *Glomus intraradices* genome sequencing project faced a bumpy road. Perhaps due to a high degree of heterozygosity and/or a larger genome than expected, assembly difficulties for this relatively small genome (estimated to be about 14–16 Mb). At the same time, the absence of a uninucleate cell stage in AM fungi's life cycle is a major obstacle to obtaining stable transgenic strains. Nevertheless, the transitory expression of an exogenous gene in *G. Biolistic* change has brought in *G. intraradices*. Due to these peculiar characteristics, transcriptomic and proteomic analyses have proved to be more competitive in Investigation of the molecular basis for AM fungal growth and function.

Although the hyphae of genetically distinct strains that anastomose and exchange genetic material, AM fungi are considered as sexual. Within this community of organisms, the very concepts of species and person are poorly described which represent a high degree of genetic and functional heterogeneity (Imran and Amanullah 2021c). Even the idea of a cell as a hyphal compartment goes against the syncytial mycelia of these fungi, in which the nucleus is continuously translocated by rapid cytoplasmic streaming. In the end, their large asexual spores contain thousands of nucleus, making classical genetic approaches unsuitable.

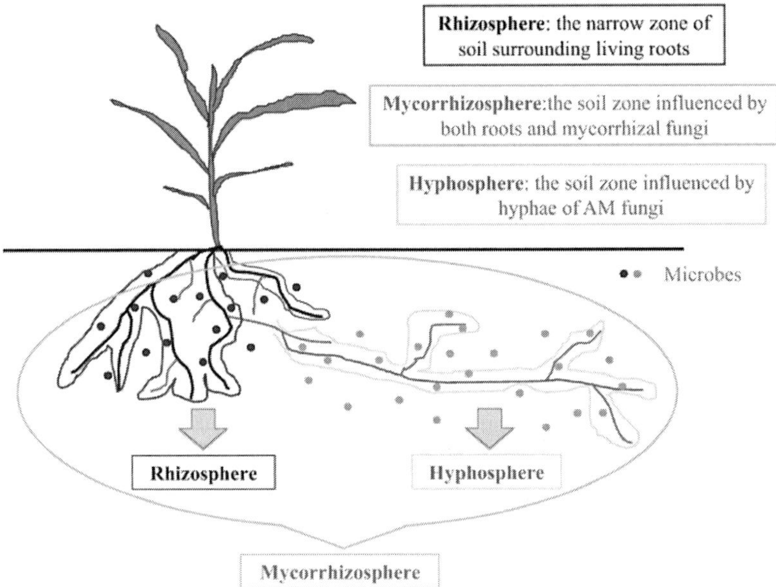

Figure 4. Mycorrhiza role in water and nutrients uptake in rhizosphere, mycorrhizosphere and hyphosphere

Unlike AM fungi, genetic methods and genomic knowledge have helped explore different aspects of AM interactions such as root colonization and symbiont-to-symbiont contact. Nonetheless, this one-way approach has most likely given us a skewed image in which the plant seems to be the key factor and decides each partner's contribution to the establishment and to the group also has trouble working.

Arbuscular Mycorrhizal Fungi's Secret to Success

Increased absorption of various mineral nutrients, including phosphorus, by mycorrhizal plants has been attributed to an increase in the surface area for absorption of nutrients (hyphea). Plant roots are the primary absorbing organ of mineral nutrients and water. In the soil, plant roots are limited to a small area for the absorption of nutrients, while arbuscular mycorrhizal fungi hyphae grow outside the depletion zone, where the available nutrients and water have already been removed from the plant roots (Ilyas et al.,

2020). Root absorptive surface area has been reported to increase up to 50 times in mycorrhizal plants compared to non-mycorrhizal plants. There are many factors that influence the nutrition and yield of plants. In addition to plant genotypes, environmental factors such as the availability of nutrients in growing media, the availability and uptake of water, and the ability of the plant to properly utilize what is available in the soil all play an integral part in plant growth (Imran et al., 2022c; Imran et al., 2022d). In order to facilitate these processes, plants are able to form a mutually beneficial relationship with arbuscular mycorrhizal fungi, also known as AMF or endomycorrhizae. Plants grown under artificial non-symbiotic conditions have shown that AM fungi contribute significantly to the absorption of soil nutrients, increase plant biomass and confer on the plant enhanced stress tolerance and pathogen resistance (Wang et al., 2022; Imran et al., 2021b; Izhar et al., 20221). AM fungi, on the other hand, have so far proved to be inculturable in the absence of a host. These fungi, being unable to consume carbohydrates even from within a plant cell, are strictly dependent on their green hosts for growth and reproduction, giving them the status of obligatory biotrophs. At an evolutionary point of view, the ecological success of AM fungi indicates the advantages of such a strict association with plants surmounted the risks of loss of saprotrophic ability.

This group of soil fungi plays a key role in plant nutrition by facilitating the absorption of nutrients and water from the soil, and is capable of forming a mutual symbiotic relationship with more than 85% of plant species, including almost all major agricultural and horticultural crops. Fossil evidence suggests that this symbiotic relationship dates back more than 400 million years and played a key role in allowing plants to migrate from the sea to dry land and use terrestrial habitats (Liu et al., 2021; Yang et al., 2022; Li et al., 2022). Without mycorhizal fungi, today's crop plants would probably not exist, nor would life on the land as we know it. This relationship begins after mycelia is born through a spore (equivalent to plant seeds) and spreads to the root of the plant. If the plant agrees that it is necessary to have a relationship, the fungus is invited. Unique to AMF, these mycelia penetrate the cell membrane and stretch into the root cortex, thus bridging the gap between root and the soil biome (Imran.2022c; Imran et al., 2020a; Ali et al., 2022). These fungi form specialized structures, namely arbuscules (which serve as sites for the exchange of nutrients within the cell) and vesicles (Fig 4).

Role Of Mycorrhizae Under Biotic and Abiotic Stress Condition

Arbuscular mycorrhizal fungi (AMFs) are soil microorganisms, capable of forming potential symbolic relationships with the roots of 90% of host plants (Aguégué et al., 2021). AMF belong to the phylum Glomeromycota, a monophyletic group that differed from the same common ancestor as Ascomycota and Basidiomycota. While AM host plants can survive if deprived of their fungal symbionts, this condition is virtually unknown in natural ecosystems, where fungi AM act as true helper microorganisms, improving overall plant fitness. Mycorrhizal mycelium can increase the absorption range of roots and improve the resistance of plants to external stresses, promotes plant growth and improve the recovery of plants (Akpınar et al., 2020). Compared with vaccination of AMF in soil, injecting AMF into the soil can improve plant growth and increase nutrient availability (Amir et al., 2021). As a biological modifier, AMF can effectively improve soil structure and promote plant growth under both normal and stressful conditions. AMF can also affect the root shape of plants, increase the capacity of root branches, increase the absorption range of roots, and thus increase the absorption of water and nutrients. In addition, AMF plants can effectively increase the absorption and utilization of mineral nutrients especially P. Mycorrhizal activity increases the concentration of phosphorus available in the rhizosphere (Anli et al., 2021). Mycorrhizae lowers the pH of the root zone by selective absorption of NH_4^+ (NH-ions) and by release of H^+ ions. Decreased soil pH increases the solubility of phosphorous precipitate. Hyphal NH_4^+ uptake also increases the flow of nitrogen to the plant as the inner surfaces of the soil absorb ammonium and disperse it by diffusion

Phosphorus to the root or through diffusion and hyphae reduces the distance required for diffusion, thus increasing uptake. The rate of phosphorus flowing into mycorrhizae can be up to six times that of root hair (Arruda et al., 2021). In some cases, the role of phosphorus uptake can be completely taken over by the mycorrhizal network, and all of the plant's phosphorus may be of hyphal origin. At the same time, compartmented boxes experiment result shown that contribution of external hyphae (*Glomus mosseae*) to the uptake of P in white clover and maize grown soil is 79% and 20% respectively (Arruda et al., 2021). The benefits of AMF are greatest in systems with low inputs. Heavy use of phosphorus fertilizer can inhibit mycorrhizal colonization and growth. Less is known about the role of nitrogen nutrition in the mycorrhizal arbuscular system and its impact on symbiosis and the community. While significant progress has been made in clarifying the mechanisms of this complex interaction, a great deal remains to be investigated. As the levels of soil phosphorus available to plants increase, the amount of phosphorus in plant tissues also increases, and the carbon leakage to the plant by AM fungi symbiosis becomes non-beneficial to the plant (Arruda et al., 2021).

The AMF is an important microbial component of farming systems, making it one of the most commonly used symbols on earth, identified as the nutrient flow between the two living plants or microbes, the symbolic relationship involved in the bilateral biological exchange (Atteya et al., 2021). The mathematical model estimated the contribution of AMF hyphae to plants through plants has shown that plants rely solely on P nutrition. You may have to rely on the pathway of mycorrhizal. Morphologically, the AMF functions as an extension of the root system via its mycelium network and allows the plant to expand outside the boundaries of the rhizosphere, and this function is likely to prevail in the case of P deficiency. In this respect, the AMF provides an additional p-optical pathway (AMF pathway).

In most situations, arbuscules are a symptomatic device, preventing the direct uptake of root outbreaks resulting in rapid absorption of nutrients from the soil solution, which immediately disrupts the creation of the depletion region. AMF hyphae have a high affinity with inorganic P and can detect irreversible soil pores and increase the displacement of inorganic P (run) due to their smaller diameter than the roots. In fact, P is absorbed by AMF hyphae in the form of polyphosphatase and then transferred by intracellular hyphae to vacuole where it is cleaved. In addition, it appears that plants have acquired specific P transporters through evolution that have been identified for various species, including rice, tomatoes, potatoes, barely, and clover (Arruda et al., 2021; Atteya et al., 2021). However, so far, and even if the ability of fungal hyphae to exclude acid phosphatase enzymes and organic acids has been confirmed, compared to that of AMFPSM. May play a direct role in improving the availability of P and the roots are still not well defined. The general consensus is that AMF is involved in P solubilisation through coordination with PGPM. In fact, recent studies have shown that PGP bacteria associated with hypophobia-known AMF hyphae may play a key role in mycorrhizal symbiosis, which strongly suggests that AMF is present. Rhizobacteria may actually benefit as a source of nutrients from fungal eukaryotes, while hypobacteria provide growth factors that promote mycorrhizal symbiosis and plant growth (Bencherif et al., 2021). In addition, hypobacteria can often form biofilm-like structures that enhance the biodynamics of nutrients. Several features of PGP, including route-side enmity, have been demonstrated.

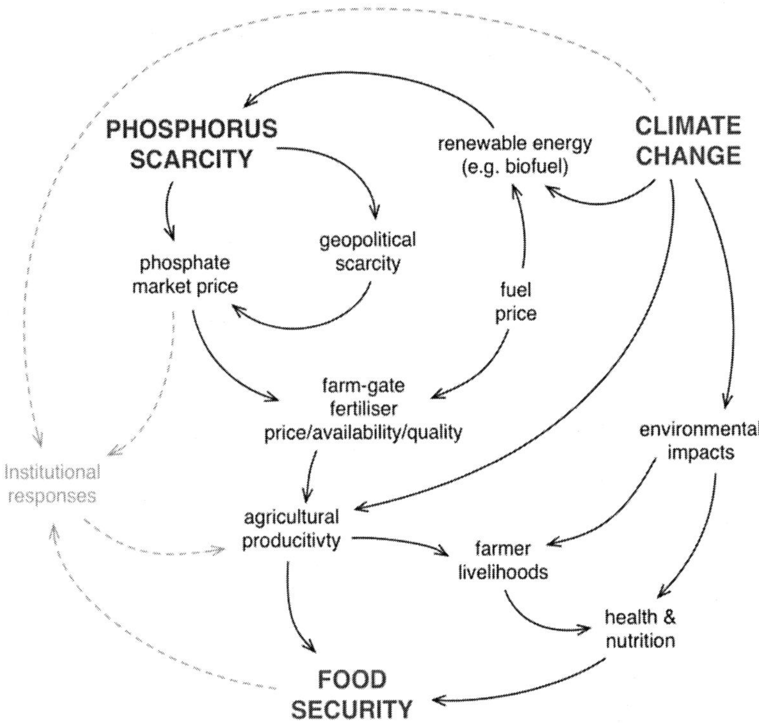

Figure 5. Depletion of plant nutrients in soil and its less availability threaten food security

The fast-changing biotechnological environment, which is affected by globalization, antagonism, financial pressures and the advancement of innovative technologies, has had an impact on the industry. It will be at the centre of interest in developed countries as demand for organic food continues to grow (Bastami et al., 2021). Growth and pathogenic resistance increases could be achieved by inoculating soil with mycorrhizae rather than by using inorganic fertilizers, pesticides and fungicides. Similarly, in developing countries, once inoculating soil with appropriate fungal isolates could eliminate the need for recurrent applications of expensive fertilizers that farmers cannot willingly afford (Bencherif et al., 2021). Due to many factors, including soil pH (Fig 6) nutrients deficiency are increasing day by day which leading to food insecurity and less biofortification of the food and its quality (Fig 5).

Mycorrhizal fungi may be one option that can immediately improve productivity while without AMF application crops may be reduce biomass production in natural and managed ecosystems without deteriorating the natural resource base and reducing the cost of fertilizers and energy needs to restore economic efficiency and environmental safety (Bhantana et al., 2021). If better mycorrhizal fungi strains were inoculated they could potentially advance the growth of almost all agronomic crops.

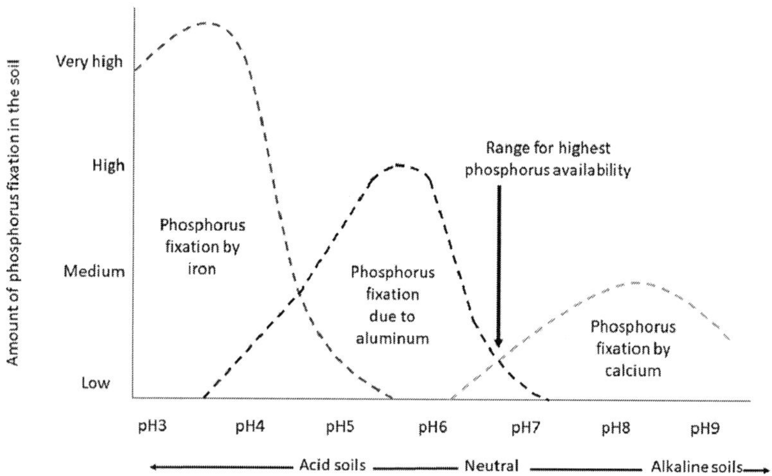

Figure 6. Phosphorus availability in various soil classes and its relation with soil pH

Impact of Mycorrhiza on Crop Production and Food Security

While mycorrhizal fungi are present in nearly 90% percent of plant families and occur for the greatest growth in many crop species, the widespread presence of these fungi in virtually all soils confides in the pressing need for inoculation with these natural connecting fungi (Bastami et al., 2021; Bhantana et al., 2021). Sophisticated scientific understanding of mycorrhizal symbioses recently enhanced commercial potential for the implementation of mycorrhizal technology in agriculture and horticulture production, forestry landscaping, phytoremediation in disturbed sites and other aspects of the plant market (Fig 7). The production of mycorrhizal inoculum is a multi-faceted procedure, as it requires professional enterprises to develop the indispensable biotechnological competence and capacity to counteract the permissible, ethical, educational and saleable requirements (Çekiç et al., 2012). There is a wide variety of formulation types, both liquid and solid, which can be tailored to the requirements. Smart and high quality commercial mycorrhizal inoculum is now available from a variety of sources. Consortium inoculum containing mixtures of species of mycorrhizal fungi regularly give the best response. Mycorrhizal inoculum comes in granular, powder, liquid and tablet forms. Dry products include dust, granules and wettable powders. Dusts have particle size ranging from 5 to 20 mm and contain <10 % of an organism by weight. Granules are distinct masses 5–10 mm 3 in size, pellets are > 10 mm 3 and briquettes are large blocks (5g) up to a few cubic centimetres (Bastami et al., 2021; Castañeda-Gómez et al., 2021; Çekiç et al., 2012). These products include an inert carrier such as carbon dioxide, lignite, clay minerals (perlite, vermiculite and bentonite), starch polymers, dry fertilizers and soil plant residues and may enhance plant biomass production (Fig 2.). Both endomycorrhizae and ectomycorrhizal are in small-scale commercial production. Mycorrhizal inoculum production systems in a gel carrier have been proposed and manufactured by some elite scientists that are highly concentrated and free from microbial contamination. Such breakthroughs are necessary for the practical use of mycorrhizal fungi (Celik et al., 2004). The greatest obstacle to the commercialization of mycorrhizal fungi appears to be the lack of large-scale fi eldest tests under typical agroecology, the adequate economics of mycorrhizal fungi technology, and a growing lack of willingness on the part of farmers to switch from an energy-intensive system to a new, but cheaper, energy-conservative system using mycorrhizal fungi.

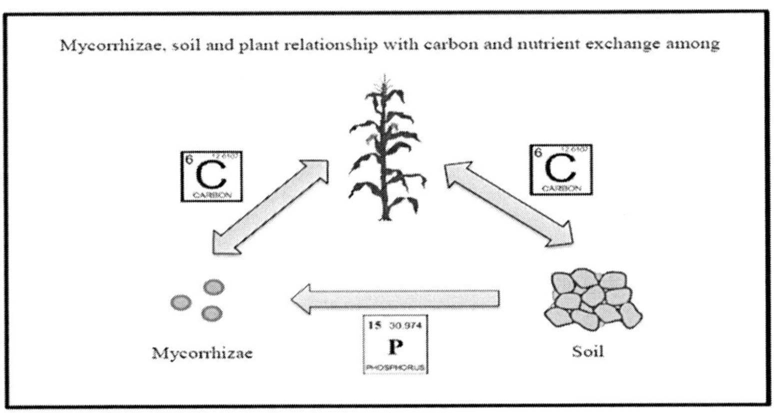

Figure 7. Mycorrhizae-soil and plant-carbon-nutrient exchange relationship

Plant and Mycorrhiza: A Symbiosis for Nutritional Security

Mycorrhizal fungi have a huge potential to be used in agriculture under a set of conditions that help plants to obtain nutrition through symbiosis (Cheng et al., 2021). Despite this, potential inoculation of crops is hardly ever seen, and endomycorrhizas are only consciously introduced in a small number of industries. Irrespective of the possible increase in yield, the money-making use of endomycorrhizas has not yet fully to take place. Several companies in USA, UK, Germany, France, India and Czech Republic establish to produce commercial inoculum, but they not find as much as expected marketing. On the other hand, forestry is one of the industries that have fully endorsed the role of mycorrhizas in plant growth (Chu et al., 2020). The bulk of commercial timber comes from trees that form ectomycorrhizal bonds, and mycorrhizal symbioses are binding in exotic trees. A number of other small industries habitually use mycorrhizal infection. Soil symbiotic microorganisms have increased the supply of nutrients and promoted the Grassland ecosystem (Demirbaş et al., 2019). Excessive soil content in mining-related soil can prevent the growth of plants, and the soil stress can be reduced by changing soil properties. Immunization of AMFs can increase resistance to plant stress.

One of the primary agricultural practices that increases crop productivity and supports robust agricultural ecosystems is beneficial microbe's application or seed inoculation and mycorrhizal soil application (Deveautour et al., 2021). It is a component that includes a variety of bio-products formed from microorganisms whosc bioavailability is essential for boosting and accelerating the biological processes of complex plant-microbe nutrition and soil health continuity. Effective bio fertilisers may be created using a variety of soil microorganisms (particularly bacteria and fungus) that display PGP characteristics. In general, there are four different forms of microbial bio formulation: (1) nitrogen-fixing bacteria (NFB); (2) phosphorus mobilizing/activating microorganisms (3) decomposing/ composting microorganisms and (4) bio-pesticides microorganism (Cheng et al., 2021; Demirbaş et al., 2019; Deveautour et al., 2021). Microbial groups that display additional PGP characteristics, such as phytohormones, siderophores, amino acids, and polysaccharides, may contribute to increased crop growth in addition to the crucial task for which they are selected. The advantages of microbial-based biofertilizers on the growth and development of significant crops can be restricted to one specific nutrient, such as N (thanks to N-fixing bacteria), but they can also be extended to a number of nutrients [i.e. owing to AMF]. Additionally, a polymicrobial compound called microorganism consortium formation, which contains a variety of microbial strains from different functional classes, can greatly encourage plant growth, productivity, and soil output.

To create effective polymicrobial formulations utilizing a variety of methods, it depends on how active, complementary, and coherent the candidate strain is. These advantageous impacts on crops when Rhizobium spp. is involved have also been documented. The development and dissemination of AMF inoculants are equally poorly understood and only inconsistently understood (Gujre et al., 2021). Utilizing fungi as inoculants may have questionable ecological implications, but phosphorus absorption is considerably more worrisome. The issue is that we don't fully understand how these fungi spread, despite the fact that we are aware that they may travel through soil, the air, and even migrating birds. Therefore, an AMF inoculant may have the potential to significantly affect local fungal populations and have disastrous effects on plant variety and the ecosystem's ability to operate if it is able to spread outside of its intended zone and become invasive. This is a stark illustration of how industrial practices clash with current scientific knowledge. The practice of AMF immunization as it currently exists either puts people in danger or endangers the environment (Gupta et al., 2021).

Long-term inoculation on how these commercial bio fertilizers are created and they may alter present fungal and plant communities will benefit in the development of appropriate practices. The primary author of this literature advise farmers, growers, and gardeners in his domain and farmers gathering to only apply inoculants when absolutely essential and to employ alternative inputs to promote the soil's general biodiversity in the interim. Symbionts are necessary for nutrient uptake and AM fungi sharing. They rely on the plant for their carbon feeding due to their low saprobic capacity.

The Mechanism of Nutrients uptake by arbuscular Mycorrhizae

Mycorrhizal fungi increase the efficiency of mineral uptake in soils or growing media where such mineral elements may be deficient or otherwise less readily available. These fungi have also been shown to moderate some nutrient uptake in soils that contain high levels of minerals, preventing plant toxicity (Gupta et al., 2021). Both of these mechanisms result in enhanced plant growth and/or increased yield. In this symbiosis, better access to and absorption of soil nutrients and water is provided to the plant. In this almost ubiquitous symbiotic relationship between plant roots and mycorrhizal fungi, it has been shown that it helps plants to allocate mineral nutrients from the soil, especially immobile elements such as phosphorus, zinc and copper, but also, under certain conditions, more mobile ions such as sulphur, calcium, sodium, potassium, iron, magnesium, manganese, chlorine, bromine and nitrogen (He et al., 2021).

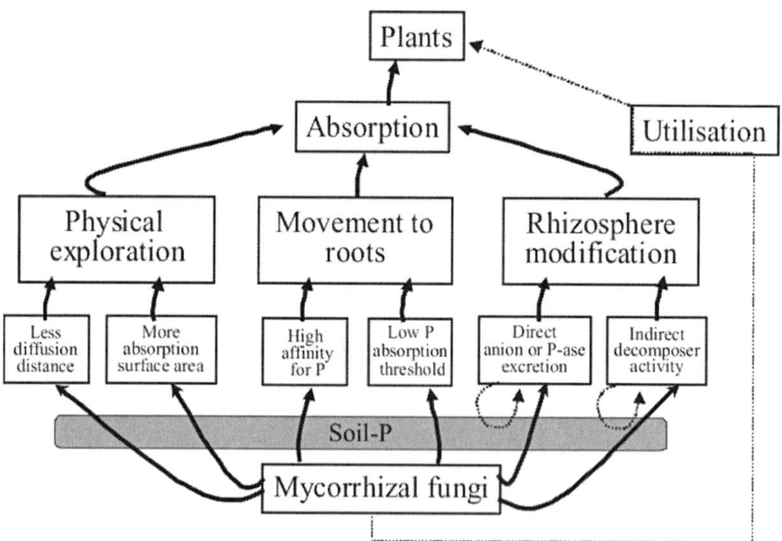

Figure 8. Complexity of nutrients availability to plants without mycorrhiza application

Mycorrhizal fungi are a common living organism in soil that facilitates plant growth, development, stress tolerance, soil remediation, C-sequestration, food safety and agricultural sustainability. Mycorrhizal fungi help the plants consume nutrients by expanding the network of mycorrhizal hyphae beyond the rhizosphere. Mycorrhizal inoculation alters the root architecture and studies have shown that the inoculated root absorption capacity of the nutrient is much better than non-inoculated root. It is believed for a long time that roots only consume nutrients through direct pathway (DP) while the role of AM fungi in nutrient absorption via mycorrhizal pathway (MP) is overlooked. But now the advancement of scientific methods and techniques has allowed the researcher to explore MP mechanisms for macro and micro nutrients, in addition to suppressing the plants' heavy metal stress (Hou et al., 2021). Plant hormones and root exudates also affect the formation and growth of pathogens, and they also point out new sites for mycorrhizal fungi and plant roots activity. MP's nutrient independence is better, and financial to DP. Understanding the exploration, mobilization and uptake of nutrients in root-mycorrhizal interactions has been discussed at molecular level. Contribution of plant and mycorrhizal transporters was addressed requiring more understanding. Also it was addressed the contribution of mycorrhizal inoculation to nutrient absorption relative to plant growth and development (Jabborova et al., 2021).

Signals of Mycorrhiza for the Exchange of Nutrients

Similar to other plant-interacting fungi, the exchange of nutrients and signals in Mycorrhizal fungi is based on the absorption of organic carbon from a living plant. Fifteen genes were annotated in the encoding putative hexose transporters (Jiang et al., 2021). The bicolour genome is upregulated during symbiosis, and much like it. Nevertheless, the genome lacks genes to encode invertases, the enzymes to glucose and fructose that hydrolyse sucrose (the most abundant sugar in plant tissues), indicating L. Bicolor relies on the glucose its green host releases. Interestingly, ustilago maydis, a biotrophic pathogen that causes smut disease on maize, has a sucrose transporter located on the plasma membrane 16 that allows the fungus to acquire sucrose directly without the need for secreted invertases (Kuyper et al., 2021). Thus, a range of mechanisms for accessing plant sugars seems to reflect some of the differences in nutritional strategies between symbiotic and pathogenic fungi. The rate of uptake of plant roots colonized by arbuscular mycorrhizal fungi is usually faster than that of non-mycorrhizal roots. Mycorrhizal fungal hyphae have a higher affinity for mineral ions and a lower absorption threshold than the roots of plants. In addition, the finest absorptive arbuscular mycorrhizal hyphae in soil are typically 2μm in diameter compared to root-hair diameters of 10 to 20μm and root-hair diameters of 100 to 500μm. Hyphae are therefore approximately 10 times more effective than root hair and approximately 100 times more effective than root hair. AM Fungi forming structures called vesicles (Liu et al., 2021a; Liu et al., 2021b; Lourdes et al., 2021). These structures act as storage sites within the roots of the plant and store absorbed minerals and lipids. These stored nutrients act as reserves and become accessible to the host plant when supplies are limited.

Mycorrhizal fungi have been shown to protect plants from saline conditions and heavy metal absorption, helping the plant to avoid nutrient lockouts. Mycorrhizal fungi have been shown to reduce the uptake of salts, balancing the toxic ion uptake of Na+ and Cl-and thus allowing the plant to absorb other ions, such as K+, Ca2+ Mn2+. Plants colonized by AMF have also been shown to have a buffer between the plant and heavy metals, preventing the absorption of elements such as copper, zinc, arsenic and aluminium (Maitra et al., 2021; Metwally et al., 2021). Accumulation of these heavy metals in plants can cause various symptoms, including: slow growth, chlorosis, root browning and death. This heavy metal resistance was found to occur through sequestration of these metals in the fungal tissue as well as through improved phosphorous nutrition of the plant.

Arbuscular Mycorrhizal Fungi: A Solution of Phosphorus Crises

Arbuscular mycorrhizal fungi have the ability to solubilize minerals otherwise insoluble nutrients cannot be available for plants. The action of mycorrhizal phosphatase enzymes turns phosphate into soluble forms and helps mycorrhizal plants to consume more phosphorus than non-mycorrhizal ones (Nouri et al., 2021). It has been shown that *Glomus intraradices'* extraradical hyphae were readily capable of hydrolyzing exogenously supplied sources of organic phosphorus and moving large amounts of phosphorus to roots.

Several mycorrhizal plants, including marigold, wheat, onion, leek and maize, may increase the activity of acid phosphatase in the roots and, therefore, organic phosphorus hydrolysis. A study on marigolds, for example, revealed that the increase of acid phosphatase activity in the association of arbuscular mycorrhizal fungi was controlled at the level of gene expression (Ortas, 2019). This provides clear evidence that molecular changes caused by arbuscular mycorrhizal fungi in the roots increase the ability of plants to convert insoluble organic P into bioavailable forms of P. In addition to their role in phosphorus nutrition, AMF plays a key role in the uptake and conversion of insoluble nitrogen into bioavailable forms. AMF increases the decomposition and subsequent capture of inorganic nitrogen from complex organic materials such as plant litter. It was also documented that the arbuscular mycorrhizal fungi species increased the mineralization of nitrogen from organic residues to different amounts, e.g., *Glomus clarium* hyphae recovered and translocated up to mineralized nitrogen phosphorus to Russian wild rye (Qu et al., 2021). AMF can increase the degradation of organic residues and the absorption of nitrogen by the host plant.

Fungi (ie, penicillium and Aspergillus), provide the release of many biological complexes to solubilize nutrients. In the form of ortho-phosphate (HPO_4) by means of special mechanisms that play a key role in the hydrolyzing of organic forms, organic acids, siderophore production, and phosphatase enzymes (Ortas et al., 2011; Poveda et al., 2019). Overall, PSM can be helpful in promoting plant growth by directly enhancing the utilization of NPK through the emission of direct organic acids and P-hydrolyzing phosphatase enzymes and thus increasing the bioavailable P pool. Or indirectly through the production of phytohormones, antifungal compounds, toxic resistant compounds, and other high value bioactive molecules that form a strong pill / root system, especially under biotic and abiotic barriers.

Despite the numerous mechanisms involved in P-solubilization, organic acids are important contributors. Saturation depends primarily on the amount and quantity of organic acids by beneficial microorganisms, but this ecosystem may also be affected by environmental characteristics. For example, N and C can directly influence the nature of the organic acids produced by the soil content, the nature of the C source can affect the process of bio-solubilization, and the high C/P ratio of organic acid production While C/N and N/P can both influence the development of microorganisms. It is also important to note that P solubilization efficiency depends more on the quantity of organic acids and on the quality rather than on the sources of P (Ortaş et al., 2017b; Qu et al., 2021). When comparing the P solubilizing properties of the two fungal strains (*Teloromyces flavus* and *Penicillium porpogeneum*), found that both strains exhibited equal solubilization potential, although *T. flavus* compared with P. purpurenum. Produced less organic acids.

Phosphorus & Nitrogen are Mediated by Arbuscular Mycorrhiza

The most influential tool in most agricultural lands which can provide clear answers to address nutritional limitations (especially N&P). Today beneficial microbes (as inoculants, biofertilizers, or bio stimulants) are commonly used to encourage plant growth and serve as biological control agents (Ortaş and Rafique, 2017). Before agriculturally beneficial microorganisms, especially AMF, were detected, farmers moved productive soil from one field to another, thus positively affecting crop productivity. It showed that clover plants can get greater quantities of nitrogen fertilizer during crop rotation than they do in other grains. The inclusion of some soil microorganisms in this phenomenon is not a novel, which began in early era (Ortas et al., 2017; Palacios et al., 2021). As a result, many phosphorus solubilizing microorganisms (PSMs) inhabit beneficial bacteria and fungi (Fig 8). That promise opened in close association with the plants.

The purpose of powerful tools is to establish sustainable agriculture by improving the use of P. A few years later, microbial inoculant based on P-solubilizing *B. megaterium* (ie, fibrobacterin) was successfully used in Russia. The introduction of efficient products aimed at enhancing efficiency of nutrients.

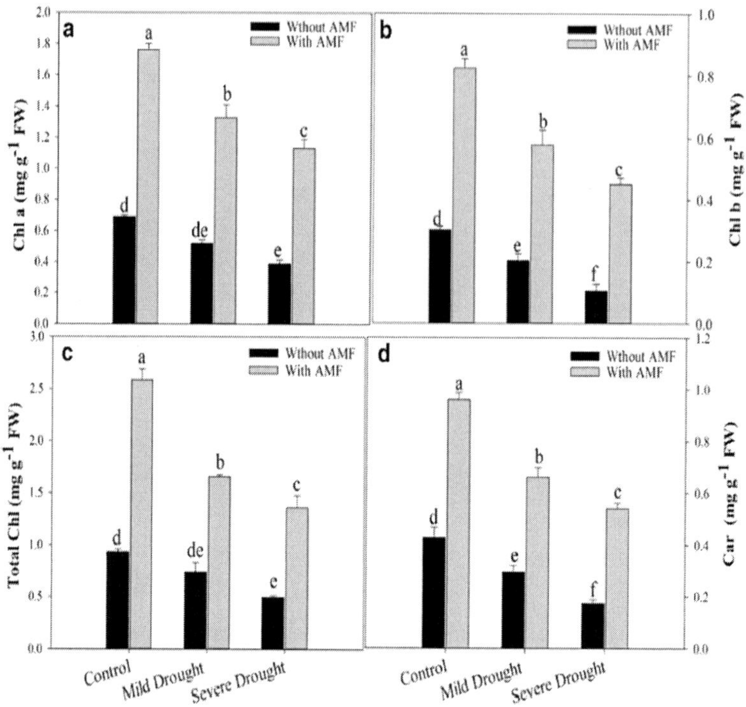

Figure 9. Effect of AMF inoculation on a chlorophyll a (Chl a), b chlorophyll b (Chl b), c total chlorophylls, and d carotenoids under drought stress.

Solubilisation and Mobilization of Phosphorus

Phosphorus is an essential mineral nutrient for agricultural and crop development and is important for physiological processes. As a matter of fact, agricultural production has increased almost threefold since agricultural revelation (after the year 1950), with the introduction of mineral fertilizers, including P fertilizers, widely attributed to agricultural systems (Quan et al., 2021). P root is known to promote root growth, rapid plant maturity, seed generation, BNF, water use efficiencies, and resistance to diseases. It also plays a central role in essential physiological processes including energy transfer and storage, light synthesis, cell division, and seed formation, all of which are energy-dependent biological processes mediated by two intracellular PK molecules (Quan et al., 2021; Riaz et al., 2021). ADP (adenosine triphosphate) and ATP (adenosine triphosphate) respectively.

According to the content of the plants and soil, P represents about 0.2% of the total dry weight of the plants and 0.05% (w/w) of the soil content, with only a small fraction of the bio-available for the plants. In soil solution, the available bioavailability of NPK rarely exceeds 10 μM and plants must have specialized transporters and complex flow systems that efficiently distribute these nutrients between all parts of the plant (Rodrigues et al., 2021; Roy-Bolduc and Hijri, 2011). The low concentration of P in the soil solution allows for plant growth to be very nutrient limited due to the nutrient production required for P. In addition, the availability of P depends on the rhizosphere pH and the occurrence of fixed and rainy conditions, thus reducing the P fertilizer performance and consequently. For example, in striking soils, the use of P fertilizers is severely hampered by rainfall and disturbance (Saboor et al., 2021). Therefore, one of the key challenges facing today's agriculture lies in finding solutions for improving or enhancing the bioavailability of nutrients while knowing productivity.

Crop, Soil, Environment and Nutrients Uptake

Agriculture resources has challenged production in addition to increasing food demand, shortage of cultivable land, and numerous environmental pressures, needs to be managed wisely through sustainable and environmentally efficient ways (Roy-Bolduc and Hijri, 2011; Sagar et al., 2021). Modern agriculture needs to be more productive, sustainable, and environmentally friendly. Microbial-based bio formulations that enhance plant efficiency are in great need, and in particular bioformulations that show the effects of complement and harmony with mineral fertilizers. Various strategies for integrating soil fertility have been used through numerous controlled and uncontrolled experiments, but to better understand the many functions of beneficial microorganisms within the soil microbial community and with plants and minerals. Mutual contact requires more effort to get better development (Sorensen et al., 2008). In fact, the common use of AMF is an emerging research area that aims to design and develop effective microbial formulations. Extremely compatible with mineral inputs, having a positive impact on both crops and the environment. This novel approach may be of global interest, especially for most N- and P-deficient agricultural ecosystems. In this chapter, we report on the importance of NF bacteria and their association with mineral N, P and K fertilization in improving crop production ability and fertilizer performance, as well as the importance of P / microorganisms resolving. The interactive and synergistic effects of multi-trophic interactions and their potential for plant mineralization, crop productivity and environmental constraints have been highlighted (Tran et al., 2021; Turan, 2021). Improving the use of mineral nutrients is essential for achieving high yields and productivity in a sustainable manner, and for developing and testing modern integrated plant nutrition management systems. The provision of primarily organic, inorganic, and chemical fertilisers is a key component in the management of crop nutrients through integrated techniques. The use of balanced fertilisers, together with the effective utilisation of organic sources and advantageous microorganisms, has, nevertheless, long been recognised as the secret to good yields and net returns. The retention, disintegration, and assimilation of plant residues, as well as the cycling of nutrients and their impact on soil structure and porosity, are all crucial functions of the soil microbial community (Tran et al., 2021; Turan, 2021; van't Padje et al., 2021a). Drought conditions limit the root system's capacity to absorb nutrients due to impaired root development and function. Drought spell in growing periods cause both carbon and oxygen to migrate to the root nodules, and a drop in nitrogen accretion inhibits legume crops nodulation and nitrogen fixation. The principal component of soil organic matter, soil organic carbon, gives the soil structure and helps it to retain more water for usage of crops in arid regions. Both direct and indirect consequences of climate change may be seen in the condition of the soil and in the actions of the soil microbes that release greenhouse gases into the atmosphere and cause global warming(Turan, 2021). Climate-driven changes in plant productivity and diversity affect soil physicochemical conditions, the supply of carbon to the soil, and the structure and activity of microbial communities involved in decomposition processes and carbon release from soil. These changes have direct effects on soil microbes and greenhouse gas production. In contrast, indirect effects are caused by changes in plant productivity and diversity.

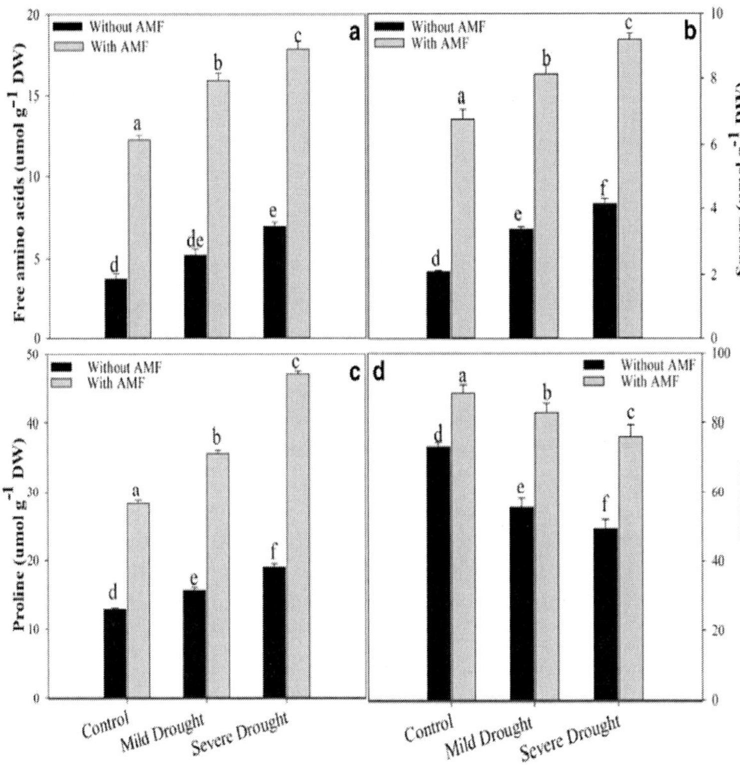

Figure 10. Effect of AMF inoculation on a free amino acids, b sugars, c proline, and d relative water content under drought stress.

Abiotic stresses, such as drought, are the key challenges for growth and production, leading to an annual of billions of dollars loss. Globally, in arid and semiarid zones, plants were affected by drought in various ways that include water shortage and transpiration reduction, which leads to oxidative stress (van't Padje et al., 2021a). Various cellular organelles generate toxic radicals that include hydroxyl ion, superoxide, and hydrogen peroxide effect on mechanical and functional constancy to oxidized the related nucleic acids, proteins, and lipids (van't Padje et al., 2021b). To combat abiotic stresses, plants adopt different physiological as well as molecular mechanisms that include the stimulation of antioxidant enzymes, i.e., catalase, peroxidase, superoxide dismutase, ascorbate peroxidase etc.

Previous studies indicated that intense drought stress always has adverse effects on the synthesis of essential oil compounds and secondary metabolites in medicinal plants. However, under the prevailing stress conditions, including drought, the input of high fertilizers does not constantly guarantee the uppermost level of secondary metabolites (Vosnjak et al., 2021; Wang and Feng, 2021). To meet the prevailing climate threats, especially drought, most plants establish various defences' techniques. Among defence techniques, the use of beneficial microbes, such as arbuscular mycorrhizal fungi (AMF), clearly proved to help the host plant through the symbiotic association of roots. AMF species belong to the subphylum Glomeromycotina of the phylum Mucoromycota. AMF are well-known for improving the association with various plant species and consequently improve the contents of secondary metabolites and thus improve plant nutrition in order to enhance growth and development. This improvement changes the synthesis of different molecule families and terpenoids, glucosinolates, and essential oil and phenolic compounds and the phytoalexins (Wang et al., 2021a; Wang et al., 2021b). Further, the potential effects of AMF on the synthesis of secondary metabolites in different crops and also in medicinal plants have already been investigated in several studies. Moreover, AMF are supposed to be the real alternative to improve the growth and development of aromatic plants which are usually found in low fertility soil. Various studies have reported the links between AMF and tobacco, while others have been shown that the inoculation with AMF improves plant growth, augments photosynthesis, and increases the concentration of mineral nutrients and antioxidant functioning. However, the potential role of AMF inoculation on the synthesis of secondary metabolites in tobacco under drought stress needs further study. Additionally, the plants' symbiotic association with root colonization through AMF may enhance the contents of phenolic compounds and also the plants' defense enzymes, e.g., phenylalanine ammonialyase and lipoxygenases, Lipoxygenases are abundant in various plant processes, e.g., plant development and growth, as well as reactions to pest and pathogen damage and response to several environmental factors (Wu et al., 2021).

Many studies documented that the activation of the lipoxygenases pathway enhances important functions in the primary and secondary metabolism in the plants. The cox enzymes are responsible for resistance against abiotic stresses, e.g., drought, higher salinity by altering the stress-responsive marker genes' expressions, abscisic acid, production of reactive oxygen species, and lipid peroxidation in Arabidopsis (Yadav et al., 2021). Phenylalanine ammonia-lyase is an enzyme extensively found in higher plants and performs its role in accomplishing primary and secondary metabolism. During AMF symbiosis, these enzymes have been investigated by researchers due to their impact on the colonization process as they produce plant biomarkers in the physiological state. Moreover, AMF creates a specific compound called nitrogen-linked glycoprotein called glomalin. Economic crop, and drought stress has a significant effect on crop quality and yield. It is pertinent to learn the mechanism and production technologies of the tobacco crop. However, little attention has been paid to AMF and its consequences on tobacco plants' drought tolerance (Yadav et al., 2021; Yazici et al., 2021). Therefore, here, the present study hypothesized that AM inoculation would stimulate tobacco plant growth to enhance mineral nutrients' uptake, regulate enzymatic activities, and defend against drought stress. The potential of AMF inoculation on crop growth, photosynthesis and chlorophyll fluorescence parameters, antioxidant enzyme activities, osmolyte metabolism, contents of glomalin, and amount of essential oil of various oilseed crops under drought stress conditions.

Indirect climate-microbe feedback is the indirect effects on soil microbial communities and their activity and hence the potential for microbial feedback to climate change-through its influence on plant growth and vegetation composition. Climate change may have stronger or weaker, permanent or periodical, favorable or unfavorable, harmful (sometimes catastrophic), primary (direct) or secondary (indirect) impact on soil processes. Among these processes, soil moisture regime and beneficial microbes play a distinguished role (Zare et al., 2021; Zhang et al., 2021). It determines the water supply of plants, influences the air and heat regimes, biological activity, and plant nutrient status of soil. In most cases, it determines the agro-ecological potential, the biomass production of various natural and agroecosystems, and the hazard of soil and water pollution. Excessive precipitation in changing climate causes a significant source of soil nutrient loss like nitrate leaching. Agricultural areas with poorly drained soils or that experience frequent or intense rainfall events can have waterlogged soils that become hypoxic. The change in soil redox status under low oxygen can lead to elemental toxicities of Mn, Fe, Al, and B that reduce crop yields and the production of phytotoxic organic solutes that impair root growth and function. Hypoxia can also result in nutrient deficiency since the active transport of ions into root cells is driven by ATP synthesized through the oxygen-dependent mitochondrial electron transport chain. Significant nitrogen losses can also occur under hypoxic conditions through denitrification as nitrate is used as an alternative electron acceptor by microorganisms in the absence of oxygen (Anli et al., 2021; Bencherif et al., 2021). Soil warming can increase nutrient uptake from 100-300% by enlarging the root surface area and increasing rates of nutrient diffusion and water influx. Since warmer temperatures increase rates of transpiration, plants tend to acquire water-soluble nutrients (nitrate, sulfate, Ca, Mg primarily move towards roots through transpiration-driven mass flow) more readily as temperature increases (Bencherif et al., 2021; Gull et al., 2004). Temperature increases in the rhizosphere can also stimulate nutrient acquisition by increasing nutrient uptake via faster ion diffusion rates and increased root metabolism. However, any positive effects of warmer temperature on nutrient capture are dependent on adequate soil moisture. If under dry conditions higher temperatures result in extreme vapor pressure deficits that trigger stomatal closure (reducing the water diffusion pathway in leaves), then nutrient acquisition driven by mass flow will decrease.

Mycorrhiza a Symbiosis for Life: Supply of Nutrients Ensure Higher Biomass & Food Security

Global food demand is increasing rapidly and in developing countries where crop lands and resources are hardly contributing to the efficient production of crops needed to meet such immediate demand for food. Sustainable agricultural production needs to be sustainable through the use of efficient agricultural biosystems that have the potential to reduce the negative effects of the entire agricultural system of biochemical diversity and low soil fertility, climate stress (Fig 10), pathogens and pests (Arruda et al., 2021; Bencherif et al., 2021; Gull et al., 2004). The issue of global food security will foster a dependence on innovation, development, and the delivery of technologies that will increase food production while ensuring sustainable agriculture. A number of advanced and effective technologies have been adopted, such as smart irrigation systems, smart fertilizers [ie, controlled release fertilizers and improved performance fertilizers, integrated fertilization, and disease biocontrol strategies as well as diverse imaging and sensing. Based technologies that provide the most valuable information to monitor and protect crop productivity (Liu et al., 2021b). Agricultural microbial biotechnology can represent a sustainable solution to improve agricultural production through the integration of microbiome and microbiome interactions.

In general, beneficial microorganisms in the rhizosphere can promote plant growth through a variety of regulatory biochemical pathways (categorized as direct and indirect mechanisms), including controlling plant hormonal signalling (Fig 9, Fig 10). Preventing pathogenic microbial stress and increasing soil-based nutrient bioavailability (Martínez-Viveros et al., 2010; Rafique et al., 2019). Direct pathways usually include resource acquisition (e.g., N, P, K and essential micronutrients), plant hormone bio-synthesis, and many molecules in the vicinity of the rhizosphere (e.g. siderophores), or antinuclear, such as aminosylcropine (Wang and Feng, 2021). Indirect mechanisms by which microorganisms of the rhizosphere can promote plant growth are primarily involved in the reduction of the inhibitory effects of various phytopathogens by acting as biocontrol agents. Biosynthesis of antimicrobial metabolites, 2,4-diacetylfluoroglucinol, etc., nutrient competition and stimulated immune system resistance or self defence systems in plant and thus promote growth, yield and biomass production under harsh environment.

The future prospect of Mycorrhiza for sustainable agriculture

A wide range of association exists in nature of which symbiotic association is the most evolutionary conserved where both the partners are mutually benefited. However, these associations of the plants with microorganisms are both harmful as well as beneficial (Wang and Feng, 2021). Hence the key to survival for plants is to promote beneficial symbiotic associations and prevent the intruding pathogens. In this chapter, we focus on the beneficial role of mycorrhiza, a heterogeneous group of fungi primarily associated with plant roots (Jansa et al., 2003). Out of these associations, one group of root obligate mycorrhiza provides the host plant with water nutrients and importantly protects from pathogens, called Arbuscular mycorrhiza fungi (AMF). This association helps in improving the nutrient supply like nitrogen and phosphorus to the plants and in turn, the fungus gets 20% of the fixed carbon from the plants. The nutrient transfer is established by the help of structures called arbuscules and the release of 'Myc factors' by fungus and strigolactones by plant root exudates induces the symbiotic reactions (Yazici et al., 2021). A coordination of these two factors, in turn, switches on the downstream signaling cascades that further strengthens the association. In the course of research over the years several genes like novel transporters or receptor kinases have been identified that favors the association and in turn helps in dissecting the cellular responses. In this chapter, we discuss the current status of several studies, which unravel the contributions of partners involved in these symbiotic associations and can be exploited for improvement of agricultural crops in diverse environmental conditions (Kuyper et al., 2021; Ma et al., 2021).

Mycorrhiza refers to the association between plant roots and biotrophic mycorrhizal fungi, these mycorrhizal fungi form a network of filaments associated with plant roots and enable roots to absorb the nutrients from the soil. Moreover, this fungus-plant alliance stimulates plant growth and accelerates root development. Mycorrhiza is classified into five groups based on their morphological characteristics as arbuscular, arbutoid, ectoericoid, monoptropoid and orchid (Poveda et al., 2019; Wu et al., 2021). Arbuscular mycorrhiza (AM) is the most common variety of mycorrhizal fungi, these are microscopic and branched forms found in the cortical cells of root. These AM fungi are found in 80% of vascular plants and involved in diverse roles from accelerating the nutrient uptake to improve plant development and improve soil health and properties thereby influencing the ecosystem. Previously, it has been shown that fungal-host interaction can produce microbe-associated molecular patterns (MAMPs), exopolysaccharides, volatile organic compounds (VOCs), Myc factors, and Nod factors (Yazici et al., 2021). Moreover, VOCs have demonstrated ability to modulate root system architecture favoring symbiotic associations and also modulate Nod factor signaling pathway enabling AMF colonization. The initiation of AMF symbiosis with host plant involves varieties of plant genes, hormones etc and of these strigolactones and fungal-derived lipochito-oligosaccharides have been identified as key players. Furthermore, it can also be said that a thorough identification and characterization of the fungal effectome can be of immense help in understanding the interaction of the fungus and the plants. Here, we have compiled and summarized the current knowledge regarding AMF symbiosis in plants (Castañeda-Gómez et al., 2021; Hussain et al., 2021). We present an overview of different phases of AM association with plant roots, the role of AM in nutrient absorption and phytohormone signaling. This chapter shades light on AMF as primary biotic soil component, which is important for efficient ecosystem functions. Information presented here can be used to improve plant AMF symbiosis and these improvements open a new avenue to employ AMF as bio-fertilizers for sustainable agriculture systems.

Conclusion

The presence of symbiotic association is a conserved phenomenon that has been documented from times immemorial. Despite the fact that the genes facilitating the symbiotic association between both the partners is conserved but the lack of knowledge about the conservation of biochemical function has made it difficult to understand their function. The present chapter clearly emphasizes the important role played by mycorrhiza in benefiting the terrestrial ecosystems and it also highlights the fact. The present findings clearly demonstrated the role played by AMF in benefiting crop plants and soil cultivation and showed that AMF considerably improved the growth mediated by drought and other various environmental factors by improving photosynthesis, mineral nutrients, glomalin contents of soil, contents of essential oil, and chemical composition, upregulating the antioxidant systems and enzymatic activity of phenylalanine ammonia-lyase and lipoxygenase. In crop cultivation, AMF improved both the quality and productivity of soil and crops that was specifically noticed in the contents of essential oil compounds. The present outcomes suggested that the symbiotic association between AMF and crop plants may help to strengthen the fabrication of defensive compounds that ultimately help to relieve abiotic and biotic stresses and may also act as biological control of crop production constraints.

Conflict of Interest

The authors declared that they have no any conflict of Interest.

References

1. Ali M. Qaisi, Abdel Rahman Al Tawaha, Imran, Moh, d Al-Rifaee.2022. Effects of chitosan and Biochar-Mended soil on growth, yield and yield components and mineral composition of Fenugreek. *Gesunde Pflanzen. doi.org/10.1007/s10343-022-00727-x*
2. Ding, W., Meng, Q., Dong, G., Qi, N., Zhao, H., Shi, S. (2022). Metabolic engineering of threonine catabolism enables Saccharomyces cerevisiae to produce propionate under aerobic conditions. Biotechnology journal, e2100579. doi: 10.1002/biot.202100579
3. Imran & Amanullah (2022a): Soybean quality and profitability improved with peach (Prunus persica L) remnants, phosphorus and beneficial microbes, *Journal of Plant Nutrition*, DOI: 10.1080/01904167.2022.2068438
4. Imran and Amanullah. (2022b). Integration of peach (*Prunus persica* L) remnants in combination with beneficial microbes and phosphorus differ phosphorus use efficiency, agronomic efficiency and partial factor productivity in soybean Vs maize crops. *Journal of Plant Nutrition.* doi.org/10.1080/01904167.2022.2099890
5. Imran and Amanullah. 2021a. Assessment of chemical and manual weed control approaches for effective weed suppression and maize productivity enhancement under maize-wheat cropping system. *Gesunde Pflanzen*, DOI: 10.1007/s10343-021-00599-7.
6. Imran and Amanullah· 2021b. Phosphorus and Boron Application Optimizing Biofortification Of phosphorus and Productivity of French Bean (*Phaseolus Vulgaris* L.). *Communication In Soil Science and Plant Analysis.* DOI:10.1080/00103624.2021.1971689
7. Imran and Amanullah. 2021c. Phosphorus biofortification and uptake in maize enhanced with integrated phosphorus management. *Phosphorus, Sulfur, and Silicon and the Related Elements*. DOI: 10.1080/10426507.2021.2022677.
8. Imran, Amanullah & Abdel Rahman Al Tawaha (2022a): Humic Acid and Sulfur Integration Enhances Growth and Yield Related Traits Of *Brassica Napus L. Journal of Plant Nutrition. d*oi.org/10.1080/01904167.2022.212882
9. Imran, Amanullah & Abdel Rahman Al Tawaha (2022b): Indigenous organic resources utilization, application methods and sowing time replenish soil nitrogen and increase maize yield and total dry biomass, *Journal of Plant Nutrition*, DOI: 10.1080/01904167.2022.2067055.

10. Imran, Amanullah and Abdel Rahman Al-Tawaha. 2021a. Carbon Sources Application Increase Wheat Yield And Soil Fertility. *Communication in soil science and plant analysis DOI: 10.1080/00103624.2020.1865397.*
11. Imran, Amanullah and Abdel Rahman M. Al Tawaha. 2021b. Management of Nano-black carbon, Phosphorous and Bio Fertilizer improve Soil organic carbon and ensilage biomass of soybean and maize. *Communication in Soil Science and Plant Analysis.* DOI.10.1080/00103624.2021.1966439
12. Imran, Amanullah and Abdel Rehman Altawaha. 2022c. Carbon Assimilation and Dry Matter Partitioning in Soybean Ameliorates with the Integration Of Nano-Black Carbon, Along With Beneficial Microbes and Phosphorus Fertilization. *Journal of Plant Nutrition, DOI:10.1080/01904167.2022.2035753*
13. Imran, Amanullah and Abdel Rehman Al-Tawaha.2020a. The Productivity of subsequent wheat enhanced with residual carbon sources and phosphorus under improved irrigation system. *Communication in soil science and plant analysis*, DOI:10.1080/00103624.2020.1763387
14. Imran, Amanullah and Ibrahim Ortash. 2022d. Agronomic Practices Improved Cucumber Productivity, Nutrients Uptake and Quality. *Gesunde Pflanzen*. 10.1007/s10343-022-00634-1.
15. Imran, Amanullah, and Abdel Rahman Al Tawaha. 2022e. Regenerating potential of dual purpose rapeseed (*Brassica napus* L.) as influenced by decapitation stress and variable rates of phosphorous. *Communication in soil science and plant analysis. Doi.org/10.1080/00103624.2022.2118297*
16. Imran, Amanullah, Asad Ali khan, Tariq Mahmood, Abdel Rahman Al Tawaha and Samia Khanum. 2021c. Adequate Fertilization, Application Method and Sowing Techniques Improve Maize yield and related traits. *Communication in soil science and plant analysis*. DOI:10.1080/00103624.2021.1925688
17. Imran, Amanullah, Muhammad Arif, Zahir Shah and Abdul Bari. 2020b. Soil application of *Trichoderma* and Peach (*Prunus persica* L.) residues possesses biocontrol potential for weeds and enhances growth and profitability of soybean (Glycine max). *Sarhad Journal of Agriculture* 36 (1) 10-20. DOI | http://dx.doi.org/10.17582/journal.sja/2020/36.1.10.20
18. Imran. 2021. The bioavailability of Phosphorus in composite vs. Hybrid maize differ with phosphorus and boron fertilization. *Phosphorus, Sulfur, and Silicon and the Related Elements. https://doi.org/10.1080/10426507.2021.1920588.*

19. Imran. 2022a. (Review) Phosphorus availability enhanced with combine application of organic amendments and beneficial microbes under soybean-wheat cropping system. *Communication in soil science and plant analysis.* doi.org/10.1080/00103624.2022.2034848
20. Imran. 2022b. Growing of Off-Season Tomato in High Tunnel and Its Nutritional Value Augmentation with Integrated Nutrients Management. *Journal of Plant Nutrition.* Doi: 10.1080/01904167.2022.2046062
21. Imran.2022c. Management of Abiotic Stresses with Nano-black carbon is a tool for Crop Production. *Journal of Plant Nutrition.* doi.org/10.1080/01904167.2022.2046085
22. Izhar Ali, Saif Ullah, Imran, Zixiong Guo 2022. Changes in Soil Properties and Microbial Communities with Biochar Amendment in Combination with Nitrogen Fertilizer in a paddy field. Frontiers Microbiology. DOI: 10.3389/fmicb.2022.834751
23. Izhar Ali, Saif Ullah, Imran, Zixiong Guo1, Shangqing Wei1 and Ligeng Jiang. 2021. Combined application of biochar and nitrogen fertilizer promotes the activity of starch metabolism enzymes and the expression of related genes in rice in a dual cropping system. BMC Plant Biology *(2021) 21:600* https://doi.org/10.1186/s12870-021-03384-w
24. Li, J., Charles, L. S., Yang, Z., Du, G., & Fu, S. (2022). Differential Mechanisms Drive Species Loss Under Artificial Shade and Fertilization in the Alpine Meadow of the Tibetan Plateau. Frontiers in Plant Science. doi: 10.3389/fpls.2022.832473
25. Li, Y., Erhunmwunsee, F., Liu, M., Yang, K., Zheng, W., Tian, J. (2022). Antimicrobial mechanisms of spice essential oils and application in food industry. Food Chemistry, 382, 132312. doi: https://doi.org/10.1016/j.foodchem.2022.132312
26. Liu, Z., Ying, H., Chen, M., Bai, J., Xue, Y., Yin, Y., Dou, Z. (2021). Optimization of China's maize and soy production can ensure feed sufficiency at lower nitrogen and carbon footprints. Nature food, 2(6), 426-433. doi: 10.1038/s43016-021-00300-1
27. Lu, L., Zhai, X., Li, X., Wang, S., Zhang, L., Wang, L., Wang, F. (2022). Met1-specific motifs conserved in OTUB subfamily of green plants enable rice OTUB1 to hydrolyse Met1 ubiquitin chains. Nature Communications, 13(1), 4672. doi: 10.1038/s41467-022-32364-3
28. Mohammad Ilyas, Gohar Ayub, Imran*, Azmat Ali Awan and Manzoor Ahmad.2020. Calcium and Boron effect on Production and Quality of Autumn Potato under Chilling Temperature. *Communication in soil science and plant analysis*. DOI: 10.1080/00103624.2020.1854286
29. Pan, C., Yang, K., Erhunmwunsee, F., Li, Y., Liu, M., Pan, S., Tian, J.

(2023). Inhibitory effect of cinnamaldehyde on Fusarium solani and its application in postharvest preservation of sweet potato. Food Chemistry, 408, 135213. doi: https://doi.org/10.1016/j.foodchem.2022.135213
30. Sun, J., Jia, Q., Li, Y., Zhang, T., Chen, J., Ren, Y., Fu, S. (2022). Effects of Arbuscular Mycorrhizal Fungi and Biochar on Growth, Nutrient Absorption, and Physiological Properties of Maize (Zea mays L.). Journal of Fungi, 8(12). doi: 10.3390/jof8121275
31. Wang, Y., Liu, S., Yang, X., Zhang, J., Zhang, Y., Liu, X., Wang, H. (2022). Effect of germination on nutritional properties and quality attributes of glutinous rice flour and dumplings. Journal of Food Composition and Analysis, 108, 104440. doi: https://doi.org/10.1016/j.jfca.2022.104440
32. Xie, Q., Liu, G., Zhang, Y., Yu, J., Wang, Y., Ma, X. (2022). Active edible films with plant extracts: a updated review of their types, preparations, reinforcing properties, and applications in muscle foods packaging and preservation. Critical Reviews in Food Science and Nutrition, 1-23. doi: 10.1080/10408398.2022.2092058
33. Yang, K., Geng, Q., Luo, Y., Xie, R., Sun, T., Wang, Z., Tian, J. (2022). Dysfunction of FadA-cAMP signalling decreases Aspergillus flavus resistance to antimicrobial natural preservative Perillaldehyde and AFB1 biosynthesis. Environmental microbiology, 24(3), 1590-1607. doi: 10.1111/1462-2920.15940
34. Yang, Y., Dou, Y., Wang, B., Xue, Z., Wang, Y., An, S., Chang, S. X. (2022). Deciphering factors driving soil microbial life-history strategies in restored grasslands. iMeta. doi: https://doi.org/10.1002/imt2.66
35. Zhang, Y., Zhang, S., Yang, X., Wang, W., Liu, X., Wang, H., Zhang, H. (2022). Enhancing the fermentation performance of frozen dough by ultrasonication: Effect of starch hierarchical structures. Journal of Cereal Science, 106, 103500. doi: https://doi.org/10.1016/j.jcs.2022.103500
36. Aguégué, M. R., Ahoyo Adjovi, N. R., Agbodjato, N. A., Noumavo, P. A., Assogba, S., Salami, H., Salako, V. K., Ramón, R., Baba-Moussa, F., and Adjanohoun, A. (2021). Efficacy of native strains of arbuscular mycorrhizal fungi on maize productivity on ferralitic soil in Benin. *Agricultural Research*, 1-15.
37. Akpınar, Ç., Demirbaş, A., and Ortaş, İ. (2020). The Effects of Different Mycorrhiza Species, Compost and Rock Phosphate Applications on the Growth of Sorghum Plant. *Turkish Journal of Agriculture-Food Science and Technology* **8**, 2583-2589.
38. Amir, H., Gensous, S., Cavaloc, Y., and Wantiez, L. (2021). Phosphorus Fertilization of an Ultramafic Soil Reduced Effects of Arbuscular

Mycorrhizal Fungi but not Mycorrhizal Colonization. *Journal of soil science and plant nutrition*, 1-11.

39. Anli, M., Symanczik, S., El Abbassi, A., Ait-El-Mokhtar, M., Boutasknit, A., Ben-Laouane, R., Toubali, S., Baslam, M., Mäder, P., and Hafidi, M. (2021). Use of arbuscular mycorrhizal fungus Rhizoglomus irregulare and compost to improve growth and physiological responses of Phoenix dactylifera 'Boufgouss'. *Plant Biosystems-An International Journal Dealing with all Aspects of Plant Biology* **155**, 763-771.
40. Arruda, B., Herrera, W. F. B., Rojas-García, J. C., Turner, C., and Pavinato, P. S. (2021). Cover crop species and mycorrhizal colonization on soil phosphorus dynamics. *Rhizosphere* **19**, 100396.
41. Atteya, A. K., Sami, R., Al-Mushhin, A. A., Ismail, K. A., and Genaidy, E. A. (2021). Response of seeds, oil yield and fatty acids percentage of Jojoba Shrub strain EAI to mycorrhizal fungi and moringa leaves extract. *Horticulturae* **7**, 395.
42. Bastami, A., Amirnia, R., Sayyed, R., and Enshasy, H. A. E. (2021). The Effect of Mycorrhizal Fungi and Organic Fertilizers on Quantitative and Qualitative Traits of Two Important Satureja Species. *Agronomy* **11**, 1285.
43. Bencherif, K., Laruelle, F., Dalpé, Y., and Lounès-Hadj Sahraoui, A. (2021). Inoculum Sources Modulate Mycorrhizal Inoculation Effect on Tamarix articulata Development and Its Associated Rhizosphere Microbiota. *Plants* **10**, 2716.
44. Bhantana, P., Rana, M. S., Sun, X.-c., Moussa, M. G., Saleem, M. H., Syaifudin, M., Shah, A., Poudel, A., Pun, A. B., and Bhat, M. A. (2021). Arbuscular mycorrhizal fungi and its major role in plant growth, zinc nutrition, phosphorous regulation and phytoremediation. *Symbiosis* **84**, 19-37.
45. Castañeda-Gómez, L., Powell, J. R., Ellsworth, D. S., Pendall, E., and Carrillo, Y. (2021). The influence of roots on mycorrhizal fungi, saprotrophic microbes and carbon dynamics in a low-phosphorus Eucalyptus forest under elevated CO2. *Functional Ecology* **35**, 2056-2071.
46. Çekiç, F. Ö., Ünyayar, S., and Ortaş, İ. (2012). Effects of arbuscular mycorrhizal inoculation on biochemical parameters in Capsicum annuum grown under long term salt stress. *Turkish Journal of Botany* **36**, 63-72.
47. Celik, I., Ortas, I., and Kilic, S. (2004). Effects of compost, mycorrhiza, manure and fertilizer on some physical properties of a Chromoxerert soil. *Soil and Tillage Research* **78**, 59-67.

48. Cheng, H.-Q., Giri, B., Wu, Q.-S., Zou, Y.-N., and Kuča, K. (2021). Arbuscular mycorrhizal fungi mitigate drought stress in citrus by modulating root microenvironment. *Archives of Agronomy and Soil Science*, 1-12.
49. Chu, Q., Zhang, L., Zhou, J., Lixing, Y., Fanjun, C., Fusuo, Z., Gu, F., and Zed, R. (2020). Soil plant-available phosphorus levels and maize genotypes determine the phosphorus acquisition efficiency and contribution of mycorrhizal pathway. *Plant and soil* **449**, 357-371.
50. Demirbaş, A., Kaya, Z., Akpınar, Ç., and Ortaş, İ. (2019). The Effects of Applications of Fertigation and Mycorrhiza on Yield and Nutrient Uptake of Pepper Plant (Capsicum annum L.) under Field Conditions. *Turkish Journal of Agriculture-Food Science and Technology* **7**, 152-161.
51. Deveautour, C., Donn, S., Bennett, A. E., Power, S., and Powell, J. R. (2021). Variability of arbuscular mycorrhizal fungal communities within the root systems of individual plants is high and influenced by host species and root phosphorus. *Pedobiologia* **84**, 150691.
52. Gujre, N., Soni, A., Rangan, L., Tsang, D. C., and Mitra, S. (2021). Sustainable improvement of soil health utilizing biochar and arbuscular mycorrhizal fungi: A review. *Environmental pollution* **268**, 115549.
53. Gull, M., Hafeez, F., Saleem, M., and Malik, K. (2004). Phosphorus uptake and growth promotion of chickpea by co-inoculation of mineral phosphate solubilising bacteria and a mixed rhizobial culture. *Australian Journal of Experimental Agriculture* **44**, 623-628.
54. Gupta, S., Thokchom, S. D., and Kapoor, R. (2021). Arbuscular mycorrhiza improves photosynthesis and restores alteration in sugar metabolism in Triticum aestivum L. grown in arsenic contaminated soil. *Frontiers in plant science* **12**, 334.
55. He, Y., Yang, R., Lei, G., Li, M., Li, T., Zhan, F., and Li, Y. (2021). Arbuscular mycorrhizal fungus–induced decrease in phosphorus loss due to leaching in red soils under simulated heavy rainfall. *Journal of Soils and Sediments* **21**, 881-889.
56. Hou, L., Zhang, X., Feng, G., Li, Z., Zhang, Y., and Cao, N. (2021). Arbuscular mycorrhizal enhancement of phosphorus uptake and yields of maize under high planting density in the black soil region of China. *Scientific reports* **11**, 1-11.
57. Hussain, S., Sharif, M., and Ahmad, W. (2021). Selection of efficient phosphorus solubilizing bacteria strains and mycorrhizea for enhanced cereal growth, root microbe status and N and P uptake in alkaline calcareous soil. *Soil Science and Plant Nutrition*, 1-10.
58. Jabborova, D., Annapurna, K., Paul, S., Kumar, S., Saad, H. A.,

Desouky, S., Ibrahim, M. F., and Elkelish, A. (2021). Beneficial features of biochar and arbuscular mycorrhiza for improving spinach plant growth, root morphological traits, physiological properties, and soil enzymatic activities. *Journal of Fungi* **7**, 571.

59. Jansa, J., Mozafar, A., Kuhn, G., Anken, T., Ruh, R., Sanders, I., and Frossard, E. (2003). Soil tillage affects the community structure of mycorrhizal fungi in maize roots. *Ecological Applications* **13**, 1164-1176.
60. Jiang, F., Zhang, L., Zhou, J., George, T. S., and Feng, G. (2021). Arbuscular mycorrhizal fungi enhance mineralisation of organic phosphorus by carrying bacteria along their extraradical hyphae. *New phytologist* **230**, 304-315.
61. Kuyper, T. W., Wang, X., and Muchane, M. N. (2021). The interplay between roots and arbuscular mycorrhizal fungi influencing water and nutrient acquisition and use efficiency. *The Root Systems in Sustainable Agricultural Intensification*, 193-220.
62. Liu, M., Shen, Y., Li, Q., Xiao, W., and Song, X. (2021a). Arbuscular mycorrhizal fungal colonization and soil pH induced by nitrogen and phosphorus additions affects leaf C: N: P stoichiometry in chinese fir (Cunninghamia lanceolata) forests. *Plant and soil* **461**, 421-440.
63. Liu, Y., Zhang, G., Luo, X., Hou, E., Zheng, M., Zhang, L., He, X., Shen, W., and Wen, D. (2021b). Mycorrhizal fungi and phosphatase involvement in rhizosphere phosphorus transformations improves plant nutrition during subtropical forest succession. *Soil Biology and Biochemistry* **153**, 108099.
64. Lourdes, G.-C. M., Stéphane, D., and Maryline, C.-S. (2021). Impact of increasing chromium (VI) concentrations on growth, phosphorus and chromium uptake of maize plants associated to the mycorrhizal fungus Rhizophagus irregularis MUCL 41833. *Heliyon* **7**, e05891.
65. Ma, X., Li, X., and Ludewig, U. (2021). Arbuscular mycorrhizal colonization outcompetes root hairs in maize under low phosphorus availability. *Annals of Botany* **127**, 155-166.
66. Maitra, P., Zheng, Y., Wang, Y.-L., Mandal, D., Lü, P.-P., Gao, C., Babalola, B. J., Ji, N.-N., Li, X.-C., and Guo, L.-D. (2021). Phosphorus fertilization rather than nitrogen fertilization, growing season and plant successional stage structures arbuscular mycorrhizal fungal community in a subtropical forest. *Biology and Fertility of Soils* **57**, 685-697.
67. Martínez-Viveros, O., Jorquera, M., Crowley, D., Gajardo, G., and Mora, M. (2010). Mechanisms and practical considerations involved in plant growth promotion by rhizobacteria. *Journal of soil science and plant nutrition* **10**, 293-319.

68. Metwally, R. A., Soliman, S. A., Latef, A. A. H. A., and Abdelhameed, R. E. (2021). The Individual and interactive role of arbuscular mycorrhizal fungi and Trichoderma viride on growth, protein content, amino acids fractionation, and phosphatases enzyme activities of onion plants amended with fish waste. *Ecotoxicology and Environmental Safety* **214**, 112072.
69. Nouri, E., Surve, R., Bapaume, L., Stumpe, M., Chen, M., Zhang, Y., Ruyter-Spira, C., Bouwmeester, H., Glauser, G., and Bruisson, S. (2021). Phosphate Suppression of Arbuscular Mycorrhizal Symbiosis Involves Gibberellic Acid Signalling. *Plant and Cell Physiology*.
70. Ortas, I. (2019). Under filed conditions, mycorrhizal inoculum effectiveness depends on plant species and phosphorus nutrition. *Journal of plant nutrition* **42**, 2349-2362.
71. Ortaş, I., and Rafique, M. (2017). The mechanisms of nutrient uptake by arbuscular mycorrhizae. *In* "Mycorrhiza-Nutrient Uptake, Biocontrol, Ecorestoration", pp. 1-19. Springer.
72. Ortaş, I., Rafique, M., and Ahmed, İ. A. (2017). Application of arbuscular mycorrhizal fungi into agriculture. *In* "Arbuscular mycorrhizas and stress tolerance of plants", pp. 305-327. Springer.
73. Ortas, İ., Rafique, M., Akpinar, C., and Kacar, Y. A. (2017). Growth media and mycorrhizal species effect on acclimatization and nutrient uptake of banana plantlets. *Scientia Horticulturae* **217**, 55-60.
74. Ortas, I., Sari, N., Akpinar, Ç., and Yetisir, H. (2011). Screening mycorrhiza species for plant growth, P and Zn uptake in pepper seedling grown under greenhouse conditions. *Scientia Horticulturae* **128**, 92-98.
75. Palacios, Y. M., Gleadow, R., Davidson, C., Gan, W., and Winfrey, B. (2021). Do mycorrhizae increase plant growth and pollutant removal in stormwater biofilters? *Water Research*, 117381.
76. Poveda, J., Hermosa, R., Monte, E., and Nicolás, C. (2019). Trichoderma harzianum favours the access of arbuscular mycorrhizal fungi to non-host Brassicaceae roots and increases plant productivity. *Scientific reports* **9**, 1-11.
77. Qu, L., Wang, M., and Biere, A. (2021). Interactive Effects of Mycorrhizae, Soil Phosphorus, and Light on Growth and Induction and Priming of Defense in Plantago lanceolata. *Frontiers in plant science* **12**.
78. Quan, L., Zhang, J., Wei, Q., Wang, Y., Qin, C., Hu, F., Chen, Y., Shen, Z., and Xia, Y. (2021). Promotion of zinc tolerance, acquisition and translocation of phosphorus in Mimosa pudica L. mediated by arbuscular mycorrhizal fungi. *Bulletin of Environmental Contamination and Toxicology* **106**, 507-515.
79. Rafique, M., Ortas, I., Rizwan, M., Sultan, T., Chaudhary, H. J., Işik,

M., and Aydin, O. (2019). Effects of Rhizophagus clarus and biochar on growth, photosynthesis, nutrients, and cadmium (Cd) concentration of maize (Zea mays) grown in Cd-spiked soil. *Environmental Science and Pollution Research* **26**, 20689-20700.

80. Riaz, M., Kamran, M., Fang, Y., Wang, Q., Cao, H., Yang, G., Deng, L., Wang, Y., Zhou, Y., and Anastopoulos, I. (2021). Arbuscular mycorrhizal fungi-induced mitigation of heavy metal phytotoxicity in metal contaminated soils: A critical review. *Journal of hazardous materials* **402**, 123919.
81. Rodrigues, M. Â., Piroli, L. B., Forcelini, D., Raimundo, S., da Silva Domingues, L., Cassol, L. C., Correia, C. M., and Arrobas, M. (2021). Use of commercial mycorrhizal fungi in stress-free growing conditions of potted olive cuttings. *Scientia Horticulturae* **275**, 109712.
82. Roy-Bolduc, A., and Hijri, M. (2011). The use of mycorrhizae to enhance phosphorus uptake: a way out the phosphorus crisis. *J. Biofertil. Biopestici* **2**, 1-5.
83. Saboor, A., Ali, M. A., Hussain, S., El Enshasy, H. A., Hussain, S., Ahmed, N., Gafur, A., Sayyed, R., Fahad, S., and Danish, S. (2021). Zinc nutrition and arbuscular mycorrhizal symbiosis effects on maize (Zea mays L.) growth and productivity. *Saudi journal of biological sciences* **28**, 6339-6351.
84. Sagar, A., Rathore, P., Ramteke, P. W., Ramakrishna, W., Reddy, M. S., and Pecoraro, L. (2021). Plant Growth Promoting Rhizobacteria, Arbuscular Mycorrhizal Fungi and Their Synergistic Interactions to Counteract the Negative Effects of Saline Soil on Agriculture: Key Macromolecules and Mechanisms. *Microorganisms* **9**, 1491.
85. Sorensen, J., Larsen, J., and Jakobsen, I. (2008). Pre-inoculation with arbuscular mycorrhizal fungi increases early nutrient concentration and growth of field-grown leeks under high productivity conditions. *Plant and soil* **307**, 135-147.
86. Tran, B. T., Cavagnaro, T. R., Jewell, N., Brien, C., Berger, B., and Watts-Williams, S. J. (2021). High-throughput phenotyping reveals growth of Medicago truncatula is positively affected by arbuscular mycorrhizal fungi even at high soil phosphorus availability. *Plants, People, Planet* **3**, 600-613.
87. Turan, V. (2021). Arbuscular mycorrhizal fungi and pistachio husk biochar combination reduces Ni distribution in mungbean plant and improves plant antioxidants and soil enzymes. *Physiologia Plantarum* **173**, 418-429.
88. van't Padje, A., Bonfante, P., Ciampi, L. T., and Kiers, E. T. (2021a). Quantifying nutrient trade in the arbuscular mycorrhizal symbiosis

under extreme weather events using quantum-dot tagged phosphorus. *Frontiers in Ecology and Evolution* **9**, 153.

89. van't Padje, A., Galvez, L. O., Klein, M., Hink, M. A., Postma, M., Shimizu, T., and Kiers, E. T. (2021b). Temporal tracking of quantum-dot apatite across in vitro mycorrhizal networks shows how host demand can influence fungal nutrient transfer strategies. *The ISME journal* **15**, 435-449.
90. Vosnjak, M., Likar, M., and Osterc, G. (2021). The Effect of Mycorrhizal Inoculum and Phosphorus Treatment on Growth and Flowering of Ajania (Ajania pacifica (Nakai) Bremer et Humphries) Plant. *Horticulturae* **7**, 178.
91. Wang, F., and Feng, G. (2021). Arbuscular mycorrhizal fungi interactions in the rhizosphere. *In* "Rhizosphere Biology: Interactions Between Microbes and Plants", pp. 217-235. Springer.
92. Wang, F., Li, K., and Shi, Z. (2021a). Phosphorus fertilization and mycorrhizal colonization change silver nanoparticle impacts on maize. *Ecotoxicology* **30**, 118-129.
93. Wang, G., Wang, L., Ma, F., Yang, D., and You, Y. (2021b). Earthworm and arbuscular mycorrhiza interactions: Strategies to motivate antioxidant responses and improve soil functionality. *Environmental pollution* **272**, 115980.
94. Wu, Y.-H., Wang, H., Liu, M., Li, B., Chen, X., Ma, Y.-T., and Yan, Z.-Y. (2021). Effects of native arbuscular mycorrhizae isolated on root biomass and secondary metabolites of Salvia miltiorrhiza Bge. *Frontiers in plant science* **12**, 66.
95. Yadav, R., Ror, P., Rathore, P., Kumar, S., and Ramakrishna, W. (2021). Bacillus subtilis CP4, isolated from native soil in combination with arbuscular mycorrhizal fungi promotes biofortification, yield and metabolite production in wheat under field conditions. *Journal of Applied Microbiology* **131**, 339-359.
96. Yazici, M. A., Asif, M., Tutus, Y., Ortas, I., Ozturk, L., Lambers, H., and Cakmak, I. (2021). Reduced root mycorrhizal colonization as affected by phosphorus fertilization is responsible for high cadmium accumulation in wheat. *Plant and soil*, 1-17.
97. Zare, M., Ganjeali, A., and Lahouti, M. (2021). Rosmarinic and caffeic acids contents in Basil (Ocimum basilicum L.) are altered by different levels of phosphorus and mycorrhiza inoculation under drought stress. *Acta Physiologiae Plantarum* **43**, 1-10.
98. Zhang, L., Chu, Q., Zhou, J., Rengel, Z., and Feng, G. (2021). Soil phosphorus availability determines the preference for direct or mycorrhizal phosphorus uptake pathway in maize. *Geoderma* **403**, 115261.

CHAPTER FIVE

SOIL SUSTAINABILITY AND MAINTENANCE: ORGANIC AMENDMENTS IS A RIGHT WAY

IMRAN[1,2,3]

[1]DEPARTMENT OF AGRONOMY, THE UNIVERSITY OF AGRICULTURE, PESHAWAR, PAKISTAN,
[2]RHIZOSPHERE LABORATORY, UNIVERSITY OF CUKUROVA, FACULTY OF AGRICULTURE, DEPARTMENT OF SOIL SCIENCE AND PLANT NUTRITION, ADANA, TURKEY,
[3]MINISTRY OF AGRICULTURE, ENTENSION WING, KHYBER PAKHTUNKHWA, PAKISTAN

Abstract

The production of major food crops in semi-arid and the Mediterranean areas is restricted by many environmental and agricultural factors. The integration of organic matter is an antique farming practices aimed at enhancing soil air, nutrient provision, soil porosity, soil quality, soil fertility and water infiltration and water retention. Combination of organic and mineral fertilizers with inoculation of beneficial microbes enhances mutual output. Organic, chemical and biological interactions not only help crop production and soil quality through successful management, but also optimize use of chemical fertilizers in various crop systems. Integrated management of P is essential to enhance crop growth, seed yield, nodulation and human health. This chapter discusses how the organic amendments can be the right way to soil sustainability.

Keywords: beneficial microbes, biological sources, crop productivity, inoculation, mineral fertilizers, Organic amendments, soil health

Introduction

Many environmental and agronomic factors limit the assembly of major food crops within the semi-arid and Mediterranean regions. In agriculture, soil fertility and plant nutrition played a significant role in growing crop yields within the 20th century. Organic matter incorporation is an ancient age agriculture practices to boost soil aeration, nutrients supply, soil porosity, soil health, soil productivity and water infiltration rate together with water holding capacity (Imran et al., 2015). It's been reported by Imran et al. (2016) and Imran and Asad (2017), that originally organic amendments was used as supplement of chemical fertilizer to produce nutrients in abundance for crop growth and maintenance. Many researchers examined that soil reclamation and soil productivity could only be achieved with soil organic matter incorporation (Baqa et al., 2015; Imran and Asad, 2015; Imran et al., 2015; Iqbal et al., 2015; Khan et al., 2015; Naveed et al., 2016). Various organic sources (manure and composted manure) bring lots of beneficial microorganisms that may act as PGPR. All manures aren't the identical manure which supplies the short response. These are low in ammonia but have good nitrate concentration. Additionally they're high is soluble phosphate and therefore the complete complement of needed nutrition including its excellent ability to provide calcium. Imran et al. (2020) reported that the utilization of a thermophilic turned leaf compost when amended with 10 to twenty castings gives season long and early growth stimulus. It also saves on application since it's useful for early and later stage nutrition without unnecessary multiple lay by applications. Plant residues are environmentally friendly materials has many favorable physicochemical characteristics, making it suitable for mixture in substrates including high porosity, good aeration, drainage and water holding capacity.

Vermicomposting

Vermicomposting, in contrast to standard compost, is that the product of an accelerated biooxydation of organic matter by the employment of high densities of earthworm populations without passing a thermophilic stage (Anwar et al., 2016; Imran et al., 2016; Iqbal et al., 2016; Khan et al., 2016; Samreen et al., 2016). Different earthworm species are able to consume a good range of organic residues like sewage sludge, animal wastes, crop residues, and industrial wastes. The earthworm-processed organic wastes are finely divided peat-like materials with high porosity, aeration, drainage, and water-holding capacity sludge Compared to standard compost which passes a thermophilic stage, vermicompost usually encompasses a much finer structure and bigger extent providing strong absorbability and retention of nutrients. Supported of these characteristics, earthworm-processed organic waste would have an excellent commercial potential within the horticultural industry as container media for growing bedding and vegetable plants. There's strong scientific evidence that vermicompost can significantly influence the expansion and productivity of plants. These characteristics of soil are referred to as soil health.

Soil Reclamation by Organic Amendments

For degraded soil reclamation organic matter is generally incorporated acts mainly as a source of nutrients, and improves some physical properties. On strongly degraded soils, mainly by acidity (pH in KCl even of three, 2, in water probably about 4) the effect of single dose of farmyard manure last even 3 years or more and consists mainly on softening of negative effects of acidity (probably complexing of exchangeable Al). It also depends on crop. it's been reported that, during 3 years the crop uptake (apparent) from farmyard manure was about of 25-40% of N, 20% P, 37-55% K and 10% of Mg introduced with (Anwar et al., 2016; Imran et al., 2016; Iqbal et al., 2016; Khan et al., 2016; Samreen et al., 2016). Organic matter application contains a central role in nutrients supply, maintaining soil functions and preventing soil degradation (Babana and Antoun, 2007; Arora et al., 2011; Bardi and Malusà, 2012). Additions of plant organic sources significantly increase the reactivity of the indigenous phosphate rock due to the various organic acids produced by the compost microflora (Lisette et al., 2003; Adesemoye et al., 2009; Vassilev et al., 2015).

It increase soil characteristics like, mobilization of nutrients by mineralization, sorption capacity, short term immobilization into soil organisms (Lugtenberg and Kamilova, 2009; Planchamp et al., 2015). Organic matter is generally utilized in various form like fresh application of the farm residues incorporated within the soil, composted style of the farm residues application and sometime partially burn by a process of pyrolysis and applying called biochar application (Van der Heijden et al., 2008; Mendes et al., 2013; Munees and Mulugeta, 2014; Verbon and Liberman, 2016). Organic sources combined application with chemical fertilizers may well be a promising technique to extend the assembly of soybean and maize. Intensive use of organic matter with chemical fertilizer for crop production increase grain yield, chemical composition of plant and soil with the advantage of maximizing economic return (Nadeem et al., 2007; Zahir et al., 2008; Tilman et al., 2011; Jacoby et al., 2017; Timmusk et al., 2017).

Organic Manures Application

The addition of organic manure improves soil fertility and increase crops yield (Alam et al., 2007; Akter et al., 2013; Akpalu et al., 2014; Imran 2018a). When fertiliser is continually applied to the soil it decrease soil bulk density and increase porosity and buffering capacity of the soil (Ankomah et al., 1996; Anchal et al., 1997; Ansari and Sukhraj, 2010; Imran et al., 2019). Soil amendments have both agronomically and economically important and help in availability of nutrients and enhance soil fertility (Arancon et al., 2005; Argaw, 2012; Imran et al., 2020b). Addition of organic manure to the soil provide the essential macro and micro nutrients and improves soil health and thereby increase the crop growth and yield (Assiouty and Sedera, 2005; Imran, 2018b). This integration can improve the yield and also sustain soil productivity (Ayoub et al., 1994; Ali et al., 2019; Karim and Imran, 2019). Fertiliser help within the enhancing soil characteristics by obtaining high yield as compared to inorganic fertilizer (Bardan, 2003; Bekere et al., 2012; Begum et al., 2015).

The organic amendments improve soil physical, chemical and biological properties (Bellore and Mall, 1975; Imran and Amanullah, 2018). Farmyard manure reduces wearing away, nutrients leaching, and losses through evaporation and had residual effects for next crops. Farmyard manure remain for extended time within the soil, and supply the nutrients over longer period upon mineralization (Berg and Lynd, 1985; Imran, 2018d). However, the inappropriate application of manure causes crop and soil problem including weakening of soil structure, water and environmental pollution (Bhattacharyya et al., 2010; Imran and Khan, 2017). Different sources of organic manure farmyard manure, manure, and poultry manure provide organic material to the soil (Brady, 2002; Carvalho et al., 2011; Imran, 2018c). Poultry manure add essential nutrient to the soil which improve soil fertility and thus improve maize productivity (Cassman et al., 1981; Chauhan et al., 1992; Chela et al., 1993; Chaturvedi 2006; Chen et al., 2006; Imran et al., 2015b; Imran et al., 2015a; Imran, et al., 2017b). Crop yield and yield component of maize was significantly increased by the appliance of PM (Chen et al., 2008; Chiezey and Odunze, 2012; Imran et al., 2017a).

Soil Health

To ensure soil quality, improved nutrients supply and potential crop production along with farmer's income there is a need of revising the current agricultural management practices with a view of recycling plant biomass and organic matter application (Cassman et al., 1981; Ahmed, et al., 2008; Azarmi et al., 2009; Aise et al., 2011; Aziz et al., 2016; Imran, 2017; Imran et al., 2018; Imran et al., 2020a). Recycling of nutrient from organic sources into the soil is a sustainable approach for improving soil chemical, physical and biological properties (Imran, 2018e; Imran, 2019). Reduction in soil quality and poor nutrients use efficiencies is the major fact consistently declining agriculture productivity. Several management practices are adopted for a variety of soil types, cropping patterns and farmers resources, to reduce the magnitude of soil fertility degradation. Regardless of soil types and crop seasons, organic amendments are generally more effective to increase soil C and organic matter status as compared to mineral fertilizers (Ashraf, 2010; Alter et al., 2015; Araújo et al., 2015; Ghani et al., 2015). To improve soil quality and fertility status sustainable fertilizer management practices have been developed by integrating them with a large number of organic nutrient sources. The application of organic amendments has been advocated to sustain soil organic matter, improve soil biological functioning, aeration and moisture retention, reduce compaction, pollutant attenuation and enhance soil nutrient supply (Boyer and Westgate, 2004; Bouranis et al., 2006; Cai et al., 2012; Chimungu et al., 2014a; Chimungu et al., 2014b; Cooper et al., 2014). Entringer et al. (2014) supported these numerous benefits associated with soil application of organic amendments depend strongly on the quantity and nature of organic wastes. Horton et al.

(2015) stated that organic amendments for such purpose are available in huge quantities and originate from diverse resources including crop residues, peach residues, fruit plants wastes, green manures, industrial wastes, animal wastes and household waste.

Nutrients Replenish with Organic Matter

Several researchers reported that the low P availability to plants is a global problem limiting crop production, especially in arid and semi-arid regions of the world (Huang et al., 2012; Jaramillo et al., 2013; Hu et al., 2015a; Hu et al., 2015). As a result, a number of agronomic practices have been proposed to enhance P availability and use by crops under diverse climatic conditions. However, no studies have focused on the use of peach wastes and beneficial microbes to increase P utilization from organic and inorganic P fertilizers (Jones, 2009; Ku et al., 2010; Ku et al., 2011; Ku et al., 2012). Phosphorus is an important element for seeds development, roots elongation, roots nodulation in legumes, N_2 fixation and its deficiency limiting crop production and proper seed development (Mano et al., 2006; Li et al., 2007; Landi et al., 2010; Luo, 2010; Lobell et al., 2011; Lynch, 2013; Lobell et al., 2014; Li et al., 2015; Martre et al. 2015). Phosphorus is needed by most of bacteria for conversion of atmospheric N (N_2) into an ammonium (NH_4). *Rhizobium* bacteria convert atmospheric N to useable form. Nitrogenase enzyme catalyzed N_2 into two molecules of ammonia (NH_3). These enzymes are synthesized by *Rhizobium*. It is well known that symbiotic N_2 fixation is a high P demanding process, hence nodule formation and N_2 fixation are generally limited by low P availability which adversely affects nodule number and mass, as well as nitrogenase activity. Symbiotic N_2 fixation has high ATP requirements for nodule development, function and need additional P for signal transduction and membrane biosynthesis (Munns et al., 2010; Meister, 2014; Ort and Long, 2014; Nuccio et al., 2015). Phosphorus is partially mobile in soil and interaction of maize and soybean with phosphorus along with beneficial microbes increase crop productivity and profitability. Combine use of phosphorus fertilizers with organic matter and beneficial microbes provide frequent supply of nutrients to plants and improve plant production and biomass with advantage of reasonable nutrients use efficiency.

Nutrients Availability and Beneficial Microbes

To ensure phosphorus and other nutrients availability to plant and minimize adsorption on soil particles, beneficial microbe's application is more productive to work as a natural chelate (Piperno et al., 2009; Postma et al., 2014). Postma and Lynch (2010) confirmed that beneficial microbes (fungi, bacteria, protozoa, algae and viruses) are containing living cells commonly used in agriculture known as bio fertilizers. Some of the biofertilizer products, used for nutrients availability in agronomic crops are Mycorrhizae, *Rhizobia*, *Azospirillum*, *Pseudomonas*, *Bacillus*, *Trichoderma*, *Streptomyces* species etc.

Postma and Lynch (2011) reported that biofertilizer is a wide term which includes a diverse category of bioinoculants such as nitrogen fixers, phosphate solubilizers, phosphate mobilizers and plant growth promoting rhizobacteria (Sakamoto et al., 2006; Saengwilai et al., 2014). Shelden and Roessner (2013) concluded that the mineral mobilizer, arbuscular mycorrhizal fungus has been found to be quite effective for growth improvement and saving in phosphorus nutrition in agronomic crops such as maize, rice, wheat and soybean under field conditions. The phosphate solubilizers (PSB, *Trichoderma*) is generally solubilize phosphate by the release of low molecular weight organic acids (formic acids, fumeric acids, acetic acids, propionic acids and succinic acids), which through their hydroxyl and carboxyl groups chelate the cations bound to phosphate, thereby converting it into soluble forms. Further, Trevisan et al. (2015) has reported that vesicular arbuscular mycorrhizal (VAM) fungi grow extensively in soil to form a well-developed hyphal network that absorbs inorganic phosphate several centimeters away from the root surface. Phosphorus is translocated rapidly to the roots overcoming the slow diffusion that occurs in the soil solution (Simons et al., 2014; Shi and Lai,

2015; Shi et al., 2015). It has been reported that soybean and wheat grain yield increases with combines application of phosphate solubilizers.

Integrated Approach and Soil Sustainability

Combined use of organic and mineral fertilizers along with beneficial microbes' inoculation increases each other's efficiency. Interaction of organic, chemical and biological sources not only sustains crop productivity and soil health by effective management but they also optimize chemical fertilizers utilization for different crops in different cropping systems. Integrated P management is essential for the improvement of crop growth, seed yield, nodulations in legumes and human health. Application of organic sources increases soil organic matter, P availability and soybean productivity. Use of proper organic and inorganic fertilizer combination, is a key factor in crop production for sustainable agriculture (Yin et al., 2014; York et al., 2015; Zhan and Lynch, 2015). Zhan et al. (2015) concluded that biochar application enhanced crop productivity, soil properties and phosphorus use efficiency (PUE) when applied with chemical fertilizer along with other organic sources.

In many parts of the globe are stricken by saturated soils from a really early time. Imran et al. (2020) reported that we continued to urge record rainfall from April, when most farmers would give some thought to planting, all the way through today. This has caused not only late-planted soybeans, but soybeans that in a very lot of cases were planted into but ideal soil conditions and are experiencing continued saturated soil conditions. This has caused soybeans to require an extended time to emerge and that they seem to own struggled to grow ever since. Recent warm weather has helped pushed beans along a touch, but with saturated soils we're still seeing soybeans with slow growth.

The most common symptoms observed are slow emergence, reduced early vigor, yellowing of lower leaves, necrosis or death, poor root systems and lack of nodule formation. the most cause for these symptoms will be attributed to saturated soils, which has caused a scarcity of oxygen and poor root development. We then saw problems like nutrient deficiencies and lack of nodulation. Determining a way to overcome these problems may require practices and products that we don't typically use.

Nitrogen

Sometimes in the aerial tissues healthy plants contain between 3 and 4 per cent nitrogen. The only other nutrients present in higher concentration are carbon, hydrogen and oxygen, nutrients that do not play an important role in most soil fertility management programmes. Nitrogen is so vital because it is a key component of chlorophyll, the combination through which plants produce water and carbon dioxide sugar (i.e. photosynthesis) with solar energy. It is also a major component of protein-building amino acids. Nitrogen is an energy transfer compound, such as ATP (adenosine triphosphate). Lastly, nitrogen is an important component of nucleic acids like DNA, the genetic material that enables cells (and possibly whole plants) to grow and reproduce.

Foliar feeding

Another idea would be an application of foliar fertilizers within the post herbicide trip. Before making a fertilizer application it's important to identify which nutrients are lacking within the plant. A tissue sample, along with a soil sample, will help determine the foremost effective route to want.

Key nutrients

A few other key nutrients of interest are sulfur (S) and molybdenum (Mo). More soybeans are receiving planned applications of sulfur than ever before. With less sulfur available through pollution we are seeing a much bigger need for sulfur applications. Sulfur helps promote nodulation in soybeans, is required for chlorophyll production and aids in seed production (Zhu et al., 2010; Zhang et al., 2014). Zurek et al. (2015) stated that sulfur is mobile within the soil, so it should are leached out with all the rain. Molybdenum is vital for fixing nitrogen by rhizobia bacteria. It going to move through the profile and may have leached out. The addition of S and Mo are likely to help soybeans that are struggling to produce nodules and begin fixing N.

Disease risk

Disease management will play an important role in protecting the soybean crop we have got. If we still receive wet weather, we are visiting likely see foliar diseases like white mold and frogeye leaf spot develop. Scout fields and be prepared to make foliar fungicide applications at R3 to safeguard soybeans from these pathogens. Overtime syndrome (SDS) goes to be something to look at for if we still have wet conditions during flowering, followed by warm, dry conditions after flowering. This disease infects the soybean during early growth but isn't recognized until later within the season. It’s more prominent in years where we've cool, wet weather at planting, which we had this year.

Aeration

One highly unpopular recommendation would be to cultivate if you are able. This might help to dry and put oxygen back within the soil. However, researchers realize that just about all growers probably don't own a rotary hoe or row crop cultivator in an era of reduced and no-till, with herbicides to manage weeds.

Conclusion

Combined use of organic and mineral fertilizers with inoculation of beneficial microbes improves each others performance. The interactions of organic, chemical and biological sources not only help crop productivity and soil health through effective management, but also maximize the use of chemical fertilizers in various crop systems. Integrated P management is important to enhance crop growth, seed yield, legume nodulation and human health.

References

1. Abdel-Ghani A. H., Kumar B., Jansen C., Gonzalez-Portilla P. J., Reyes-Matamoros J., San Martin J. P. Lee, M., & Lubberstedt, Th. Association analysis of genes involved in maize (*Zea mays* L.) root development with seedling and agronomic traits under contrasting nitrogen levels. *Plant Mol. Biol.* 2015, 88, 133–147.
2. Abebe, G., Assefa, T., Harrun, H., Mesfine, T., & Al-Tawaha, A. M. Participatory selection of drought tolerant maize varieties using mother and baby methodology: a case study in the semi-arid zones of the central rift valley of Ethiopia. *World Journal of Agricultural Sciences*, 2005d, *1*(1), 22-27.
3. Abebe, G., Hattar, B., & Al-Tawaha, A. M. Nutrient availability as affected by manure application in cowpea [*Vigna unguiculata* (L.) Walp.] on calacarious soils. *Journal of Agriculture and Social Science*, 2005c, *1*(1), 1-6.

4. Abebe, G., Sahile, G., & Al-Tawaha, A. M. Effect of soil solarization on orobanche soil seed bank and tomato yield in the central rift valley of Ethiopia. *World Journal of Agricultural Sciences*, 2005b, *1*(2), 143-147.
5. Abebe, G., Sahile, G., & Al-Tawaha, A. M. Evaluation of potential trap crops on Orobanche soil seed bank and tomato yield in the central rift valley of Ethiopia. *World Journal of Agricultural Sciences*, 2005a, *1*(2), 148-151.
6. Abera, T., Feyisa, D., Yusuf, H., Nikus, O., & Al-Tawaha, A. M. Grain yield of maize as affected by biogas slurry and N-P fertilizer rate at Bako, Western Oromiya, Ethiopia. *Bioscience Research*, 2005, *2*(1), 31-38.
7. Adesemoye, A. O., & Kloepper, J. W. Plant-microbes interactions in enhanced fertilizer-use efficiency. *Appl. Microbiol. Biotechnol.* 2009, 85, 1–12.
8. Ahmed, M. F., Kennedy, I. R., Choudhury, A. T. M. A., Kecskés, M. L., & Deaker, R. Phosphorus adsorption in some Australian soils and influence of bacteria on the desorption of phosphorus. *Comm. Soil Sci. Plant Anal.* 2008, 39, 1269-1294.
9. Aise, D., Erdal, S., Hasanand, A., & Ahment, M. Effects of different water, phosphorus and magnesium doses on the quality and yield factors of soybean in Harran plain conditions. *Int. J. Phys. Sci.* 2011, *6*(6), 1484-1495.
10. Akpalu, M. M., Siewobr. H., Oppong, D., & Akpalu, S. E. Phosphorus application and rhizobia inoculation on growth and yield of soybean (*Glycine max* L. Merrill). *American J. Exp. Agric.*, 2014, *4*(6), 674-685.
11. Akter, F., Islam, Md. N., Shamsuddoha, A. T. M., Bhuiyan, M. S. I., & Shilpi, S. Effect of phosphorus and sulphur on growth and yield of soybean (*Glycine max* L.). *Intel. J. Biores. Strs. Manag.*, 2013, *4*(4), 555-560.
12. Alam, M. N., Jahan, M. S., Ali, M. K., Ashraf, M. A., & Islam, M. K. Effect of vermicompost and chemical fertilizers on growth, yield and yield components of potato in barind soils of Bangladesh. *J. App. Sci. Res.*, 2007, 3, 1879–1888.
13. Ali, I., Khan, A. A., Imran, inamullah, Khan, A., Asim, M., Ali, I., B., Zib, Khan, I., Rab, A., Sadiq, G., Ahmad, N., & Iqbal, B. Humic acids and Nitrogen levels optimizing productivity of green gram (*Vigna radiata* L.). *Russian Agriculture Sciences*, 2019, *45*(1), 43-47.
14. Alter, S., Bader, K. C., Spannagl, M., Wang, Y., Bauer, E., Schön, C. C., & Mayer, K. F. X. DroughtDB: an expert-curated compilation of

plant drought stress genes and their homologs in nine species. *Database*, 2015, 1-7.

15. Anchal, D., Kharwara, P. C., Rana, S. S. & Dass, A. Response of gram varieties to sowing dates and phosphorus levels under on farm condition. *Himachal J. Agric. Res.*, 1997, 23, 112-5.
16. Ankomah, A. B., Zapata, F., & Hardarson, G. Yield, nodulation and N2 fixation by cowpea cultivars at different phosphorus levels. *Biol. Fert. Soils*, 1996, 22, 10-15.
17. Ansari, A. A., & Sukhraj, K. K. Effect of vermiwash and vermicompost on soil parameters and productivity of okra in Guyana. *Afr. J. Agric. Res*, 2010, 5, 1794-1798.
18. Anwar, S., Israeel, Iqbal, B., Khan, A. A., Imran, Shah, W. A., Islam, M., Khattak, W. A., Ikramullah, Akram, W., & Abbas, W. Nitrogen and phosphorus fertilization of improved varieties for enhancing phenological traits of wheat. *Pure and Applied Biology*, 2016, 5(3), 511-519.
19. Arancon, N. Q., Edwards, C. A., Bierman, P., Metzger, J. D., & Lucht, C. Effects of vermin composts produced from cattle manure, food waste and paper waste on the growth and yield of peppers in the field. *Pedobio*. 2005, 49, 297-306.
20. Araújo, S. S., Beebe, S., Crespi, M., Delbreil, B., González, E. M., Gruber, V., Lejeune-Henaut, I., Link, W., Monteros, M.J., E., Prats, Rao, I., Vadez, V., & Vaz Patto, M. C. Abiotic stress responses in legumes: strategies used to cope with environmental challenges. *Crit. Rev. Plant Sci.*, 2015, 34, 237–280.
21. Argaw, A. Evaluation of co-inoculation of *Bradyrhizobium japonicum* and phosphate solubilizing *Pseudomonas* spp. effect on soybean (*Glycine max* L.) in assossa area. *J. Agric. Sci. Tech.* 2012, 14, 213-224.
22. Arora, N. K., Khare, E., & Maheshwari, D. K. Plant growth promoting rhizobacteria: constraints in bioformulation, commercialization, and future strategies. In: D. K. Maheshwari (ed.), *Plant Growth and Health Promoting Bacteria, Microbiology Monographs* (pp. 97-112). Springer-Verlag Berlin Heidelberg, 2010.
23. Ashraf, M. Inducing drought tolerance in plants: recent advances. *Biotechnol. Adv*, 2010, 28, 169–183.
24. Assaf, T. A., Hameed, K. M., Turk, M. A. & Tawaha, A. M. Effect of soil amendment with olive mill by-products under soil solarization on growth and productivity of faba bean and their symbiosis with mycorrhizal fungi. *World Journal of Agricultural Sciences*, 2006, *2*(1), 21-28.

25. Assefa, T., Abebe, G., Fininsa, C., Tesso, B & Al-Tawaha, A. M. Participatory bean breeding with women and small holder farmers in eastern Ethiopia. *World Journal of Agricultural Sciences*, 2005, *1*(1), 28-35.
26. Assiouty, F. M., & Abo-Sedera, S. A. Effect of bio and chemical fertilizers on seed production and quality of spinach. *Intel. J. Agric. Biol.* 2005, 6, 947-952.
27. Ayoub, M., Guertin, S., Lussier, S., & Smith, D. L. Timing and level of nitrogen fertilizer effects on spring wheat yield in eastern Canada. *Crop Sci.*, 1994, 34, 748–756.
28. Azarmi, R., Giglou, M. T., & Taleshmikail, R. D. Influence of vermicompost on soil chemical and physical properties in tomato field. *Afr. J. Biol.*, 2009, 7, 2397–2401.
29. Aziz, A. L. A., Ahiabor, B. D. K., Opoku, A. & Abaidoo, R. C. Contributions of rhizobium inoculants and phosphorus fertilizer to biological nitrogen fixation, growth and grain yield of three soybean varieties on a fluvic luvisol. *American J. Exp. Agric.*, 2016, *10*(2), 1-11.
30. Babana, A. H., & Antoun, H. Effect of Tilemsi phosphate rock-solubilizing microorganisms on phosphorus uptake and yield of field-grown wheat (*Triticum aestivum* L.) in Mali. In: Velázquez, E., & Rodríguez-Barrueco, C., (eds.), First International Meeting on Microbial Phosphate Solubilization (pp. 51-58). Dordrecht, Springer, 2006.
31. Baqa, S., Khan, A. Z., Inamullah, Imran, Khan, A. A., Anwar, S., Iqbal, B., Khan, S., Usman, A. Influence of farm yard manure and phosphorus application on yield and yield components of wheat. *J. Pure Appl Biol.*, 2015, *4*(4), 458- 464.
32. Bardan, M. M. Effect of nitrogenous and phosphetic fertilization on some economical characters of soybean Crawford cultivar under calcareous soil conditions. *Egyptian J. Agric. Res.*, 2003, *81*(2), 433-440.
33. Bardi, L., & Malusà, E. Drought and nutritional stresses in plant: alleviating role of rhizospheric microorganisms. In: N. Haryana and S. Punj (eds.), *Abiotic Stress: New Research* (pp. 1-57). Hauppauge: Nova Science Publishers Inc., 2012.
34. Begum, A., Islam, M. D. A., Ahmed, Q. M., & Rahman, M. Effect of nitrogen and phosphorus on the growth and yield performance of soybean. *Res. Agric. Livest. Fish.* 2015, *2*(1), 35-42.
35. Bekere, W., Endalkachew, W., & Tesfu, K. Growth and nodulation response of soybean (*Glycine max* L.) to bradyrhizobium inoculation and phosphorus levels under controlled condition in south western Ethiopia. *Afric. J. Agric. Res.*, 2012, *7*(30), 4266-4270.

36. Bellore, S. K., & Mall, L. R. Chlorophyll content as an ecological index of dry matter production. *Indian J. Bot. Soc.*, 1975, *54*(1–2), 75–77.
37. Berg, R. K., & Lynd, J. Q. Soil fertility effects on growth, yield, nodulation and nitrogenase activity of Australian winter pea. *J. Plant Nutr.*, 1985, 8, 131-145.
38. Bhattacharyya, R. P. S. C., Chandra, S., Kundu, S., Supradip, S., Mina, B. L., Srivastva, A. K., & Gupta, H. Fertilization effects on yield sustainability and soil properties under irrigated wheat-soybean rotation of an Indian Himalayan upper valley. *Nutr. Cyc. Agroeco.*, 2010, 86, 255-268.
39. Bouranis, D., Chorianopoulou, S. N., Kollias, C., Maniou, P., Protonotarios, V. E., Siyiannis, V.F., & Hawkesford, M. J. Dynamics of aerenchyma distribution in the cortex of sulfate-deprived adventitious roots of maize. *Ann. Bot.*, 2006, 97, 695–704.
40. Boyer, J., & Westgate, M. Grain yields with limited water. *J. Exp. Bot.* 2004, 55, 2385–2394.
41. Brady, N. Phosphorus and potassium. The nature and properties of soils. Prentice-Hall of India Pvt. Limited, New Delhi, 2002, p 352.
42. Cai, H. G., Chen, F., Mi G., Zhang, F., Maurer, H. P., Liu, W., Reif, J. C., & Yuan, L. Mapping QTL for root system architecture of maize (*Zea may* L.) in the field at different developmental stages. *Theor. Appl. Genet.*, 2012, 125, 1313–1324.
43. Carsky, R. J., Singh, B. B., & Oyewole, R. Contribution of early–season cowpea to late season maize in the savanna zone of West Africa. *Biol. Agric. Hort.*, 2001, 18, 303-315.
44. Carvalho, E. R., Rezende, P. M., Andrade, M. J. B., Passos, A. M. A., & Oliveira, J. A. Fertilizante mineral: The nature and properties of soils. *Revista Ciência Agro.*, 2011, 42, 930-939.
45. Cassman, K. G., Munns, D. N., & Beck, D. P. Phosphorus nutrition of *Rhizobium japonicum*: Strain differences in phosphate storage and utilization. *Soil Sci. Amer. J.*, 1981, 45, 517-520.
46. Cassman, K. G., Whitney, A. S., & Fox, R. L. Phosphorus requirements of soybean and cowpea as affected by mode of nitrogen nutrition. *Agric. J.*, 1981, 73, 17-22.
47. Chaturvedi, I. Effects of phosphorus levels alone or in combination with phosphate-solubilizing bacteria and farmyard manure on growth, yield and nutrient up-take of wheat (*Triticum aestivum*). *J. Agric. Soc. Sci.*, 2006, *2*(2), 96-100.
48. Chauhan, Y. S., Johanson, C., & Venkataratnam, N. Effect of phosphorus deficiency on phenology and yield components of short duration pigeonpea. *Trop. Agric.*, 1992, *69*(3), 235-238.

49. Chela, G. S., Tiwana, M. S., Thind, I. S., Puri, K. P., & Kaur, K. Effect of bacterial cultures and nitrogen fertility on the yield and quality of maize fodder (*Zea mays* L.). *Ann. Biol.*, 1993, 9, 83-86.
50. Chen, Y. P., Rekha, P. D., Arunshen, A. B., Lai, W. A., & Young, C. C. PSB from subtropical soil and their tri-calcium phosphate solubilizing abilities. *Appl. Soil Ecol.*, 2006, 34, 33-41.
51. Chen, Z., Shengwu, M., Lily, L. Studies on phosphorus solubilizing activity of a strain of hosphobacteria isolated from chestnut type soil in China. *Bio Tech*, 2008, 99, 6702-6707.
52. Chiezey, U. F., & Odunze, A. C. Soybean response to application of poultry manure and phosphorus fertilizer in the sub-humid Savanna of Nigeria. *J. Ecol. Nat. Env.*, 2012, *1*(2), 25-31.
53. Chimungu, J. G., & Brown, K. M., Lynch, J. P. Reduced root cortical cell file number improves drought tolerance in maize. *Plant Physiol.*, 2014a, 166, 1943–1955.
54. Chimungu, J. G., & Brown, K. M., Lynch, J. P. Large root cortical cell size improves drought tolerance in maize. *Plant Physiol.*, 2014b, 166, 2166–2178.
55. Cooper, M., Gho, C., Leafgren, R., Tang, T., & Messina, C. Breeding drought-tolerant maize hybrids for the US corn-belt: discovery to product. *J. Exp. Bot.*, 2014, 65, 6196–6204.
56. de Souza, T. C., de Castro, E. M., Magalhães, P. C., Lino, L. D. O., Alves, E. T., & de Albuquerque, E. P. Morphophysiology, morphoanatomy, and grain yield under field conditions for two maize hybrids with contrasting response to drought stress. *Acta Physiol. Plant*, 2013, 35, 3201–3211.
57. Entringer, G. C., Guedes, F. L., Oliveira, A. A., Nascimento, J. P., & Souza, J. C. Genetic control of leaf curl in maize. *Genet. Mol. Res.*, 2014, 13, 1672–1678.
58. Frey, F. P., Urbany, C., Hüttel, B., Reinhardt, R., & Stich, B. Genome-wide expression profiling and phenotypic evaluation of European maize inbreeds at seedling stage in response to heat stress. *BMC Genomics*, 2015, 16, 123.
59. Gong, F. P., Yang, L., Tai, F. J., Hu, X. L., Wang, W. "Omics" of maize stress response for sustainable food production: opportunities and challenges. *Omics*, 2014, 18, 711–729.
60. Hachez, C., Veselov, D., Ye, Q., Reinhardt, H., Knipfer, T., Fricke, W., Chaumont, F. Short-term control of maize cell and root water permeability through plasma membrane aquaporin isoforms. *Plant Cell Environ.*, 2012, 35, 185–198.

61. Horton, D. E., Johnson, N. C., Singh, D., Swain, D. L., Rajaratnam, B., Diffenbaugh, N. S. Contribution of changes in atmospheric circulation patterns to extreme temperature trends. *Nature*, 2015, 522, 465–469.
62. Hu, X. L., Wu, L. J., Zhao, F. Y., Zhang, D. Y., Wang, W., Zhu, G., Li, Ch., & Wang, W. Phosphoproteomic analysis of the response of maize leaves to drought, heat and their combination stress. *Front. Plant Sci.*, 2015a, 6, 298.
63. Hu, X. L., Yang, Y. F., Gong, F. P., Zhang, D. Y., Wang, W., Wu, L., Li, C., & Wang, W. Protein sHSP26 improves chloroplast performance under heat stress by interacting with specific chloroplast proteins in maize (*Zea mays*). *J. Proteomics* 2015b, *115*, 81–92.
64. Huang, H., Mølle, I. M., & Song, S. Q. Proteomics of desiccation tolerance during development and germination of maize embryos. *J. Proteomics* 2012, *75*, 1247–1262.
65. Imran, & Amanullah. Global impact of climate change on water, soil resources and threat towards food security: evidence from Pakistan. *Adv Plants Agric Res.*, 2018, *8*(5), 350–355.
66. Imran, & Khan, A. A. Phenological charateristics of *Brassica napus* L. as influenced by biochar application and shoot cutting duration (days). *Civi and Environ Res.*, 2015, *7*(3), 104-107.
67. Imran, & Khan, A. A. Biochar application and shoot cutting duration (days) influenced growth, yield and yield contributing parameters of *Brassica napus* L. *J. Bio. Agric H Care*, 2015, *5*(5), 1-6.
68. Imran, & Khan, A. A. Canola yield and quality enhanced with sulphur fertilization. *Russ. Agricult. Sci.*, 2017, 43, 113-110.
69. Imran, & Khan, A. A. Effect of transplanting dates on yield and yield components of various rice genotypes in hilly area cold climatic region of Khyber Pakhtunlhwa-Pakistan. *J. Bio, Agric H. Care*, 2015, *5*(7), 1-10.
70. Imran, & Khan, A. A. Grain yield and phenology of maize cultivars influenced by various phosphorus sources. *Food Science and Quality Management*, 2015, 37, 74-78.
71. Imran, & Khan, A. A. Influence of compost application and seed rates on production potential of late sown maize on high elevation in swat - Pakistan. *J. Environ Earth Sci.*, 2015, *5*(5), 36- 40.
72. Imran, Amanullah, Arif, M., Shah, Z., & Bari, A. Soil application of *Trichoderma* and Peach (*Prunus persica* L.) residues possesses biocontrol potential for weeds and enhances growth and profitability of soybean (*Glycine max*). *Sarhad Journal of Agriculture*, 2020a, *36*(1), 10-20.

73. Imran, Amanullah, Bari, A., & Ali, R. Peach sources, phosphorous and beneficial microbes enhance productivity of soybean. *Soy. Res.*, 2018, *16*(2), 39-48.
74. Imran, Amanullah, Bari, A., Khan, H., & Ali. R. *Climatic Variability and Agronomic Cropping Pattern* (pp 33-44). Springer Nature, Singapore, 2019.
75. Imran, Amanullah, Khan, A. A., Bari, A., Din, R.U, Ali, R., Ahmad, N., Ali, A., Maula, F., Ahmad, F., Naveed, Sh., Ullah, I., Zada, H., & Khan, G. R. Production statistics and modern technology of maize cultivation in Khyber Pakhtunkhwa Pakistan. *Plant Sci. Arc.* 2019, *2*(2), 1-12.
76. Imran, Bari, A., Ali, R., Ahmad, N., Ahmad, Z., Khattak, M. I., Ali, A., Ahmad, F., Khan, I., & Naveed, S. traditional rice farming accelerate CH_4 and N_2O emissions functioning as a stronger contributors of climate change. *Int J Environ Sci Nat Res.*, 2017b, *9*(3), 89-92.
77. Imran, Hussain, I., Khattak, I., Rehman, A., Ahmad, F., & Zada, H., & Naveed, Sh. Roots nodulation, yield and yield contributing parameters of mungbean cultivars as influenced by different phosphorous level in Swat-Pakistan. *J. Pure Appl Biol.*, 2015, *4*(4), 557-567.
78. Imran, Jamal, N., Alam, A., & Khan, A. A. Grain yield, yield attributes of wheat and soil physio-chemical characteristics influenced by biochar, compost and inorganic fertilizer application. *Agric Res Tech.*, 2017, *10*(4), 103-108.
79. Imran, Khan, A. A., & Ahmad, F. Phenology of various rice genotypes as affected by different transplanting dates under cold climatic region of Khyber Pakhtunkhwa-Pakistan. *J. Environ E. Science*, 2015a, *5*(3), 119-112.
80. Imran, Khan, A. A., Ahmad, F. & Irfanullah. Influence of hydrated calcium sulphate ($CaSO_4.2H_2O$) and nitrogen levels on water infiltration rate and maize varieties productivity in rainfed area of Swat, Pakistan. *Chem Environ Res.*, 2015, *7*(3), 15-20.
81. Imran, Khan, A. A., Ahmad, F., & Irfanullah. Nitrogen levels, tillage practices and irrigation timing influenced yield, yield components and oil contents of Canola. *Civil Environ Res.*, 2015, *7*(5), 74-77.
82. Imran, Khan, A. A., Akhtar, K., Zaheer, S., Faisal, S., & Ali, S. Rice seedling characteristics of various genotypes influenced by different sowing dates in Swat-Pakistan. *J. Environ E. Science*, 2015, *5*(1), 121-125.
83. Imran, Khan, A. A., Inam, I., & Ahmad, F. Yield and yield attributes of mung bean (*Vigna radiata* L.) cultivars as affected by phosphorous

levels under different tillage systems, *Cogent Food and Agriculture*, 2016, 2, 1151982.

84. Imran, Khan, A. A., Inamullah, Zada, H., Ahmad, F., Shah, S. T., Usman, A., & Irfanullah. Yield and yield attributes of rapeseed cultivars as influence by sulfur level under Swat valley conditions. *J. Pure and Applied Biology*, 2015, *4*(3), 296-301.
85. Imran, Khan, A. A., Irfanullah, & Ahmad, F. Production potential of rapeseed (*Brassica napus* L.) as influenced by different nitrogen levels and decapitation stress under the rainfed agro-climatic condition of Swat Pakistan. *J. Glob. Innov. Agric. Soc. Sci.*, 2014, *2*(3), 112-115.
86. Imran, Khan, A. A., Zada, H., Ahmad, F., & Irfanullah. Graine yield and yield components of wheat cultivar "siran 2010" as affected by phosphorous levels under rain fed condition. *J. N. Sci. Res*, 2015, *5*(5), 139-143.
87. Imran, Khan, A.A., Inamullah, & Luqman. Weeding stages and their effect on yield and yield components of rice in upper Swat, Pakistan. *Pak. J. Weed Sci. Res*. 2015, *21*(4), 555-563.
88. Imran, Khan, A.A., Khan, I.U., & Naveed, S. Weeds density and late sown maize productivity influenced by compost application and seed rates under temperate environment. *Pak. J. Weed Sci. Res*., 2016, *22*(1), 169-181.
89. Imran, Khattak, I., Hussain, I., Rehman, A., Anwar, S., Ahmad, F., Khan, A. A., & Zada, H. Growth and yield of maize hybrids as effected by different sowing dates in Swat Pakistan. *J. Pure Appl Biol*., 2015, 4(4).
90. Imran, Maula, F., Uzair, M., & Zada, H. Farmers income enhancement through off-season vegetables production under natural environment in Swat-Pakistan. *J. Environ. E. Science*, 2015, *5*(5), 58- 64.
91. Imran, Naveed, S., Khan, A. A., & Khattak, I. impact of phosphorus levels and seed rates on growth and yield of late sown maize on high elevation in Swat, Pakistan. *Pak. J. Agric. Res*., 2015, *28*(4), 406-413.
92. Imran, Uzair, M., Maula, F., Vacirca, M., Farfaglia, S., & Khan, N. M. Introduction and promotion of off-season vegetables production under natural environment in hilly area of upper Swat-Pakistan. *J. Bio. Agric. H. Care*, 2015, 5(11), 42-48.
93. Imran, Zada, H., Naveed, S., Khattak, I., & Ahmad, S. Variable rates of phosphorous application influenced phenological traits of green gram (*Vigna radiata* L.). *J. Agric. Res*., 2016, 1(3), 000113.
94. Imran. *Advance production technologies of legume crops* (pp. 297-311). Springer Nature, Singapore, 2019.

95. Imran. Climate change is a real fact confronting to agricultural productivity. *Int. J. Environ. Sci. Nat. Res.*, 2017, *3*(3), 65-73.
96. Imran. Climate change is threat toward agronomy (base of food, fiber system), and food security. *Food Nutr. J.*, 2018a, *3*(1), 1-8.
97. Imran. Ecological environmental variability influence growth and yield potential of rice under northern climatic scenario. *Russian Agricultural Sciences*, 2018c, *44*(1), 18–24.
98. Imran. Influence of nitrogen levels and decapitation stress on biological potential of rapeseed (*Brassica napus* L) under water deficit condition of Swat-Pakistan. *J. N. Sci. Res.*, 2015, *4*(3), 296-301.
99. Imran. Organic matter amendments improve soil health, productivity and profitability of maize and soybean. *Ann Rev Res.*, 2018e, *1*(3), 555564.
100. Imran. Phosphorous fertilization influenced weeds attributes and phenological characteristics of mungbean cultivars (*Vigna radiata* L.). *Russian Agricultural Sciences*, 2018, *44*(3), 229–234.
101. Imran. Physiological and morphological traits of agronomic crops influenced by climate change. *Mod Concep Dev Agrono*, 2018b, *1*(4), 95-98.
102. Imran. *Plant microbe interaction in Agronomic crops* (pp. 445-459). Springer, 2019.
103. Imran. 2015. Effect of germination on proximate composition of two maize cultivars. *J. Bio Agric H. Care*, 2015, *5*(3), 123-128.
104. Iqbal, B., Ahmad, B., Ullah, I., Imran, Khan, A. A., Anwar, S., Ali, A., Shahzadand, K., & Khan, S. Effect of phosphorus, sulphur and different irrigation levels on phenological traits of triticale. *Pure and Applied Biology* 2016, 5(2), 303-310.
105. Iqbal, B., Jan, M. T., Inamullah, Imran, Khan, A. A., Muhammad, Z., Anwar, S., Khan, S., Baqa, S., & Usman, A. Integrated management of compost type and fertilizer-n in maize. *J. Pure Appl Biol.*, 2015, *4*(4), 453-457.
106. Iqbal, B., Jan, M. T., Muhammad, Z., Khan, A. A., Anwar, S., Imran & Shahzad, K. Phenological traits of maize influenced by integrated management of compost and fertilizer nitrogen. *J. Pure Appl Biol.*, 2016, *5*(1), 58-63.
107. Iqbal, B., Jan, M.T., Imran, Zar, M., Khan, A. A., Anwar, S., & Shahzad, K. Growth and phenology of maize as affected by integrated management of compost and fertilizers nitrogen. *J. Pure Appl Biol.*, 2015, *4*(4), 58-63.

108. Islam, M., Anwar, S., Bashir, S., Khattak, W. A., Imran, Ali, M., & Khan, N. Growth and yield components of wheat varieties as affected by dual purpose practices. *J. Pure Appl Biol.*, 2015, *4*(4), 491-946.
109. Jacoby, R., Peukert, M., Succurro, A., Koprivova, A., & Kopriva, S. The role of soil microorganisms in plant mineral nutrition-current knowledge and future directions. *Front. Plant Sci*, 2017, *8*, 1-19.
110. Jaramillo, R. E., Nord, E. A., Chimungu, J. G., Brown, K. M., & Lynch, J. P. Root cortical burden influences drought tolerance in maize. *Ann. Bot.*, 2013, 112, 429–437.
111. Jones T. J. Maize tissue culture and transformation: the first 20 years. In: Kriz, A. L., & Larkins, B. A. (eds.), *Molecular Genetic Approaches to Maize Improvement* (pp. 7–27). Springer, Heidelberg, 2009.
112. Karim, S. A., & Imran. Reduction in substrate moisture content reduce final yield of wheat. *Biomed J. Sci Tec Res.*, 2019, 1-5.
113. Khan, A. A., Imran, Shah, F. A., Inamullah, Muhammad, L. Z., Muhammad, N., & Khan, N. Phenological traits of rice as influenced by seedling age and number of seedling per hill under temperate region. *J. Bio, Agric and H. Care*, 2015, *5*(3), 145- 149.
114. Khan, A. A., Khan, M. N., Uallah, I., Shah, S., Arshad, I. R., Zeb, I. M. A., & Imran. Effect of potash application on growth, yield and yield components of spring maize hybrids. *Pure and Applied Biology*, 2015, *4*(2), 195-203.
115. Khan, A. Z., Imran, Muhammad, A., Khalil, A., Gul, H., Akbar, H., & Wahab, S. Impact of fertilizer priming on seed germination behavior and vigor of maize. *Pure and Applied Biology*, 2016, *5*(4), 744-751.
116. Ku, L. X., Wei, X. M., Zhang, S. F., Zhang, J., Guo S. L., & Chen, Y. Cloning and characterization of a putative TAC1 ortholog associated with leaf angle in maize (*Zea mays* L.). *PLoS ONE*, 2011, *6*(6), e20621.
117. Ku, L. X., Zhang, J., Guo, S. L., Liu, H. Y., Zhao, R. F., & Chen, Y. H. Integrated multiple population analysis of leaf architecture traits in maize (*Zea mays* L.). *J. Exp. Bot.*, 2012, 63, 261–274.
118. Ku, L. X., Zhao, W. M., Zhang, J., Wu, L. C., Wang, C. L., Wang P. A., Zhang, W. Q., Chen, Y. H. Quantitative trait loci mapping of leaf angle and leaf orientation value in maize (*Zea mays* L.). *Theor. Appl. Genet.*, 2010, 121, 951–959.
119. Landi, P., Giuliani, S., Salvi, S., Ferri, M., Tuberosa, R., Sanguineti, M. C. Characterization of root-yield-1.06, a major constitutive QTL for root and agronomic traits in maize across water regimes. *J. Exp. Bot.*, 2010, 61, 3553–3562.

120. Lee, K. D., Sulpanjani, Tawaha, A. M. & Min Yang, S. Effect of Phosphorus application on yield, mineral contents and active components of *Chrysanthemum coronarium* L. *Bioscience Research*, 2005a, *2*(3), 118-124.
121. Lee, K. D., Tawaha A. R. M. & Supanjani. Antioxidant status, stomatal resistance and mineral composition of hot pepper under salinity and boron stress. *Bioscience Research*, 2005d, *2*(3), 148-154.
122. Lee, K. D., Turk, M. A & Tawaha, A. M. Nitrogen fixation in rice based farming system. *Bioscience Research*, 2005b, *2*(3), 130-138.
123. Li, C. H., Li, Y. X., Shi, Y. S., Song, Y. C., Zhang, D. F., Zhang, D., Buckler, E. S., Zhang, Z., Wang, T., Li, Y. Genetic control of the leaf angle and leaf orientation value as revealed by ultra-high density maps in three connected maize populations. *PLoS ONE*, 2015, 1-13.
124. Li, K., Xu, C., Zhang, K., Yang, A., & Zhang, J. Proteomic analysis of roots growth and metabolic changes under phosphorus deficit in maize (*Zea mays* L.) plants. *Proteomics*, 2007, 7, 1501–1512.
125. Lisette, J., Xavier, C., & Germida, J. J. Selective interactions between arbuscular mycorrhizal fungi and *Rhizobium leguminosarum* bv, Viceae enhance pea yield and nutrition. *Biol. Fertil. Soil*, 2003, 37, 261–267.
126. Lobell, D. B., Bänziger, M., Magorokosho, C., & Vivek, B. Nonlinear heat effects on African maize as evidenced by historical yield trials. *Nat. Clim. Chang.*, 2011, 1, 42–45.
127. Lobell, D. B., Roberts, M. J., Schlenker, W., Braun, N., & Little, B. B. Rejesus, R. M., Hammer, G.L. Greater sensitivity to drought accompanies maize yield increase in the US Midwest. *Science* 344, 2014, 516–519.
128. Lugtenberg, B., & Kamilova, F. Plant-growth-promoting rhizobacteria. *Annu. Rev. Microbiol.*, 2009, 63, 541–556.
129. Luo L. J. Breeding for water-saving and drought-resistance rice (WDR) in China. *J. Exp. Bot.* 2010, 61, 3509–3517.
130. Lynch, J. P. Steep, cheap and deep: an ideotype to optimize water and N acquisition by maize root systems. *Ann. Bot.*, 2013, 112, 347–357.
131. Mano Y., Omori F., Takamizo T., Kindiger B., Bird R. M., & Loaisiga C. H. Variation for root aerenchyma formation in flooded and non-flooded maize and teosinte seedlings. *Plant Soil* 281, 2006, 269–279.
132. Martre, P., Quilot-Turion, B., Luquet, D., Ould-Sidi Memmah, M. M., Chenu, K., Debaeke, P. Model-assisted phenotyping and ideotype design. In: Sadras, V., & calderini, D. (eds.), *Crop Physiology* (349-373). London: Academic Press, 2015.

133. Meister, R., Rajani, M. S., Ruzicka, D., & Schachtman, D. P. Challenges of modifying root traits in crops for agriculture. *Trends Plant Sci.*, 2014, 19, 779–788.
134. Mendes, R., Garbeva, P., & Raaijmakers, J. M. The rhizosphere microbiome: significance of plant beneficial, plant pathogenic, and human pathogenic microorganisms. *FEMS Microbiol. Rev.* 2013, 37, 634–663.
135. Mesfine, T., Abebe, G., & Al-Tawaha, A. M. Effect of reduced tillage and crop residue ground cover on yield and water use efficiency of sorghum (*Sorghum bicolor* (L.) Moench) under semi-arid conditions of Ethiopia. *World Journal of Agricultural Sciences*, 2005, *1*(2), 152-160.
136. Munees, A., & Mulugeta, K. Mechanisms and applications of plant growth promoting rhizobacteria: current perspective. *J. King Saud Univ. Sci.*, 2014, 26, 1–20.
137. Munns, R., James, R. A., Sirault, X. R., Furbank, R. T., & Jones, H. G. New phenotyping methods for screening wheat and barley for beneficial responses to water deficit. *J. Exp. Bot.*, 2010, 61, 3499–3507.
138. Nadeem, S. M., Zahir, Z. A., Naveed, M., & Arshad, M. Preliminary investigations on inducing salt tolerance in maize through inoculation with rhizobacteria containing ACC deaminase activity. *Can. J. Microbiol.*, 2007, 53, 1141–1149.
139. Naveed, S., Ibrar, M., Khattak, I., Khan, I., Imran, M. I., Khan & Khan, H. Anthelmintic, antilice, insecticidal, cytotoxic and phytotoxic potential of ethanolic extracts of two wild medicinal plants *Iphiona Grantioides* and *Pluchea arguta*. *J. Woulfenia*, 2016, 23 (11), 13-25.
140. Nikus, O., Abebe, G., Takele, A., Harrun, H., Chanyalew, S., Al Tawaha A.M., & Mesfin, T. Yield response of tef (*Eragrostis tef* (Zucc.) Trotter) to NP fertilization in the semi arid zones of the central rift valley in Ethiopia. *European Journal of Scientific Research*, 2005b, 4(4), 49-60.
141. Nikus, O., Nigussie, M., & Al Tawaha, A. M. Agronomic performance of maize varieties under irrigation in Awash valley, Ethiopia. *Bioscience Research*, 2005a, *2*(1), 26-30.
142. Nuccio, M. L., Wu, J., Mowers, R., Zhou, H. P., Meghji, M., Primavesi, L. F., Paul, M. J., Chen, X., Gao, Y., Haque, E., Basu, Sh. S., & Lagrimini, L. M. Expression of trehalose-6-phosphate phosphatase in maize ears improves yield in well-watered and drought conditions. *Nat. Biotechnol.*, 2015, *33*(8), 862–869.

143. Ort, D. R., & Long, S. P. Limits on yields in the corn belt. *Science*, 2014, 344, 484–485.
144. Piperno, D. R., Ranereb, A. J., Holstb, I., Iriarted, J., & Dickauc, R. Starch grain and phytolith evidence for early ninth millennium B.P. maize from the central balsas river valley, Mexico. *Proc. Natl. Acad. Sci. USA*, 2009, 106, 5019–5024.
145. Planchamp, C., Glauser, G., & Mauch-Mani, B. Root inoculation with Pseudomonas putida KT2440 induces transcriptional and metabolic changes and systemic resistance in maize plants. *Front. Plant Sci.*, 2015, 5, 719.
146. Postma, J. A., & Lynch, J. P. Root cortical aerenchyma enhances the growth of maize on soils with suboptimal availability of nitrogen, phosphorus, and potassium. *Plant Physiol.*, 2011, 156, 1190–1201.
147. Postma, J. A., & Lynch, J. P. Theoretical evidence for the functional benefit of root cortical aerenchyma in soils with low phosphorus availability. *Ann. Bot.*, 2010, 107, 829–841.
148. Postma, J. A., Dathe, A., & Lynch, J. P. The optimal lateral root branching density for maize depends on nitrogen and phosphorus availability. *Plant Physiol.*, 2014,166, 590–602.
149. Saengwilai, P. P., Tian, X. L., & Lynch, J. P. Low crown root number enhances nitrogen acquisition from low-nitrogen soils in maize. *Plant Physiol.*, 2014, 166, 581–589.
150. Sakamoto, T., Morinaka, Y., Ohnishi, T., Sunohara, H., Fujioka, S., Ueguchi-Tanaka, M., Mizutani, M., Sakata, K., Takatsuto, S., Yoshida, Sh., Tanaka, H., Kitano, H., & Matsuoka, M. Erect leaves caused by brassinosteroid deficiency increase biomass production and grain yield in rice. *Nat. Biotechnol.*, 2006, *24*(1), 105–109.
151. Samreen, U., Ibrar, M., Lalbadshah, Naveed, S., Imran & Khatak, I. Ethnobotanical study of subtropical hills of darazinda, Takht-e-Suleman range F.R D.I. Khan, Pakistan. *Pure and Applied Biology*, 2016, *5*(1), 149-164.
152. Shelden, M. C., & Roessner, U. Advances in functional genomics for investigating salinity stress tolerance mechanisms in cereals. *Front. Plant Sci.*, 2013, 4, 123.
153. Shi, J. R., Habben, J. E., Archibald, R. L., Drunnond, B., Chamberlin, M. A., Williams, R Lafitte, H. R., & Weers, B. P. Overexpression of ARGOS genes modifies plant sensitivity to ethylene, leading to improved drought tolerance in both *Arabidopsis* and maize. *Plant Physiol.*, 2015, 169, 266–282.
154. Shi, J., & Lai, J. Patterns of genomic changes with crop domestication and breeding. *Curr. Opin. Plant Biol.*, 2015, 24, 47–53.

155. Simons, M., Saha, R., Guillard, L., Clément, G., Armengaud, P., Cañas, R., Maranas, C. D., Lea, P., J., & Hirel, B. Nitrogen-use efficiency in maize (*Zea mays* L.): from 'omics' studies to metabolic modelling. *J. Exp. Bot., 2014*, 65, 5657–5671.
156. Sulpanjani, A., Yang, M. S., Tawaha, A. R. M. & Lee, K. D. Effect of magnesium application on yield, mineral contents and active components of *Chrysanthemum coronarium* L. under hydroponics conditions. *Bioscience Research*, 2005a, *2*(2), 73-79.
157. Supanjani, Tawaha, A. M., Yang, M. S., & Lee, Y. D. Calcium effects on yield, mineral uptake and terpene components of hydroponic *Chrysanthemum coronarium* L. *Research Journal of Agriculture and Biological Science*, 2005c, *1*(1), 146-151
158. Supanjani, Tawaha, A. M., Min Yang, M. S., & Lee, K. D. Role of calcium in yield and medicinal quality of *Chrysanthemum coronarium* L. *Journal of Agronomy*, 2005b, *4*(3), 188-192.
159. Tawaha, A. M. & Turk, M. A. Effects of dates and rates of sowing on yield and yield components of Narbon vetch under semi-arid condition. *Acta Agronomica Hungarica*, 2001, *49*(1), 103-105.
160. Tawaha, A. M. Turk, M. A., & Lee, K. D., Supanjani, Nikus, O., Al-Rifaee M., & Sen, R. Awnless barley response to Crop Management under Jordanian Environment. *Bioscience Research*, 2005a, *2*(3), 125-129.
161. Tawaha, A. M., & Turk, M. A. Lentil (*Lens culinaris* Medic.) productivity as influenced by rate and method of phosphate placement in a Mediterranean environment. *Acta Agronomica Hungarica*, 2002, *50*(2), 197-201.
162. Tawaha, A. R. M., Turk, M. A., & Lee, K.D. Adaptation of chickpea to cultural practices in a Mediterranean type environment. *Research Journal of Agriculture and Biological Science*, 2005b, *1*(2), 152-157.
163. Tilman, D., Balzer, C., Hill, J., & Befort, B. L. Global food demand and the sustainable intensification of agriculture. Proc. *Natl. Acad. Sci. USA*, 2011, 105, 20260–20264.
164. Timmusk, S., Behers, L., Muthoni, J., Muraya, A., & Aronsson, A. C. Perspectives and challenges of microbial application for crop improvement. *Front. Plant Sci.*, 2017, 8, 49.
165. Trevisan, S., Manoli, A., Ravazzolo, L., Botton, A., Pivato, M., Masi, A., & Quaggiotti, S. Nitrate sensing by the maize root apex transition zone: a merged transcriptomic and proteomic survey. *J. Exp. Bot.*, 2015, 66, 3699–3715.

166. Turk, A. M. & Tawaha, A. M. Response of winter wheat to applied n with or without ethrel spray under irrigation planted in semi-arid environments. *Asian Journal of Plant Sciences*, 2002b, *1*(4), 464–466.
167. Turk, A.M. & Tawaha, A. M. Impact of seeding rate, seeding date, rate and method of phosphorus application in faba (*Vicia faba* L. Minor) in the absence of moisture stress. *Biotechnology, Agronomy, Society and Environment*, 2002a, 6(3), 171-178.
168. Turk, M. A., Tawaha A. M., Taifor, H., Al-Ghzawi, A., Musallam, I. W., Maghaireh, G. A., & Al-Omari, Y. I. Two row barley response to plant density, date of seeding and rate and method of phosphorus application in the absence of moisture stress. *Asian Journal of Plant Sciences*, 2003a, *2*(2), 180-183.
169. Turk, M. A., & Tawaha, A. M. Influence of rate and method of phosphorus placement to Garlic (*Allium sativum* L.) in a Mediterranean environment. *Journal of Applied Horticulture*, 2001, *3*(2), 115-116.
170. Turk, M. A., Hameed, K. M., Aqeel, A. M., & Tawaha, A. M. Nutritional status of durum wheat grown in soil supplemented with olive mills by products. *Agrochimica*, 2003c, 209-219.
171. Turk, M. A., Tawaha A. M., Samara, N., & Latifa, N. The response of six row barley (*Hordeum vulgare* L.) to nitrogen fertilizer application and weed control methods in the absence of moisture stress. *Pakistan Journal of Agronomy*, 2003b, *2*(2), 101-108.
172. Van der Heijden, M. G. A., Bardgett, R. D., & Van Straalen, N. M. The unseen majority: soil microbes as drivers of plant diversity and productivity in terrestrial ecosystems. *Ecol. Lett.*, 2008, 11, 296–310.
173. Vassilev, N., Vassileva, M., Lopez, A., Martos, V., Reyes, A., Maksimovic, I., Eichler-Löbermann, B. & Malusà, E. Unexploited potential of some biotechnological techniques for biofertilizer production and formulation. *Appl. Microbiol. Biotechnol.*, 2015, 99, 4983–4996.
174. Verbon, E. H., & Liberman, L. M. Beneficial microbes affect endogenous mechanisms controlling root development. *Trends Plant Sci.*, 2016, 21, 218–229.
175. Yin, H., Chen, C. J., Yang, J., Weston, D. J., & Chen, J. G. Functional genomics of drought tolerance in bioenergy crops. *Crit. Rev. Plant Sci.*, 2014, 33, 205–224.
176. York, L. M., Galindo-Castañeda, T., Schussler, J. R., & Lynch J. P. Evolution of US maize (*Zea mays* L.) root architectural and anatomical phenes over the past 100 years corresponds to increased tolerance of nitrogen stress. *J. Exp. Bot.*, 2015, 66, 2347–2358.

177. Zahir, Z. A., Munir, A., Asghar, H. N., Shaharoona, B., & Arshad, M. Effectiveness of rhizobacteria containing ACC-deaminase for growth promotion of pea (*Pisum sativum*) under drought conditions. *J. Microbiol. Biotechnol.*, 2008, 18, 958–963.
178. Zhan, A., & Lynch, J. P. Reduced frequency of lateral root branching improves N capture from Low-N soils in maize. *J. Exp. Bot.*, 2015, 66, 2055–2065.
179. Zhan, A., Schneider, H., & Lynch, J. P. Reduced lateral root branching density improves drought tolerance in maize. *Plant Physiol.*, 2015, 168, 1603–1615.
180. Zhang, J., Ku, L. X., Han, Z. P., Guo, S. L., Liu, H. J., Zhang, Z. Z., Cao, L. R., Cui, X. J., & Chen, Y. H. The ZmCLA4 gene in the qLA4-1 QTL controls leaf angle in maize (*Zea mays* L.). *J. Exp. Bot.*, 2014, 65, 5063–5076.
181. Zhu, J., Brown, K. M., & Lynch, J. P. Root cortical aerenchyma improves the drought tolerance of maize (*Zea mays* L.). *Plant Cell Environ.*, 2010, 33, 740–749.
182. Zurek, P. R., Topp, C. N., & Benfey, P. N. Quantitative trait locus mapping reveals regions of the maize genome controlling root system architecture. *Plant Physiol.*, 2015, 167, 1487–1496.
183. Imran, Amanullah & Abdel Rahman Al Tawaha (2022d): Indigenous organic resources utilization, application methods and sowing time replenish soil nitrogen and increase maize yield and total dry biomass, *Journal of Plant Nutrition*, DOI: 10.1080/01904167.2022.2067055.
184. Imran, Amanullah and Ibrahim Ortash. 2022b. Agronomic Practices Improved Cucumber Productivity, Nutrients Uptake and Quality. *Gesunde Pflanzen*. 10.1007/s10343-022-00634-1.
185. Imran, Amanullah, and Abdel Rahman Al Tawaha. 2022c. Regenerating potential of dual purpose rapeseed (*Brassica napus* L.) as influenced by decapitation stress and variable rates of phosphorous. *Communication in soil science and plant analysis. Doi.org/10.1080/00103624.2022.2118297*
186. Imran and Amanullah. (2022b). Integration of peach (*Prunus persica* L) remnants in combination with beneficial microbes and phosphorus differ phosphorus use efficiency, agronomic efficiency and partial factor productivity in soybean Vs maize crops. *Journal of Plant Nutrition.* doi.org/10.1080/01904167.2022.2099890
187. Imran, & Abdel Rahman Al Tawaha (2022): Humic Acid and Sulfur Integration Enhances Growth and Yield Related Traits of *Brassica Napus L. Journal of Plant Nutrition.* doi.org/10.1080/01904167.2022.212882

188. Imran. 2022a. Growing of Off-Season Tomato in High Tunnel and Its Nutritional Value Augmentation with Integrated Nutrients Management. *Journal of Plant Nutrition*. Doi: 10.1080/01904167.2022.2046062
189. Imran and Amanullah. 2021a. Assessment of chemical and manual weed control approaches for effective weed suppression and maize productivity enhancement under maize-wheat cropping system. *Gesunde Pflanzen*, DOI: 10.1007/s10343-021-00599-7.
190. Imran, Amanullah and Abdel Rehman Altawaha. 2022c. Carbon Assimilation and Dry Matter Partitioning in Soybean Ameliorates with the Integration of Nano-Black Carbon, Along with Beneficial Microbes and Phosphorus Fertilization. *Journal of Plant Nutrition, DOI:10.1080/01904167.2022.2035753*
191. Imran. 2022b. (Review) Phosphorus availability enhanced with combine application of organic amendments and beneficial microbes under soybean-wheat cropping system. *Communication in soil science and plant analysis.* doi.org/10.1080/00103624.2022.2034848
192. Imran.2022c. Management of Abiotic Stresses with Nano-black carbon is a tool for Crop Production. *Journal of Plant Nutrition.* doi.org/10.1080/01904167.2022.2046085
193. Imran and Amanullah. 2021b. Phosphorus biofortification and uptake in maize enhanced with integrated phosphorus management. *Phosphorus, Sulfur, and Silicon and the Related Elements*. DOI: 10.1080/10426507.2021.2022677.
194. Imran. 2021. The bioavailability of Phosphorus in composite vs. Hybrid maize differ with phosphorus and boron fertilization. *Phosphorus, Sulfur, and Silicon and the Related Elements. https://doi.org/10.1080/10426507.2021.1920588.*
195. Imran, Amanullah, Asad Ali khan, Tariq Mahmood, Abdel Rahman Al Tawaha and Samia Khanum. 2021a. Adequate Fertilization, Application Method and Sowing Techniques Improve Maize yield and related traits. *Communication in soil science and plant analysis*. DOI:10.1080/00103624.2021.1925688
196. Imran, Amanullah and Abdel Rahman M. Al Tawaha. 2021b. Management of Nano-black carbon, Phosphorous and Bio Fertilizer improve Soil organic carbon and ensilage biomass of soybean and maize. *Communication in Soil Science and Plant Analysis.* DOI.10.1080/00103624.2021.1966439
197. Imran and Amanullah· 2021c. Phosphorus and Boron Application Optimizing Biofortification Of phosphorus and Productivity of French

Bean (*Phaseolus Vulgaris* L.). *Communication In Soil Science and Plant Analysis.* DOI:10.1080/00103624.2021.1971689

198. Imran, Amanullah and Abdel Rahman Al-Tawaha. 2021c. Carbon Sources Application Increase Wheat Yield And Soil Fertility. *Communication in soil science and plant analysis DOI: 10.1080/00103624.2020.1865397.*
199. Imran, Amanullah and Abdel Rehman Al-Tawaha.2020a. The Productivity of subsequent wheat enhanced with residual carbon sources and phosphorus under improved irrigation system. *Communication in soil science and plant analysis*, DOI:10.1080/00103624.2020.1763387
200. Imran, Amanullah, Muhammad Arif, Zahir Shah and Abdul Bari. 2020b. Soil application of *Trichoderma* and Peach (*Prunus persica* L.) residues possesses biocontrol potential for weeds and enhances growth and profitability of soybean (Glycine max). *Sarhad Journal of Agriculture* 36 (1) 10-20. DOI | http://dx.doi.org/10.17582/journal.sja/2020/36.1.10.20

Chapter Six

Agronomic and Molecular Approaches: A Solution to the Water and Environmental Crisis

Waqas Liaqat[1], Samina Yasmin[2], Celaleddin Barutçular[1], Muhammad Tanveer Altaf[3], Muhammad Faheem Jan[4] and Imran[5,6]

[1]Department of Field Crops, Faculty of Agriculture, Institute of Natural and Applied Sciences, Çukurova University, Adana, 01330, Türkiye

[2]Department of Plant Protection, Faculty of Agriculture, Institute of Natural and Applied Sciences, Çukurova University, Adana, 01330, Türkiye

[3]Faculty of Agricultural Sciences and Technology, Department of Plant Protection, Sivas University of Science and Technology, Sivas 58140, Türkiye

[4]Northeast Agricultural University, College of Agriculture, Department of Agronomy, Heilongjiang-Harbin 150030, China

[5]Department of Agronomy, The University of agriculture, Peshawar, Pakistan

[6]Rhizosphere laboratory, Department of soil and Plant nutrition, Cukurova University, Adana Turkey

Abstract

Anthropogenic activities in the past and present have resulted in global warming and a variety of biotic and abiotic stressors. Among these, drought stress is regarded as one of the most detrimental to crops. Plants are more susceptible to drought stress since they are sessile. Drought stress has a deleterious impact on several biological processes in plants, from seedling to maturity stages. Drought stress has an impact on plants' morphological, physiological, biochemical, and metabolic pathways, resulting in decreased plant productivity. It is expected that drought will continue to increase in the future as a result of changing climatic patterns. It is critical to know the plant response mechanisms to drought stress in order to mitigate the negative impacts of drought stress on plants. It is vital to employ diverse agronomical, molecular, and biotechnological strategies to compensate for yield losses resulting from drought. Several agronomic techniques such as mulching, biochar application, and the use of plant growth-promoting rhizobacteria, etc., are gaining popularity in this context. To better understand and increase drought tolerance in plants, quantitative trait loci, trangenic methods, and genome-wide association studies, among others, must be employed in a variety of major crops. This book chapter integrates various possible applicative agronomical and molecular approaches that could help in mitigating the negative effects of drought on crop growth and development.

Keywords: Global warming, mulching, biochar, quantitative trait loci, genome wide association study

Introduction

Agriculture consumes 70% of the world's water, accounting for 70% of overall usage. According to Chen et al. (2018), rainfed agriculture accounts for 80% of worldwide farmed area and supplies 60-70% of the world's food. As a result of growing drought conditions, rainfed agriculture is becoming more popular around the world for assisting in food production. Climate change or varying rainfall patterns affect agricultural production in dry and semi-arid regions due to water constraints (Li et al., 2017). Hence, water management and conservation in the agricultural industry have become a difficulty. Water scarcity is a global issue affecting both developing and developed economies (Hanna et al., 2016). Anthropogenic activities disrupt the equilibrium between incoming solar radiation and outgoing radiation, increasing the earth's energy in the form of heat and eventually contributing to global warming (IPCC Climate Change, 2014). Global warming increases water evaporation, which in turn exacerbates drought conditions. Owing to a rise in temperature, major changes in atmospheric conditions may occur, including a rise in summertime temperatures and an increase in drought (Stocker, 2014; IPCC Climate change, 2013). These climate variations have an impact on trophic interactions, species distribution, and the functioning of ecosystems. High temperatures and water scarcity are two interrelated stresses that have a significant impact on crop development, productivity, and ultimately food safety (Zandalinas et al., 2017). Determining the morpho-physiological mechanisms and molecular signaling pathways responsible for greater drought tolerance is vital for future drought management.

Plants face continuous changes in environmental conditions and are frequently subjected to abiotic stresses such as lack of adequate water, salt, excess light, high/low temperatures, and nutritional imbalance, all of which result in a decline in plant performance (Verslues et al., 2006). Water deficit occurs when the plant's water demand cannot be completely met, which occurs when the level of transpired water exceeds the amount of water absorbed by the roots which is caused by insufficient precipitation, decreasing ground water level, or the retention of water by soil particles (Salehi-Lisar and Bakhshayeshan-Agdam, 2016). In dry and semi-arid regions, inadequate water accessibility, availability, and precipitation are the primary factors limiting agricultural production (Salehi-Lisar and Bakhshayeshan-Agdam, 2016). A range of factors including infrequent yearly precipitation and uneven temporal distribution, high evaporation, and water scarcity contribute to the global problem of drought which is restricting crop production and reducing agricultural development (Peng et al., 2016). As a result of the severe effects of global climate change, these concerns are growing more serious (El-Beltagi et al., 2020; El-Beltagi et al., 2022). Water availability for agricultural farmers is continuously reducing since the water requirements of urban populations are substantially rising. Farmers are seeking creative methods to increase soil moisture in order to combat water scarcity (Mohamed et al., 2021; Shalaby et al., 2022). Water scarcity caused mostly by low yearly precipitation and is extremely detrimental to plant growth and eventually to the sustainability of food production. Stress losses can range from 50-80 percent depending on the duration of the stress and the plant species (Christensen and Christensen, 2007). The impacts of drought on plant growth has been studied in wheat, rice, barley, and corn (Rampino et al., 2006; Khan and Bano, 2018; Lafitte et al., 2006) and these studies reported the negative impacts of drouhgt on the performance of these plants. Drought stress negatively affetcs the nutrient uptake and translocation as the soil nutrients are transported to the roots via water. Hence, drought stress decreases nutritional absorption and mass-flux of water-soluble nutrients such as calcium, sulfate, nitrate, magnesium, and silicon (Atouei et al., 2019). Drought stress increases formation of free radicals that harm plant defense system causing an increase in reactive oxygen species (ROS) like superoxide radicals, hydroxyl radicals, and hydrogen peroxide induces oxidative stress. Plants as sessile organisms are subjected to a variety of adverse environmental influences; as a result, they have developed several defense strategies and resistance mechanisms to deal with abiotic stresses (Ballhorn et al., 2009). Photosynthesis activity is reduced principally by stomatal closure, membrane damage, and altered functioning of numerous enzymes particularly those involved in ATP synthesis (Kumawat and Sharma, 2018; Sharma et al., 2020).

Drought stress causes an increase in the production of ROS and reactive nitrogen species which disrupts the redox regulating function of the cell (Laxa et al., 2019; Sharma et al., 2020). Other physiological functions like rapid stomata closure and water use efficiency and biochemical like synthesis of osmolytes, aquaporins, and a potent antioxidant apparatus responses may contribute to the drought tolerance of certain plant individuals thereby supporting the use of drought-tolerant genotypes (Farooq et al., 2009). The consequences of drought on agriculture are exacerbated by the loss of water resources and the alarming increase in global food demand (O'Connell, 2017). The unpredictability of the drought is influenced by a number of variables, including the uneven and unreliable distribution of precipitation, evapotranspiration, and water-holding capacity of the rhizosphere (Devincentis, 2020). Drought stress impairs various molecular, biochemical, physiological, morphological, and ecological features and processes in plants. Plant productivity and quality suffer in water-stressed areas (Battaglia et al., 2018). The primary elements that determine plant responses to drought are growth stages, age, plant species, and drought intensity and duration (Gray and Brady, 2016). Hence, the drought resistance mechanism varies among plant species. Plants are able to lower their resource use and adapt their growth to adverse environmental situations such as drought (Bielach et al., 2017). Molecular networks, such as those involved in signal transduction are responsible for the enhancement of these responses to drought stress (Zandalinas et al., 2020). The molecular mechanisms of plant response to drought stress also include stomatal control via increased ion transport, transcription factor activity, and abscisic acid signaling (Kumar et al., 2019). With changing conditions, it is necessary to increase the plant's drought tolerance. To fulfill future food demands, it will be of the utmost importance to encourage additional research on drought-resistant plants and to employ affordable and beneficial agricultural practices (Raza et al., 2019).

Drought and climate change

Global climate change is predicted to intensify in the future as air temperatures and atmospheric CO2 levels continue to rise altering rainfall patterns and distribution (Yang et al., 2019; Yin et al., 2018). Although a lack of water input from rainfall is usually the primary cause of drought stress, evaporation from soils caused by high temperature events, high light intensity, and dry winds can intensify an already existing drought stress event (Cohen et al., 2021). On a global scale, climate change often causes widespread drought stress conditions across huge areas. Together with drought, salt stress is regarded as a major cause of water deficit in plants (Mostofa et al., 2018; Adnan et al., 2020). Furthermore, the melting of glaciers is shrinking water reservoirs, limiting agriculture water availability. Moreover, as a result of global warming, annual accumulated precipitation has reduced in many rain-fed agricultural areas around the world (Warner and Afifi, 2014). Water loss due to global warming is occurring not only in the soil but also at the plant level. Internal water in plants is lost to the atmosphere to a large extent as a result of higher temperatures caused by global warming, a phenomenon that exacerbates already existing water deficiency problems in various agricultural systems around the world (Sultan et al., 2019). If predicted increases in air temperature of around 2 degrees Celsius above current levels occur by the end of the century, almost one-fifth of the world's population will face acute water shortages (Ray et al., 2019).

A vast number of drought metrics have been used to estimate drought in the future nevertheless, some are more acceptable than others when analyzing the effects of drought on plants (Dai, 2011). Knowing that potential evapotranspiration (PET) rises as a function of temperature, drought metrics are using PET to estimate water demand demonstrated a large-scale change towards drier conditions as temperatures rise (Zhao et al., 2015; Cook et al., 2014). Similarly, assessing increases in aridity using the aridity index, which is a function of PET shows considerable projected trends toward drier conditions (Fu, 2014; Scheff and Frierson, 2015; Berg et al., 2016). However, because actual evapotranspiration (ET) does not increase as quickly as PET (Milly and Dunne, 2016), this indicated dryness may not reflect actual plant stress. While metrics that measure atmospheric demand ("atmosphere-centric") reveal significant increases in both drought and aridity (Zhao and Dai, 2015; Cook et al., 2014; Huang et al., 2016; Lin et al., 2015). Yet, the actual water accessible to plants is also affected by a number of factors that change over time or are difficult to measure at global scales (depth of the rooting zone, root biomass, and soil properties including texture and depth (Stephenson, 1998). In addition, our ability to validate simulations of soil moisture particularly at large scales is observationally constrained (Swann, 2018). Model consensus indicates that surface soil layers will become drier in the future particularly in regions where precipitation is anticipated to decrease substantially (Cook et al., 2015; Berg et al., 2017). Yet, surface soil moisture may not be the most significant for plants whose roots levels extend considerably below the top 10 cm, the depth of soil moisture often recorded (Collins et al., 2013). Deep soil layers do not dry as quickly as surface soil layers.

Impact of drought on agriculture productivity

Depending on the dynamics of the environmental conditions, plants may be subjected to a variety of stresses that can have a negative impact on their growth and development (Bukhari et al., 2019). Various metabolic modifications and gene expressions enable plants to thrive under these conditions (Ahanger et al., 2017). Drought stress could have a significant impact on grain quality and productivity (Figure 1). Consequently, studying plants' ability to cope with water scarcity is important and should continue to attract attention in the near future, particularly in arid and semi-arid regions (Sobhanian et al., 2020). The appropriate level of water availability is required for plant growth and development; soil moisture fluctuations that exceed the optimal level might impair grain yield and quality. Inadequate water availability in the rhizosphere, on the other hand, inhibits plant nutrient uptake by impeding plant growth (Elemike et al., 2019). Common indications of drought stress in plants include leaf rolling, stunning plants, leaf yellowing, leaf blistering, and permanent wilting (Corso et al., 2020). In addition, the plant response to a given water shortage is highly reliant on the occurrence and severity of previous drought stress events (Hafez et al., 2015; Battaglia et al., 2020). Plant growth is inhibited under drought conditions due to a loss in cell wall extensibility and turgor (Seleiman et al., 2019). Respiration can decrease when drought conditions are severe however, increases in respiration have been observed under modest stress (Birami et al., 2018). To compensate for the water deficit, stressed plants maintain their osmotic adjustment by increasing the sugar content of their roots and leaves and plants subjected to drought stress in the past have shown considerably larger development in roots compared to shoots (Miranda et al., 2020).

Environmental factors such as drought duration, intensity, and frequency, soil properties, growth conditions and stages, and plant species have a significant impact on the severity and duration of drought-related symptoms in plants (Zoghi et al., 2019). Increases in the rate of leaves senescence and drooping, burning and limp leaves, leaf rolling and brittleness, closed flowers, and flower sagging, etiolation, wilting, turgidity, premature fall, senescence, and yellowing of leaves are common indicators of drought stress in plants (Ruehr et al., 2019; Khan et al., 2018). Other symptoms presented by plants under dry circumstances include twig cracks, branch dieback, necrosis, stunted growth, bark crack, shrub canopy, and tree thinning (Toscano et al., 2019). In some circumstances, plants may die under high drought stress. The lack of water in the soil hinders the germination of seeds which is necessary for their survival (Cardoso, 2013). Drought stress affects seed germination, plant growth and yield in general, and transpiration rate, net photosynthetic rate, stomatal conductance, leaf relative water content, and water potential in particular (Ullah et al., 2017). Significant changes in seed germination have been documented in a variety of plant species including some of the most intensively cultivated crops such as maize (Queiroz et al., 2019), sorghum (Patanè et al., 2013) and wheat (Qayyum et al., 2011). Drought is also responsible for less/late emergence and seedling establishment (Yan, 2015). Drought stress has a negative impact on productivity because of abnormal physiological processes such as loss of turgidity, rate of carbon assimilation, gaseous exchange, and oxidative damage (Hussain et al., 2018; Ullah et al., 2018). Due to the fact that reactive oxygen species are extremely reactive and poisonous, their concentration may be damaging or beneficial to cells (Miller et al., 2010). Nutrients, particularly macronutrients are essential for the growth and production of crops and the translocation of nutrients is influenced by drought stress (Aroca, 2012). Moreover, drought stress influences the composition of minerals, proteins, and antioxidants (Singh et al., 2012). Drought is a multidimensional stress for plants therefore, it can affect various aspects of plant growth and development (Bhargava and Sawant, 2013; Farooq et al., 2009). Dehydration reduces plant water potential and turgor hence, plant cells cannot perform their normal functions (Rahdari and Hoseini, 2012). A decrease in turgor inhibits cell expansion and growth. Cell division and development are essential for the initiation of plant growth and establishment (Bhargava and Sawant, 2013). Leaf wilting, a decrease in plant height, and an interruption in the creation of buds and flowers are visible indications of plants subjected to water constraint during the initial vegetative stage (Bhatt and Rao, 2005). Shoot length was also decreased under water deficit conditions in *Lathyrus sativus* L (Gheidary et al., 2017). Under conditions of water scarcity, the root-to-shoot ratio of plants generally increases resulting in a large reduction in plant biomass (Akhtar and Nazir, 2013). The leaf is the plant organ where the majority of photosynthetic products are produced. The number of leaves reduced when *Andrographis paniculate* was subjected to water stress (Bhargavi et al., 2017). Water stress reduces leaf area which leads to less photosynthesis and as a result lower agricultural output. A decrease in soil moisture causes a decrease in leaf water content which leads to a decrease in turgor pressure of guard cells due to stomata closure (Deka et al., 2018). Yield reduction in various major crops due to drought has been shown in table 1.

Table 1. Yield losses (%) in important crops under drought stress.

Crops	**Yield reduction**	**References**
Wheat	43.64%	Zhao et al. (2020)
Wheat	36.49%,	Pour-Aboughadareh et al. (2020)
Wheat	20.6%	Daryanto et al. (2016)
Wheat	37%	Shamsi et al. (2010)
Wheat	35%	Shamsi and Kobraee (2011)
Wheat	57%	Balla et al. (2011)
Wheat	44%	Prasad et al. (2011)
Wheat	18–53%	Majid et al. (2007)
Wheat	43–51%	Gupta et al. (2001)
Rice	22.6%	Carrijo et al. (2017)
Rice	53–92%	Lafitte et al. (2007)
Rice	>50%	Daryanto et al. (2017)
Rice	50.6%	Guan et al. (2010)

Rice	21%	Sarvestani et al. (2008)
Rice	>70%	Shamsudin et al. (2016)
Rice	70%	Dixit et al. (2014)
Rice	51–60%	Swamy et al. (2017)
Maize	29.1%	Rabbani and Emam (2011)
Maize	41.6–46.6%	Mi et al. (2018)
Maize	22.5%	Ghooshchi et al. (2008)
Sorghum	50.1%	Batista et al. (2019)
Soybean	43-44%.	Kobraei et al. (2011)
Soyabean	42%	Maleki et al. (2013)
Mashbean	31–57%	Baroowa and Gogoi (2014)
Chickpea	27–40%	Mafakheri et al. (2010)
Chickpea	49–54%	Samarah et al. (2009)
Common bean	58–87%	Martınez et al. (2007)
Common bean	40%	Ghanbari et al. (2013)
Cowpea	34–66%	Ahmed and Suliman (2010)
Cowpea	60%	Ogbonnaya et al. (2003)
Mung bean	31–57%	Ahmad et al. (2015)
Lentil	70%	Shrestha et al. (2006)

The primary effect of water scarcity in plants is a reduction or suppression of photosynthesis (Nezhadahmadi et al., 2013). Among the key restrictions for photosynthesis are lower leaf area, increased stomata closure, and consequent reduced leaf cooling by evapotranspiration which raises osmotic stress and destroys the photosynthetic machinery (Bhargava and Sawant,

2013). Among these, the reduction in photosynthetic process in drought-stressed plants is mostly due to a decrease in CO2 conductivity via stomata and mesophyll constraints (Singh and Thakur, 2018). Reduction in photosynthetic activity caused by drought may also be attributable to impaired stomatal mobility (Chaves et al., 2009). The loss of CO2 uptake causes a decline in photosynthesis (Deepak et al., 2019), which has been demonstrated to reduce the activity of Rubisco, nitrate reductase, sucrose phosphate synthase, and ribulose bisphosphate (RuBP) synthesis. Chlorophyll concentration, which is of paramount significance for photosynthesis is another photosynthetic characteristic substantially influenced by water deficiency that has been characterized as a distinguishing indicator of photo oxidation and chlorophyll degradation ((Rahdari et al., 2012; Anjum et al., 2011). For instance, drought stress modifies leaf chlorophyll production and chlorophyll a/b ratio in soybean (Chowdhury et al., 2017). The reduction in photosynthetic activity, chlorophyll content, photosystem II photochemical efficiency, variation in stomatal movement, and disturbance of water status in plants led to a reduction in plant productivity (Xiang et al., 2013). Moreover, a key cause for reduction in amount of chlorophyll due to drought stress is the drought-promoted O2 − and H2O2 which lead to lipid peroxidation and eventually chlorophyll degradation (Karimpour et al., 2019). Drought stress reduces net photosynthesis and alters plant carbon allocation and metabolism resulting in energy dissipation and decreased yield (Cuellar-Ortiz et al., 2008). Acute drought stress conditions also harm the Rubisco enzyme and other photosynthesis-related enzymes resulting in a reduction of photosynthetic pigment content (Brito et al., 2019). In the presence of water shortage; soil nutrient accessibility, root nutrient translocation, and eventually ion content in various plant tissues are typically reduced (Kheradmand et al., 2014). Water scarcity conditions reduced potassium (K) absorption by plants (Hu and Schmidhalter, 2005). This decrease in K was caused by decreased K mobility, decreased transpiration rate, and impaired root membrane transporter activity (Hu et al., 2013). Water deficit suppressed genes encoding K transporters and inner K channels are triggered by CIPK23, a protein kinase that collaborates with calcineurin B-like calcium sensors (Li et al., 2009). Under water-stress conditions, there was a considerable decrease in leaf P amount in *Ocimum gratissimum* (Osuagwu et al., 2010) and a decrease in K level in *Thymus daenensis* (Bahreininejad et al., 2013). K content also decreased in *Ocimum basilicum* and *Ocimum americanum* plants when water supply was limited (Khalid, 2006).

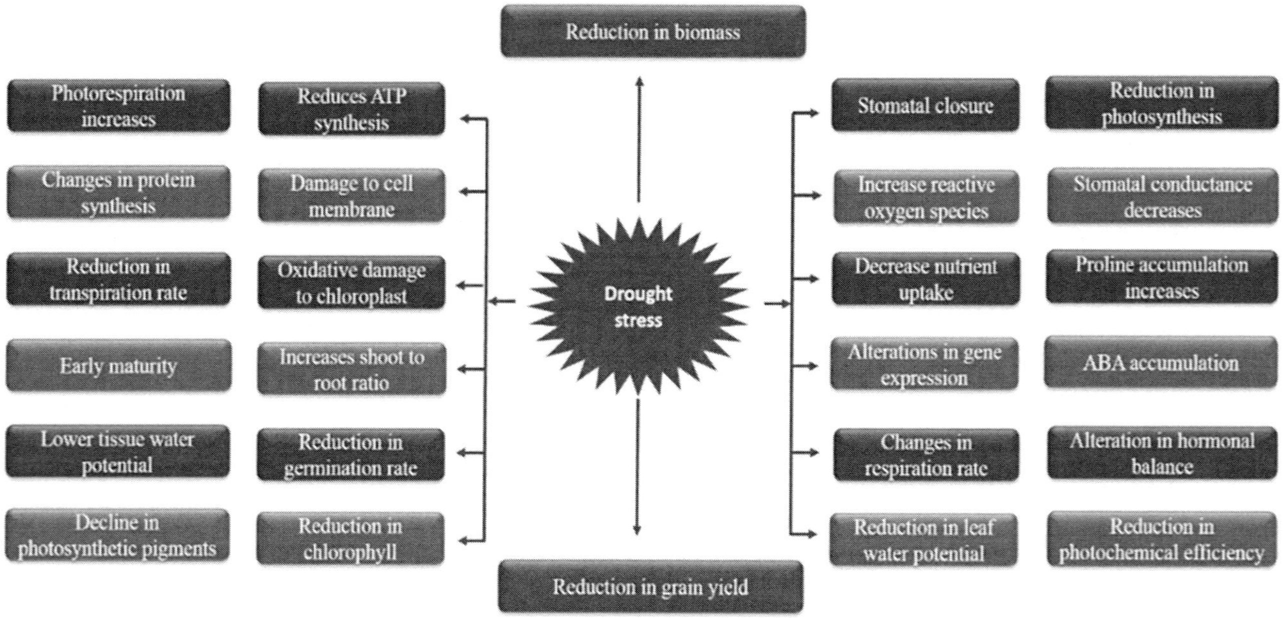

Figure 1. Various effects of drought stess on plants.

Agronomical strategies to combat the negative impacts of drought

Seed priming

Seed priming is a pre-sowing, controlled hydration method that enables germination and metabolic activities to begin without actual germination (Farooq et al., 2006). Seed priming hydrates the seed to initiate pre-germinative metabolic and biochemical activity during phase II of seed germination without radical protrusion (Paparella et al., 2015). Compared to non-primed seeds, primed seeds promote uniform germination through enzyme activation, cell repairing processes, protein synthesis, and enhanced antioxidant defense mechanisms (Jafar et al., 2012). Several priming treatments such as water-based, PGR-based (plant growth regulator), chemical-based, osmotic solution-based, etc., are commonly used to improve drought tolerance in numerous agricultural plants. Although all of these approaches share a similar characteristic i.e., partial pre-hydration and earlier activation of germination events in seeds, their effectiveness is greatly dependent on the treated plant species and priming technique employed. Seed priming induces quick imbibition of seeds with a restricted amount of water in order to initiate the pre-germination metabolic activities without radical protrusion. Due to the quick absorption of water, the germination, emergence, and establishment of primed seeds occur earlier than that of control seeds (Mustafa et al., 2017). Seed priming causes numerous metabolic changes in the seed and initiates the imbibition process in a physiological state (Paparella et al., 2015). According to (Meena et al., 2013), hydropriming of wheat seeds increased WUE, which in turn increased yield. It was reported that hydropriming treatment hydrolyzing the endosperm tissues of Solanum lycocarpum and also increasing the seed germination, seedling growth, and development by endo-β-mannanase activity, favored breaking the mechanical resistance for cell elongation in embryo and embryonic cover. Aquaporins, an important intrinsic protein, control the passage of water through the cell membrane via the intercellular region in both monocot and dicot groups (Borges et al., 2014; Gomes et al., 2009). The deep study of the positive role of water on germination and the profile of aquaporin expression genes were reported during imbibition and primary embryo growth stages in many species such as Arabidopsis (Alleva et al., 2010), Brassica napus (Vander et al., 2006), and Oryza sativa (Ge et al., 2014). These investigations demonstrated the possible effect of aquaporins on seed germination in response to abiotic stresses.

The seed priming techniques stimulate the physiological systems that boost crop output during drought stress (Gholami et al., 2013). It was revealed that chemical priming is an effective remedy for drought. Current research has focused on exploiting the priming-induced mechanisms of drought tolerance in several crops. Moreover, seed priming could improve tolerance mechanisms to subsequent stresses via the mechanism of "stress memory," in which a plant exposed to a particular stress can generate a stress memory for improved and more rapid response to subsequent stress events (Pastor et al., 2013). Plants that have previously been subjected to one type of stress (stress priming/hardening) may develop tolerance/protection to another type of stress via the generation of secondary metabolites (Tabassum et al., 2018). Despite the brief time interval between priming and subsequent stress, the molecular mechanism of seed priming-induced drought stress resistance to subsequent stress continues for a longer period of time due to stress memory (Xiao et al., 2017). Several studies have found that priming can activate the tolerance mechanism during the jointing stage (Li et al., 2015), for instance, tolerance to water deficit conditions at the grain filling stage in wheat (Wang et al., 2015). Similarly, osmoprimed and hydroprimed seedlings exhibited enhanced transgenerational drought resistance via modifying water relations, osmolyte accumulation, malondialdehyde levels, and lipid peroxidation during the terminal growth stage (Tabassum et al., 2018). When compared to non-primed wheat plants, priming before anthesis improved tolerance to a subsequent drought stress during grain filling, resulting in increased photosynthetic rate, ascorbate peroxidase activity, and greater yield (Hayat et al., 2012).

Plant hormones also respond to drought by long distance signaling in root and shoot and organize the water transport. Among plant hormones, ABA possesses a critical mechanism for inducing drought resistance (Lata and Prasad, 2011). A number of hormones are produced in response to drought stress. These hormones may be implicated in the organization of physiological events as gesture molecules in signaling networks (Verma et al., 2016). Under conditions of water deficiency, plants detect stress via sensors involved in response signaling which can be transduced via multiple metabolic cycles, signaling pathways, and transcription factors (Hirayama and Shinozaki, 2010). Arabidopsis seeds primed with the non-protein amino acid b-aminobutyric acid demonstrated drought tolerance via ABA accumulation, stress-responsive gene expression, and stomatal closure (Jakab et al., 2005). Under drought stress, the parental line of rice treated with osmopriming chemicals aids in the buildup of a high proline content via proline expression genes as well as the removal of a methyl group from nucleotides in DNA in the preceding progeny (Marthandan et al., 2020). Seed priming has been identified as the most essential short-term strategy for mitigating the negative effects of drought on plants (Khan et al., 2019). Seed priming is used in crops such as wheat, maize, and chickpea to mitigate the negative effects of drought stress (Khan et al., 2019; Nawaz et al., 2013). Primed seeds produced better germination and stands in water-stressed areas (Nawaz et al., 2016). Seed priming is associated with optimal stand, increased yield, and drought tolerance, early and synchronous germination followed by emergence. Osmopriming with KNO3 and hydropriming increased the productivity of specific crops during periods of water shortage (Delac et al., 2018).

Mulching

Mulching is a common technique in which materials such as plastic, agricultural leftovers, livestock manure, sands, rocks, and cement are applied to the field before, during, or immediately after sowing to support and spread over the soil surface (Gan et al., 2008). Mulching's primary goals are to reduce evaporation or water erosion (Gan et al., 2013), raise soil temperature, increase soil water supply capacity (Zhou et al., 2011), and inhibit weeds (Chalker-Scott, 2007). Mulching increases agricultural yield, promotes plant development, and reduces water consumption (Yu et al., 2018). Mulching improves soil properties, soil moisture availability, and soil productivity (Kader et al., 2017). It also reduces the pH of the soil and boosting nutrients availability. After decomposition, organic mulch provides nutrients to the soil and increases the soil's nutrient availability for a more prolonged time (Yu et al., 2018). Plastic mulches have the potential to greatly improve soil health and insect management (Chalker-Scott, 2007). As a result, it helps to minimize fertilizer leaching and retains nutrients close to the roots of the plants allowing them to be utilized properly. When organic mulch decomposes in the soil, the organic content of the soil increases quickly, improving the soil's ability to store water (Yang et al., 2015). Due to the fact that mulches reduce evaporation, more moisture is available near the roots of plants hence, lengthening the time required for plants to absorb water. Thus, mulched regions need less water (Kader et al., 2017). Mulches regulate the root zone temperature causing the soil to become cooler in summer and warmer in winter (Wang et al., 2016). Mulch reduces runoff or gives soil more time to absorb rainwater by reducing rain's kinetic energy or by decreasing the movement of rainwater (El-Beltagi et al., 2022).

Mulching is considered to benefit stressful environments (like drought, heat and salinity) because it modifies the evaporation and transpiration rates (Jimenez et al., 2017). The effect of mulching depends on the climate and amount of precipitation. Mulches enhance soil moisture retention and structure while preventing weed growth (Mutetwa and Mtaita, 2014). However, the soil moisture variance under diverse mulching materials depends on the soil types and meteorological conditions that influence the effectiveness of various mulching materials to conserve moisture. In general, mulching treatments retain more soil moisture than bare soil (Zhao et al., 2014). When organic mulch is added to the topsoil, it inhibits the growth of weeds, promotes the infiltration of rainwater, and decreases evaporation (Kader et al., 2019). Mulch greatly enhanced overall soil water holding capacity, soil moisture retention, soil porosity, and hence water-use efficiency (Thakur and Kumar, 2021). According to McMillen (2013), mulching with grass clippings, wheat straw, or leaf litter at a depth of 5–10 cm increased soil moisture by 10% compared to bare soil. Plastic mulch treatments retain the most soil moisture compared to organic mulch treatments, which retain more moisture than bare soil (Ogundare et al., 2015). After 90 days, soil under transparent plastic mulch had the highest moisture content (21.1%), followed by black (20.4%) and blue plastic mulch (19.2%), respectively, while the control had the lowest moisture level (14.6%). In semi-arid agriculture, two of the most critical processes that govern soil water availability to crops are infiltration and evaporation. According to Abu-Awwad (1999), covering the surface of the soil decreased the amount of irrigation water required by a pepper crop by 14–29% and by an onion crop by 70%.

Water loss from unmulched land is higher than that from ground coated with plastic. This is because of prolonged contact with water-evaporating elements such the sun, wind, and heat (El-Ganainy et al., 2022). When compared to the conventional tillage method, plastic mulching is superior in terms of its impact on plant yield and water-use efficiency (WUE). When compared to bare soil, black and white plastic mulch increases WUE in potato plants by 31% (Li et al., 2018). According to Goodman (2020), the use of straw mulch reduced evaporation by around 35%. The appropriate mulch can reduce the frequency of irrigation and in certain situations, totally eliminate it. Mulch can also protect trees and plants from drought and cold damage (El-Beltagi et al., 2022). Surface mulching is a common water-saving practice in rainfed agricultural systems (Zribi et al., 2015; Qin et al., 2016). Out of all natural resources, water is the scarcest in the farming system. Nonetheless, it is well understood that different plant species have varying water requirements (Zhou et al. 2009). Mulching increases yield and WUE by decreasing evaporation and boosting soil transpiration, according to Zhou et al. (2009). The use of plastic mulch enhances WUE by 20–60% and reduces evaporation (Lee et al., 2019), which improves soil water retention and infiltration and creates a suitable environment for root proliferation and seed germination (Liu et al., 2010). Mulch treatment has been demonstrated to boost water retention and grain yield in rice farming. A study of 36 rice farming sites found that mulching rice fields with mulch increased yields by 18%. Black plastic mulch has the potential to increase water productivity, rice yield, and quality while also preserving soil moisture (Jabran et al., 2015). The various types of mulches are dispayed in figure 2.

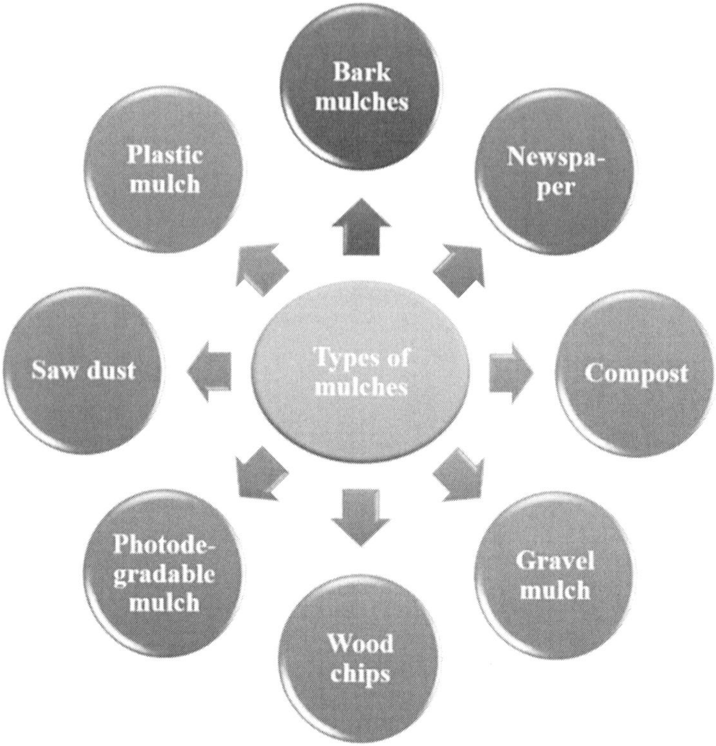

Figure 2. Different types of mulches.

Plant Growth Promoting Bacteria

Bacteria are essential rhizosphere microorganisms that influence plant physiological and biochemical activity. Plant growth promoting bacteria (PGPB) play a crucial role in plant growth and development by providing plant growth enhancing chemicals to their host plants in exchange for protection and nourishment in a mutualistic relationship. These bacteria secrete a variety of chemicals such as osmolytes, antioxidants, phytohormones, and others that improve root osmotic potential during drought stress (Ullah et al. 2017b, 2019a). Also, it regulates plant growth under drought stress, either directly (improved production of phytohormones, increased nutrient availability) or indirectly (induce systematic resistance, pathogen suppression, lytic enzyme synthesis) (Glick 2014; Ullah et al. 2015a). Azospirillum brasilense, for example, enhanced

ABA levels in Arabidopsis thaliana during drought stress (Cohen et al., 2015). ABA, on the other hand, improves drought tolerance by stimulating transcription of water-deficit-related genes and increasing root hydraulic conductivity. PGPB also helps plants in drought-stressed conditions by increasing root surface area and the number of root tips (Vacheron et al. 2013). PGP rhizobacteria are also involved in the improvement of the ROS scavenging mechanism during drought stress. The inoculation of PGPR reduces the accumulation of antioxidant enzymes such as SOD, APX, POD, and CAT, hence reducing oxidative damage during drought stress. Under drought stress, Bacillus pumilus and B. frmus boosted the levels of APX and CAT and decreased oxidative damage (Gururani et al. 2013).

The use of PGPR and commensal microbes particularly arbuscular mycorrhiza (AM) fungi may aid in the development of ways to improve plant water retention capability. Pseudomonas sp. in soil, in particular is of significant importance due to its versatility in catabolism, remarkable root colonization potential, and ability to create a number of enzymes and metabolites that contribute to plant abiotic stress tolerance (Pallai et al., 2012). Tolerance mechanisms like increased osmotic adjustment and hydration of the leaves reduced oxidative damage, improved nutritional status, and enhanced the efficiency of intrinsic water usage have been suggested to elaborate the contribution of PGPR in improving stress tolerance (Kasim et al., 2012). Certain PGPR such as Azospirillum can conserve water by forming cysts around their roots and generating polyhydroxybutyrate and melanin (Dos Reis et al., 2016). Free-moving soil bacteria maintain interactions with plant roots, assisting plant defense against a variety of challenges like drought, heavy metal toxicity, infections, and salt (Khan et al., 2019; Ikram et al., 2019). The higher hydration generated by the PGPR strain could be attributed to improved water efficiency and/or enzymatic reduction of plant ethylene concentration, reducing the inhibitory effect of ethylene on seedling root biomass (Pagnani et al., 2019). Co-inoculation of arbuscular mycorrhizal fungus (Glomus intraradicers or Glomus mosseae) and PGPR Pseudomonas mendocina with Lactuca sativa L. increased antioxidant catalase under extreme drought circumstances, highlighting that inoculants can be employed to reduce drought-induced oxidative damage (Rezaei-Chiyaneh et al., 2020; Rubin et al., 2017). The PGPR species indicated above were also found to be beneficial in mitigating drought-induced oxidative damage in plants. PGPB strain Pseudomonas fluorescens Pf1 increased catalase and peroxidase activity in green grams under water stress (Kenneth et al., 2019). It can also act as a storage molecule for protein synthesis. Under stressful conditions, starch production is inhibited and proline accumulation is exploited as a carbon and nitrogen source for plant survival (Swamy, et al., 2016; Kenneth et al., 2019). By promoting root formation, Rhizobium assists plants in maintaining a favorable water balance in their tissues under water-deficient environments. Rhizosphere bacteria that support plant growth penetrate roots and sustain symbiotic associations in order to stimulate plant growth and provide stress resistance (Saikia et al., 2018). Several strains of PGPRs such as Azospirillum sp., Azotobacter sp., and P. fluorescens, are well-known for their beneficial effects on plant growth and aiding in water stress (Zhu et al., 2020).

Root biomass was high in plants inoculated with Phyllobacterium brassicacearum STM196 and modifying root architecture boosted water absorption. STM196 resulted in a larger contact surface with the soil resulting in increased water transfer from the roots to the shoot. Certain rhizosphere bacteria assist plants in maintaining a desirable moisture content in their tissues under water scarcity by promoting root development (Kalam et al., 2017). Inoculation with PGPR strains enhanced plant development by improving root architecture and as a result, nutrient uptake (Primo et al., 2015). Despite a lack of experimental evidence, it has also been proposed that PGPR promotes plant absorption of water and mineral ions through stimulating proton pump ATPase (Khan et al., 2018). Nutrient availability to the plant is severely reduced under drought environments; however, the use of suitable microbes restores nutrient bioavailability in drought-stressed circumstances. A number of PGPR have been identified and proposed for application in sustainable agriculture under water shortage conditions (Selvakumar et al., 2015).

Application of silicon

Treatment of plants with silicon (Si) has been found to reduce some of the harmful effects of drought. Si can alter plant growth, relative water content, photosynthetic rate, chlorophyll concentration, and other physiological responses during drought (Helaly et al., 2017). Although drought and its relationship to Si have been extensively researched, many plant mechanisms and responses remain unknown. It has been widely documented that Si can alleviate stress induced by a variety of abiotic conditions including drought (Zargar et al., 2019). Stress can be reduced through a variety of mechanisms. Physically, Si taken up by the plant might deposit in the form of SiO2 on the leaf apoplast, reducing evapotranspiration and osmotic stress (Rizwan et al., 2015). Si activates antioxidant defense systems and reducing oxidative damage. It was discovered that Si boosts sugarcane (*Saccharum officinarum* L.) proline concentration during drought stress and ascorbate acid peroxidase activity after 30 days of stress (Bezerra et al., 2019). This shows that Si boosts the activity of ascorbate acid peroxidase to reduce ROS generation (Bezerra et al., 2019). In contrast, peroxidase, catalase, and superoxide dismutase activities were lower in mango plants (*Mangifera indica* L.) with Si addition under drought stress compared to stressed plants without Si addition (Helaly et al., 2017).

Si application to water-stressed plants increases total soluble sugars, proline, ascorbic acid, and glutathione content in wheat which may provide antioxidant defense against drought (Alzahrani et al., 2018). In a research on rice (*Oryza sativa* L.), proline concentration was found to be lower in Si-treated plants exposed to drought (Mauad et al., 2016). Addition of Si triggered diverse proline accumulation responses in plants which may be advantageous for drought-exposed plants. Si promotes nutrient uptake and alters gas exchange characteristics, resulting in an increase in photosynthetic rate on physiological and biochemical levels. The addition of Si to water-stressed tomato plants and Kentucky bluegrass (*Poa pratensis* L.) has a favorable effect on photosynthesis as compared to those with no added Si (Zhang et al., 2018; Chen et al., 2014). Application of Si restrained chlorophyll degradation and increased optimal photosynthetic efficiency of PSII (Fv/Fm) and the electron transport rate which contributed to improvement of the net photosynthetic rate (Cao et al., 2015). Furthermore, the expression of the photosynthesis-related PetF, PetE, PsbQ, PsbP, PsbW, and Psb28 genes were down-regulated under water deficit, but up-regulated by Si (Zhang et al., 2018). Si increased the net photosynthetic rates of wheat and tomato plants under drought stress (Cao et al., 2020; Alzahrani et al., 2018). Exogenous Si treatment improved maize (*Zea mays* L.) tolerance to water shortage by increasing photosynthetic efficiency, stomatal conductance, and cell membrane integrity (Abd El-Mageed, et al., 2020). By increasing growth and the number of leaves per plant, silicon treatment improves drought tolerance in maize and cantaloupe (Abd El-Mageed, et al. 2020; Alam et al., 2020). In mango, the addition of Si enhanced relative growth rate, net assimilation rate, and relative water content in comparison to water-stressed plants and control plants (Helaly et al., 2017). Likewise, the addition of Si to water-stressed Kentucky bluegrass boosted net photosynthesis, leaf water content, and turf quality (Chen et al., 2014). Silicon also enhances the water potential and relative water content of leaves during drought stress in wheat and sugarcane (Bezerra et al., 2019; Alzahrani et al., 2018). In addition, Si boosts the chlorophyll and carotenoid content of water-stressed mango and tomato plants relative to control plants (Helaly et al., 2017; Zhang et al., 2018). Treatment of plants with Si during drought stress increases Si concentrations in high and low Si-accumulating wheat cultivars (Thorne et al., 2021). Si can be accumulated in above-ground plant tissues even in nonaccumulator species (e.g., oilseed rape (Brassica napus L.) if present at high levels in the soil and can facilitate water uptake under drought conditions (Saja-Garbarz et al., 2021). Foliar application of Si reduces the accumulation of Cd2+, Ni2+ and Cr3+ in maize leaves and grain if the plants are under drought stress by metal-contaminated irrigation water (Abd El-Mageed et al., 2020). Si may also stimulate the transcription of genes involved in antioxidant defense, photosynthesis, osmotic adjustment, lignin, and suberin metabolism, however the exact nature of these connections remains unknown (Zhu and Gong, 2014).

Biochar application

Biochar application has been frequently recognized to have favorable effects in limited water situations (Akhtar et al. 2014; Batool et al. 2015; Paneque et al. 2016). The use of biochar boosted the growth and biomass of drought-stressed plants (Imran 2022b,c). Biochar, for example, boosted corn plant height and leaf area under drought conditions (Haider et al., 2015). Similarly, biochar treatment enhanced field-grown wheat biomass in semiarid Mediterranean conditions (Olmo et al., 2014). Biochar also boosted tomato plants' fruit yield and quality, total soluble solids, and titratable acidity under drought stress (Akhtar et al., 2014). In addition, the application of biochar to silt clay soil improved tomato growth relative to the control (Vaccari et al., 2015). Githinji (2014) observed that application of biochar to sandy loam soil enhanced tomato leaf quality in comparison to the control. According to studies, biochar treatment enhanced photosynthesis in drought-stressed plants (Lyu et al. 2016; Paneque et al. 2016; Xiao et al. 2016). Akhtar et al. (2014) reported that biochar considerably improved the chlorophyll contents, photosynthetic rate (Pn), stomatal conductance (Gs), relative water contents (RWC), water use efficiency (WUE), and increased stomatal density of tomato leaves subjected to drought stress (Imran 2021). Biochar application also increased the leaf RWC, osmotic potential, and transpiration rate of maize under drought stress as compared to the control (Haider et al., 2015).

Furthermore, biochar improved the WUE of maize grown in sandy soil (Uzoma et al., 2011). Similarly, biochar improved photosynthesis, WUE, and Gs of drought-stressed okra (*Abelmoschus esculentus* L. Moench) relative to the control (Batool et al., 2015; Imran et al., 2021b). The increases in gas exchange characteristics, WUE, RWC, and Pn caused by biochar reflect an improvement in the water status of plants under drought stress. The response of biochar to drought stress varies depending on the soil, biochar type, and plant species. When combined with microorganisms, biochar has been found in studies to reduce water stress in plants (Mickan et al., 2016; Liu et al., 2017). In drought-stressed lupin (*Lupinus angustifolius* L.) seedlings, inoculation of biochar with Bradyrhizobium sp. boosted growth, biomass, nitrogen and phosphorus intake, and nodulation compared to alone microbial inoculation (Egamberdieva et al., 2017). In another study, Liu et al. (2017) reported that inoculation of Rhizophagus irregularis with biochar (birch wood, 500 °C) decreased potato leaf area, WUE, N, and P while having no effect on root biomass and soil pH under limited root zone water compared to the respective treatments without biochar application.

However, with limited irrigation, wood-derived biochar soil amendment showed no significant influence on soil biota groups (nematodes, protozoa, fungus, bacteria, and arthropods) (Pressler et al., 2017). This demonstrated that biochar application with other amendments like microbial inoculation might be useful in reducing drought stress in plants depending upon type of biochar and other factors. Numerous studies have found that biochar enhanced soil physical characteristics when subjected to drought stress (Baronti et al., 2014; Bruun et al., 2014). Biochar-mediated decreases in bulk density and increases in soil water contents have been documented in a wide range of soil textures (Jeffery et al., 2015; Igalavithana et al., 2017). Biochar application boosted the aggregate stability of sandy-clay soil, according to Baiamonte et al. (2015). Increased aggregate stability may be useful in boosting soil water retention particularly under conditions of limited water availability (Imran et al., 2022b, d). Generally, biochar application improves soil physical qualities under limited water availability and will aid plant growth in semiarid environments. In addition, biochar aging may impact soil properties therefore, research is required to determine the effect of biochar aging on soil properties under drought conditions (Arthur et al., 2015, Imran et al., 2021a,c). Moreover, soil water retention may vary based on biochar sources and application rate (Brantley et al., 2015). Additional research utilizing a variety of biochars, pyrolysis settings, and application rates may assist to clarify the impact of biochar on soil water holding capacity during drought stress (Lee et al., 2015). The incorporation of biochar raises the soil's water-holding capacity which is a crucial aspect in boosting plant growth and produce during drought (Akhtar et al., 2014; Foster et al., 2016). Increased WHC of soil during drought stress may be due to biochar's increased CEC and porous structure (Artiola et al., 2012). The application of biochar to a vineyard field improved the soil water content measured at various time intervals (Baronti et al., 2014). In a four-year field research, the use of biochar increased the soil moisture content without impacting the maize production (Haider et al., 2017). Ajayi and Horn (2016) examined the capacity of clay and biochar to enhance the water retention and mechanical strength of sandy soil. Results indicated that soil water retention capacity was greater with Na-bentonite at more negative matric potentials whereas biochar was more successful at improving soil water retention capacity at saturation possibly due to the increase in soil porosity (Ajayi and Horn, 2016). The addition of biochar to sandy loam soil enhanced the WHC of the rice-growing soil (de Melo Carvalho et al., 2014). Another study found that biochar increased the WHC of silty sand under maize cultivation in pots (Haider et al., 2015). Under semiarid Mediterranean circumstances, the application of biochar to vertisol soil enhanced the soil's ability to retain water and promote wheat growth (Olmo et al., 2014). Andrenelli et al. (2016) reported that biochar may improve the

WHC of soils and the majority of the research were conducted on sandy soils. However, additional research is necessary to determine the impact of biochar in enhancing the WHC of soils with varied textures and environmental circumstances (Imran et al., 2022a,c).

Molecular strategies to overcome drought impacts

Drought tolerance is a complex process due to the interactions between stress factors and various molecular, biochemical and physiological phenomena affecting plant growth at different developmental stages (Imran and amanullah 2022a,b). Drought tolerance is highly complex process and related with the regulation of multiple genes expression induced during drought stress (Lang and Buu, 2008). Achieving high yield of crops under drought stress is the target of crop breeders (Farooq et al., 2020). Over the years, molecular information has been generated on the response of plants to various abiotic stresses. Plants respond to abiotic stresses such as drought stress by the induction of both regulatory and functional sets of genes. Most common stress signaling pathways have been distinguished into abscisic acid (ABA) dependent and abscisic acid independent pathway (Ramzan et al., 2018; Zlatev and Lidon, 2012). Genetically, stress is an environmental condition that inhibit a plant from obtaining complete genetic expression. Environmental stresses are severe threats to agriculture production that are currently intensified by industrialization and global warming. Once genetic markers have been developed for a particular trait, the markers can be used to evaluate the variation available for the breeders (Kang et al., 2017). DNA markers have broad applications for improving a plant's genetic makeup such as genetic identification of parents, the assessment of variation on a genetic level, and the identification, genetic confirmation, and development of linkage groups with high resolution. A vast range of molecular markers is available for genetic analysis of crops (Varshney et al., 2005). These molecular markers are classified on the basis of how markers are utilized.

DNA Markers and abiotic stress tolerance

Rapid amplified polymorphic DNAs (RAPD) are PCR-based markers, widely used due to facilitate experimental approach and excellent genetic screening of intra and inter-specific. Preliminary sequence data of samples are not necessary in RAPD analysis. Several loci from many individuals can be analyzed for screening purpose using limited resource hybrids (Shasany et al., 2005). PCR-based markers have been very useful for identifying the drought stress tolerant genes in many crops. Various mechanisms are available in plants for tolerance against stress and such responses are regulated genetically. Molecular markers can help to identify and categories drought resistant genotype and RAPD is a basic tool for detecting resistant approach. Experiments are conducted to evaluate the genetic diversity of drought-resistant genotypes growing under drought regime. Molecular markers or DNA markers effectively differentiate drought-resistant from drought-sensitive genotypes. Polymorphic primer pairing between tolerant and sensitive genotypes confirmed genetic variation. On the basis of available genetic information, resistant wheat varieties might be developed through cross of drought-resistant and drought-susceptible genotypes (Shao et al., 2005; Kirigwi et al., 2007). Field evaluation is laborious and time consuming as compared to DNA screening for classifying quality parameters, crop productivity and stress resistance. Molecular techniques are sensitive, simple, rapid, and requires very small amount of genomic material (DNA) and can process large number of DNA samples in vitro conditions. Several primers were linked to drought resistance sequences that provide valuable information for breeding drought-resistant progeny and these progenies can be further screened by marker-assisted selection. Wheat (*Triticum aestivum* L.) quality and yield in many regions are severely effected by various stresses including drought stress (Shao et al., 2005; Kirigwi et al., 2007). Changes induced in the genome due to drought stress contributes directly to stress resistance (Zhou et al., 2010). DNA primers were designed and utilized against drought resistance in hybrid varieties of wheat in early DNA fingerprinting for RAPD analysis. A RAPD primer P6 (TCGGCGGTTC) amplified a 920 base pair (bp) fragment in semi-drought resistant genotypes and drought-resistance genotypes that were absent in drought-sensitive varieties. Furthermore, a RAPD fragment of 717bps (Dreb-B1gene) was located on B genome derived from drought-resistant "Baraktali-95" (Huseynova and Rustamova et al., 2010). Experimental findings showed two negative and four positive RAPD markers that confirmed RAPD reliability for identifying the drought-resistance wheat genotypes (Rashed et al., 2010).

Quantitave trati loci and drought tolerance

Drought tolerance is controlled by polygenes that have additive effects in their expression. The loci on chromosomes residing such genes are known as quantitative trait loci (QTL) (Ashraf et al., 2008). Natural genetic variation of a crop can be exploited either through direct selection under stressful conditions whether natural, stimulated or QTL mapping and subsequent molecular marker assisted selection (Vinh and Paterson, 2005). QTL mapping identifies the location, numbers and magnitude of phenotypic effects and pattern of gene action. The function of polygenes in controlling a single trait has been widely studied by traditional means but the use of molecular markers and QTL mapping has made it more convenient to dissect the complex trait (Humphreys and Humphreys, 2005). For qualitative analysis, phenotypic evaluation is carried out of a large population segregating for variety of genetic markers then a part or the whole population is genotyped and finally appropriate analysis is carried out to pinpoint the loci controlling a trait. Due to complexity of abiotic stress tolerance and problems encountered in phenotypic based selection, QTL mapping has been considered crucial to the use of molecular markers for improving stress tolerance (Ashraf et al., 2008). Variety of DNA markers e.g. AFLPs, CAPs, RAPDs, RFLPs, SNPs and SSRs has been used to assess inheritance of stress tolerance. The data that provide molecular basis of drought tolerance is mostly based on segregation mapping and QTL analysis. In several crops like wheat, maize, cotton, barley, sorghum, and rice QTL mapping for the drought tolerance trait has been perfomed (Bernier, 2008). Guo et al. (2013) identified 51 QTLs for 10 traits on 10 different chromosomes under water deficit condition in which 22 QTL were responsible for various root characteristic associated with drought stress tolerance in rice. Likewise, Zhang et al. (2003) identified 36 QTLs for root trait and 5 QTL for osmotic adjustment in rice under various water limited regimes of lowland and upland environments.

In cotton, cross between *Gossypium barbadense* (cv. F177) and *Gossypium hirsutum* (cv. Siv'on) using F3 families identified a subset of 33 QTLs under drought stress i.e. 17 QTL for fiber quality, 11 for plant productivity and 5 for some physiological traits (Saranga et al., 2001). Through markers-assisted, near isogenic lines were produced through exchanging QTL for yield and some drought-related traits. For most of trait examined, near-isogenic lines showed a marked adaptation to water stress but not for yield. For example, *G. barbadense* near-isogenic lines showed a steady photosynthetic efficiency under varying water limited conditions (Levi et al., 2009). In barley, 187 recombinant inbred lines resulting from a cross between two Mediterranean cultivars, Tadmor and Fr/Apm, QTLs related to some key growth and water relation characteristics were examined. QTLs involved in variation in number of tillers (NT), relative water content (RWC), total shoot fresh mass were found to be located on a RFLP-RAPD genetic map. It was observed that different region of genome (DNA) mediate in constitutive water stress responses. It is important to note that epistatic interaction among many QTLs and between QTLs and other markers were noted only under water stress which suggest that some chromosomal regions are definitely involved in controlling the expression of traits under water limited conditions. From previous studies it is concluded that molecular mapping and a number of QTLs associated with drought tolerance identified in different crops can be effectively used in plant breeding program for improving drought tolerance in plants. QTL mapping is an effective way of detecting specific components that permit direct appraisal of stress tolerance. In 2005, a first attempt was made to clone QTL, which is an important milestone in molecular breeding having substantial role to understand and manipulate the traits responsible for drought tolerance (Ashraf, 2010).

Transgenic approaches for drought tolerance

Transgenic approaches involves modification of qualitative and quantitative traits in an organism by cloning desired gene from one specie to another. Genetic engineering involves the incorporation of only specific cloned genes into an organism and restricts the undesirable genes from donor organism (Gosal et al., 2009). Modern techniques in recombinant-DNA technology has allowed the efficient transformation and generation of transgenic lines in number of crop species (Ashraf et al., 2008). In saline soil, plant responses to drought stress differently and complicatedly because it involve many genes in additive effects. Despite this, efforts have been made to generate transgenic lines of different crops. The major emphasis of genetic engineers has been to transfer genes that encode antioxidants, compatible organic osmolytes, plant growth regulators, heat shock and late embryogenesis abundant proteins and transcription factors involved in gene expression (Ashraf and Foolad, 2007). It has been proved through experiments that compatible organic solutes play central role in drought tolerance. Overproduction of compatible organic osmolytes is one of the prominent response of plants to drought stress (Serraj and Sinclair, 2002). Transfer of such genes in transgenic plants is the goal of genetic engineers. Among many organic osmolytes, glycine betaine (GB), a quaternary ammonium compound occur abundantly in plants under drought stress (Mansour, 2000; Mohanty et al., 2002). Two key enzyme, choline monoxygenase (CMO) and beta aldehyde dehydrogenase (BADH) are responsible for the production of GB. These enzymes have been engineered in different crops. For example, transgenic *Nicotina Tobaccum* have been produced with over expression of CMO and higher accumulation of GB under drought stress condition (Shen et al., 2002; Zhang et al., 2008). Beside GB, some other compatible organic osmolytes like proline plays an important role under drought stress. Pyrroline-5-carboxylate synthetase (P5CR) encodes for biosynthesis of proline, gene for (P5CR) has been successfully engineered in soybean (De Rondhe et al., 2004), petunia (Yamada et al., 2005) and tobacco (Gubis et al., 2007). All these transgenic plants showed enhance production of proline as a response to drought stress condition. Trehalose, a non-reducing sugar is an organic osmolyte which plays a key role in plants against stress tolerance. Gene encoding for trehalose-6-phosphate synthetase has been genetically engineered in tobacco (Karim et al., 2007) and rice plants (Jhang et al., 2003). Studies demonstrated that engineering constitutive over expression of genes encoding TPP or TPS (Trehalose-6-phosphate phosphatase) resulted in enhanced production of trehalose under stress condition (Wu and Garg, 2003). Mannitol is an important osmoprotectant and plays a crucial role in plants under stress. Gene mannitol-1-phosphate-dehydrogenase (mt1D) involved in the biosynthesis of mannitol have been genetically transformed in plants e.g. wheat and tobacco. In transformed wheat mt1D gene overexpress in both drought and salt stress while in genetically improved tobacco over production of mannitol did not affect drought tolerance. In such cases there is need to ascertain whether or not mannitol over-expression is related to drought in different plant species (Abebe et al., 2003).

Plants under drought stress leads to enhanced accumulation of reactive oxygen specie (ROS) thus causing oxidative stress. To counteract these effects plants intrinsically develop different types of antioxidant. Overproduction of antioxidant has been found to be associated with drought stress tolerance (Sunkar et al., 2006). Different plant species have been transformed with different antioxidant encoding genes in order to have enhanced drought tolerance. For example, gene SOD encoding superoxide dismutase in alfalfa, potato and rice, ascorbate peroxidase from *Arabidopsis* in tobacco chloroplasts (Wang et al., 2005; Badawi et al., 2004). Transgenic lines showed enhanced drought tolerance. Degree of drought tolerance in transgenic plants varies with growth and developmental stages. For example transgenic lines at early stage of growth may not show the same level of tolerance as at advanced stage. Most of genetically engineered crops are based on a single gene transformation and manipulation of number of different genes against drought tolerance seems to be a plausible approach (Ashraf and Akram, 2009).

Genome-wide association study

Advances in modern techniques in bioinformatics, genetics, molecular biology, genome-wide association studies (GWASs) based on natural population and linkage disequilibrium can detect genetic variation polymorphisms in the whole genome and conduct population-level statistical analysis of the genotype and phenotype which can more accurately detect genes associated with specific traits. Thus, making it more reliable to detect genes that control the drought-tolerance traits of crops and identify the excellent genetic variation materials (Guo et al., 2018; Hoang et al., 2019). With the completion of most gene sequencing, GWASs have been widely used in drought-tolerance research of cotton (Hou et al., 2018), maize (Wang et al., 2016; Wu et al., 2021), rice (Ma et al., 2016; Guo et al., 2018; Hoang et al., 2019), soybean (Sun et al., 2022; Chen et al., 2020), sorghum (Spindel et al., 2018; Maina et al., 2022), wheat (Mei et al., 2022; Zheng et al., 2022; Mathew et al.,2019) and some other crops and many significant marker-trait association loci and candidate genes related to drought tolerance have been detected, some of which are related to transcription factors (TFs) (Mei et al., 2022; Sun et al., 2022; Spindel et al., 2018). Some other trait loci identified by GWAS contain genes related to osmotic stress regulation such as basic leucine zipper (bZIP) protein gene OsbZIP80, which plays an important role in tolerance to abiotic stress and is considered a rice-dehydration stress inducible gene (Hoang et al., 2019). A dehydration response element-binding protein factor detected by GWAS plays a transcriptional activation role in abscisic acid (ABA)-independent drought stress responses by binding to the dehydration response element (Mei et al., 2022). Drought tolerance is a complex trait regulated by many genes and hence it involve variety of drought-response signaling pathways and metabolic networks (Guan et al., 2019).

Drought resistance is a complex and intrinsic feature and the enhancement of plant drought resistance via molecular breeding is a rapidly expanding area of basic and applied study. By altering the expression of drought-responsive genes, plant drought tolerance has been improved by using knowledge gained from model plants (Qin et al., 2011; Osakabe et al., 2014). In crops however, limited allelic variation governing drought tolerance has been identified (Wang and Qin, 2017). GWAS has contributed to the identification of favorable alleles for drought tolerance particularly in maize. Setter et al. (2011) revealed that a candidate gene association study found loci involved in the accumulation of carbohydrates and ABA metabolites under drought, eight of which were confirmed to be strongly linked with drought tolerance. Later, each genotype of this association panel was crossed with a common tester (CML312), and the resulting F1 plants were phenotyped for nine attributes in seven environments under well-watered and water-stressed conditions. An additional 42 related SNPs were discovered as a result of this investigation (Xue et al., 2013). Liu et al.

(2013) analyzed all the functional (Dehydration Responsive Element Binding protein) DREB genes in maize and studied their relations with the natural variation in drought tolerance among 368 maize varieties collected from subtropical/ tropical and temperate regions (Fu et al., 2013; Li et al., 2013). Thirunavukkarasu et al. (2014) documented that by using 240 accessions of subtropical maize with a high density marker set, 61 significant SNPs were recognized under water stress condition; 48% of which were stress tolerance associated genes. Furthermore, the genetic basis of drought resistance in barley, chickpea, A. thaliana, Medicago truncatula were also studied by GWAS (Verslues et al., 2014; Thudi et al., 2014; Wehner et al., 2015; Kang et al., 2015). A combined data from 15 bi-parental populations of maize were developed under the Water-Efficient Maize for Africa (WEMA) project to identify drought resistance genes by GWAS. The study identified several hundred genetic variants that are related with flowering time and plant height under various water deficit conditions (Wallace et al., 2016).

Regulation of gene expression under drought stress

Several regulatory genes for drought response and osmotic stress have been found through molecular and genetic studies (Xiong et al., 2002). There are two primary pathways involved in water stress response: ABA dependent and ABA in-dependent pathways (Abe et al. 2003). Because of the RD29A/COR78/LTI78 (responsive to dehydration/cold-regulated/low-temperature-induced) gene research, the transcriptional factors that lead to the activation of stress responsive genes are well understood. The promoter region of this gene contains both an ABRE (abscisic acid responsive element) and a dehydration responsive element/C-repeat) (Shinozaki and Yamaguchi-Shinozaki 2007). Both ABRE and DRE/CRT are misacting elements with ABRE regulating ABA-dependent gene expression under stress and DRE/CRT regulating ABA-independent gene expression under stress. Dehydration stress induced transcription factors (DREB2A and DREB2B) which promote the expression of several genes involved in drought stress tolerance. Transcription factors that are grouped in ERF/ AP2 (ethylene-responsive element binding factor/petal 2) bind to DRE/CRT has been isolated and named as DREB1A/CBF3 (DRE-binding protein/C-repeat-binding factor), DREB1B/CBF1 and DREB1C/CBF2 (Shinozaki and Yamaguchi-Shinozaki 2007). Because the DREB1/CBF pathway is linked to ABA-independent gene expression in response to drought and cold stress, it plays a key role in stress tolerance. The transcriptional factors DREB1A/CBF3 play a significant role in the expression of the galactinol synthase gene in transgenic Arabidopsis plants with overexpression leading to the formation and accumulation of galactinol and raffinose making the plant drought-tolerant by slowing transpiration (Park et al. 2006). A list of genes involoved in tolerance against drought stress in various crops has been shown in table 2.

Table 2. A list of key genes involved in drought tolerance in various crops.

Crop	Gene	Function	References
Wheat	TaNAC69	Enhanced drought tolerance	Budak et al. (2013)
Maize	NF-YB2	Enhanced yield and photosynthetic rate under drought	Vandenbroucke and Metzlaff (2013)
Rice	OsNAC5	Enhanced drought tolerance	Kim (2014)
Rice	ONAC045	Enhanced drought tolerance	Zheng et al. (2009)
Rice	OsbZIP46	Enhanced drought tolerance	Tang et al. (2012)
Rice	OsNAC10	Enhanced grain yields and drought tolerance	Jeong et al. (2010)
Sugarcane	SodERF3	Enhanced drought tolerance	Trujillo et al. (2008)
Soybean	P5CS	Enhanced drought tolerance	Valliyodan and Nguyen (2006)
Tobacco	HSP70-1	Contributes to drought stress tolerance	Cho and Hong (2006)

Transgenic tobacco	RD22	Responsive to dehydration	Jamoussi et al. (2014)
Alfalfa	AVP1	Improved drought tolerance	Bao et al. (2009)
Cotton	TsVP	Improved drought resistance	Lv et al. (2009)
Transgenic rice	ZAT10	Enhanced grain yield under drought stress	Xiao et al. (2009)
Arabidopsis	WRKY20	Regulate ABA signaling	Luo et al. (2013)

Conclusion and future outlooks

Our capacity to feed the globe is being challenged by the ever-increasing global population and climate change-induced drought stress. Due to the tremendous losses in crop output caused by drought stress during the past few decades, there is a global interest in enhancing plant drought tolerance and yield. Generally, plants respond to drought stress via morphological, physiological, biochemical, and molecular mechanisms. It is obvious that like today, dryness due to climate change will be the primary factor restricting crop production in the future. Drought stress negatively affects plant growth and yield. Unquestionably, the timing, duration, and severity of stress play a crucial role in defining a plant's response to a lack of water. Yet, drought is difficult to manage under natural circumstances. The reaction of plants to stress at various growth stages is a critical factor for the development of high-stress tolerant cultivars. The response of plants to stress is the consequence of the co-operation of morphological, molecular, biochemical, and physiological mechanisms. In order to be prepared for the coming food and shelter crises, high yielding drought tolerant crops should be produced by combining traditional and modern technologies. To counteract the negative impact of drought stress, plants have developed mechanisms that increase their drought tolerance. Hence, there is a need to investigate unexplored adaption traits in various plant species and incorporate them into genotypes that can survive the negative effects of drought stress without affecting their productivity. Breeding approaches have more potential for enhancing plant performance and yield in water-deficient environments. Certain breeding strategies, molecular and genomics approaches, and an emphasis on omics technologies including metabolomics, proteomics, phenomics, genomics, transcriptomics, and glyomics are of great importance for enhancing drought stress tolerance in plants. Emerging technologies including gene selection, microarrays, next-generation sequencing, RT-PCR, and CRISPR will enable scientists to comprehend and enhance drought tolerance in important crops. Biotechnological approaches should also be considered in order to create transgenic plants that can withstand water scarcity. In addition, practices including growth hormones, seed priming, osmoprotectants, selenium, potassium and silicon application are useful for plants in water deficit conditions. Similarly, the beneficial effect of nanoparticle applications, microbes, hydrogel and metabolic engineering techniques also regulates the antioxidant enzymes activity in plants for adaptation to drought stress, enhancing plant tolerance through cell homeostasis maintenance, and ameliorating the adverse effects of water stress in plants. These novel approaches improve understanding and could increase plant yield in arid environoment, hence reducing the danger to global food security.

References

1. Abd El-Mageed, T.A.; Shaaban, A.; Abd El-Mageed, S.A.; Semida, W.M.; Rady, M.O.A. Silicon defensive role in maize (*Zea mays* L.) against drought stress and metals-contaminated irrigation water. Silicon 2020, 13, 2165–2176.
2. Abe H, Urao T, Ito T, Seki M, Shinozaki K, Yamaguchi-Shinozaki K (2003) Arabidopsis AtMYC2 (bHLH) and AtMYB2 (MYB) function as transcriptional activators in abscisic acid signaling. Plant Cell Online 15(1):63–78
3. Abebe T, Guenzi AC, Martin B, Cushman JC. Tolerance of mannitol-accumulating transgenic wheat to water stress and salinity. Plant Physiol 2003;131:1748–55
4. Abu-Awwad, A.M. Irrigation water management for efficiency water use in mulched onion. J. Agron. Crop Sci. 1999, 183, 1–7
5. Adnan, M.; Fahad, S.; Zamin, M.; Shah, S.; Mian, I.A.; Danish, S.; Zafar-ul-Hye, M.; Battaglia, M.L.; Naz, R.M.M.; Saeed, B.; et al. Coupling Phosphate-Solubilizing Bacteria with Phosphorus Supplements Improve Maize Phosphorus Acquisition and Growth under Lime Induced Salinity Stress. Plants 2020, 9, 900.
6. Ahanger, M.A.; Tomar, N.S.; Tittal, M.; Argal, S.; Agarwal, R.M. Plant growth under water/salt stress: ROS production; antioxidants and significance of added potassium under such conditions. Physiol. Mol. Biol. Plant 2017, 23, 731–744.

7. Ahmad, A.; Selim, M.M.; Alderfasi, A.A.; Afzal, M. Effect of drought stress on mungbean (*Vigna radiata* L.) under arid climatic conditions of Saudi Arabia. Ecosyst. Sustain. Dev. 2015, 192, 185–193.
8. Ahmed, F. E., and A. S. H. Suliman, 2010: Effect of water stress applied at different stages of growth on seed yield and water-use efficiency of cowpea. Agric. Biol. J. North Amer. 1, 534–540
9. Ajayi AE, Horn R (2016) Comparing the potentials of clay and biochar in improving water retention and mechanical resilience of sandy soil. Int Agrophys 30:391–399
10. Akhtar SS, Li G, Andersen MN, Liu F (2014) Biochar enhances yield and quality oftomato under reducedirrigation.AgriWaterManag 138:37–44
11. Akhtar, I.; Nazir, N. Effect of waterlogging and drought stress in plants. Int. J. Water Res. Environ. Eng. 2013, 2, 34–40
12. Alam, A.; Hariyanto, B.; Ullah, H.; Salin, K.R.; Datta, A. Effects of silicon on growth, yield and fruit quality of cantaloupe under drought stress. Silicon 2020, 13, 3153–3162.
13. Alleva, K.; Chara, O.; Amodeo, G. Aquaporins: Another piece in the osmotic puzzle. FEBS Lett. 2010, 586, 2991–2999.
14. Alzahrani, Y.; Ku¸svuran, A.; Alharby, H.F.; Ku¸svuran, S.; Rady, M.M. The defensive role of silicon in wheat against stress conditions induced by drought, salinity or cadmium. Ecotoxicol. Environ. Saf. 2018, 154, 187–196.
15. Andrenelli MC, Maienza A, Genesio L, Miglietta F, Pellegrini S, Vaccari FP, Vignozzi N (2016) Field application of pelletized biochar: short term effect on the hydrological properties of a silty clay loam soil. Agric. Water Manag 163:190–196
16. Anjum, S.A.; Xie, X.-y.; Wang, L.-c.; Saleem, M.F.; Man, C.; Lei, W. Morphological, physiological and biochemical responses of plants to drought stress. Afr. J. Agric. Res. 2011, 6, 2026–2032
17. Aroca R (2012) Plant responses to drought stress. From morphological to molecular features. Springer, Berlin
18. Arthur E, Tuller M, Moldrup P, de Jonge LW (2015) Effects of biochar and manure amendments on water vapor sorption in a sandy loam soil. Geoderma 243:175–182
19. Artiola JF, Rasmussen C, Freitas R (2012) Effects of a biochar-amended alkaline soil on the growth of romaine lettuce and bermudagrass. Soil Sci 177:561–570
20. Ashraf M, Akram NA. Improving salinity tolerance of plants through conventional breeding and genetic engineering: an analytical comparison. Biotechnol Adv 2009;27:744–52.

21. Ashraf M, Athar HR, Harris PJC, Kwon TR. Some prospective strategies for improving crop salt tolerance. Adv Agron 2008;97:45-110.
22. Ashraf M, Foolad MR. Roles of glycinebetaine and proline in improving plant abiotic stress resistance. Environ Exp Bot 2007;59:206–16.
23. Ashraf M. Inducing drought tolerance in plants: Recent advances. Biotechnology Advances 28 (2010) 169–18.
24. Atouei, M.T.; Pourbabaee, A.A.; Shorafa, M. Alleviation of Salinity Stress on Some Growth Parameters of Wheat by Exopolysaccharide-Producing Bacteria. Iran. J. Sci. Technol. Trans. A Sci. 2019, 43, 2725–2733
25. Badawi GH, Kawano N, Yamauchi Y. Over-expression of ascorbate peroxidase in tobacco chloroplasts enhances the tolerance to salt stress and water deficit. Physiol Plant 2004;121:231–8.
26. Bahreininejad, B.; Razmjou, J.; Mirza, M. Influence of water stress on morpho-physiological and phytochemical traits in Thymus daenensis. Int. J. Plant Prod. 2013, 7, 151–166
27. Baiamonte G, De Pasquale C, Marsala V, Cimò G, Alonzo G, Crescimanno G, Conte P (2015) Structure alteration of a sandyclay soil by biochar amendments. J Soils Sedim 15:816–824
28. Balla, K., Rakszegi, M., Li, Z., Bek´ es, F., Bencze, S., and Veisz, O. 2011. Quality ´ of winter wheat in relation to heat and drought shock after anthesis. Czech J. Food Sci. 29: 117–128
29. Ballhorn, D.J.; Kautz, S.; Heil, M.; Hegeman, A.D. Cyanogenesis of wild lima bean (*Phaseolus lunatus* L.) is an efficient direct defence in nature. PLoS ONE 2009, 4, e5450.
30. Bao AK, Wang SM, Wu GQ, Xi JJ, Zhang JL, Wang CM (2009) Overexpression of the arabidopsis H+–PPase enhanced resistance to salt and drought stress in transgenic alfalfa (*Medicago sativa* L.). Plant Sci 176(2):232–240
31. Imran, Amanullah & Abdel Rahman Al Tawaha (2022d): Indigenous organic resources utilization, application methods and sowing time replenish soil nitrogen and increase maize yield and total dry biomass, *Journal of Plant Nutrition*, DOI: 10.1080/01904167.2022.2067055.
32. Imran, Amanullah and Ibrahim Ortash. 2022b. Agronomic Practices Improved Cucumber Productivity, Nutrients Uptake and Quality. *Gesunde Pflanzen*. 10.1007/s10343-022-00634-1.
33. Imran, Amanullah, and Abdel Rahman Al Tawaha. 2022c. Regenerating potential of dual purpose rapeseed (*Brassica napus* L.) as influenced by decapitation stress and variable rates of phosphorous. *Communication in soil science and plant analysis.*

Doi.org/10.1080/00103624.2022.2118297

34. Imran and Amanullah. (2022b). Integration of peach (*Prunus persica* L) remnants in combination with beneficial microbes and phosphorus differ phosphorus use efficiency, agronomic efficiency and partial factor productivity in soybean Vs maize crops. *Journal of Plant Nutrition.* doi.org/10.1080/01904167.2022.2099890
35. Imran, & Abdel Rahman Al Tawaha (2022): Humic Acid and Sulfur Integration Enhances Growth and Yield Related Traits of *Brassica Napus L. Journal of Plant Nutrition.* *d*oi.org/10.1080/01904167.2022.212882
36. Imran. 2022a. Growing of Off-Season Tomato in High Tunnel and Its Nutritional Value Augmentation with Integrated Nutrients Management. *Journal of Plant Nutrition.* Doi: 10.1080/01904167.2022.2046062
37. Imran and Amanullah. 2021a. Assessment of chemical and manual weed control approaches for effective weed suppression and maize productivity enhancement under maize-wheat cropping system. *Gesunde Pflanzen*, DOI: 10.1007/s10343-021-00599-7.
38. Imran, Amanullah and Abdel Rehman Altawaha. 2022c. Carbon Assimilation and Dry Matter Partitioning in Soybean Ameliorates with the Integration of Nano-Black Carbon, Along with Beneficial Microbes and Phosphorus Fertilization. *Journal of Plant Nutrition, DOI:10.1080/01904167.2022.2035753*
39. Imran. 2022b. (Review) Phosphorus availability enhanced with combine application of organic amendments and beneficial microbes under soybean-wheat cropping system. *Communication in soil science and plant analysis.* doi.org/10.1080/00103624.2022.2034848
40. Imran.2022c. Management of Abiotic Stresses with Nano-black carbon is a tool for Crop Production. *Journal of Plant Nutrition.* doi.org/10.1080/01904167.2022.2046085
41. Imran and Amanullah. 2021b. Phosphorus biofortification and uptake in maize enhanced with integrated phosphorus management. *Phosphorus, Sulfur, and Silicon and the Related Elements*. DOI: 10.1080/10426507.2021.2022677.
42. Imran. 2021. The bioavailability of Phosphorus in composite vs. Hybrid maize differ with phosphorus and boron fertilization. *Phosphorus, Sulfur, and Silicon and the Related Elements. https://doi.org/10.1080/10426507.2021.1920588.*
43. Imran, Amanullah, Asad Ali khan, Tariq Mahmood, Abdel Rahman Al Tawaha and Samia Khanum. 2021a. Adequate Fertilization,

Application Method and Sowing Techniques Improve Maize yield and related traits. *Communication in soil science and plant analysis*. DOI:10.1080/00103624.2021.1925688
44. Imran, Amanullah and Abdel Rahman M. Al Tawaha. 2021b. Management of Nano-black carbon, Phosphorous and Bio Fertilizer improve Soil organic carbon and ensilage biomass of soybean and maize. *Communication in Soil Science and Plant Analysis*. DOI.10.1080/00103624.2021.1966439
45. Imran and Amanullah· 2021c. Phosphorus and Boron Application Optimizing Biofortification Of phosphorus and Productivity of French Bean (*Phaseolus Vulgaris* L.). *Communication In Soil Science and Plant Analysis*. DOI:10.1080/00103624.2021.1971689
46. Imran, Amanullah and Abdel Rahman Al-Tawaha. 2021c. Carbon Sources Application Increase Wheat Yield And Soil Fertility. *Communication in soil science and plant analysis DOI: 10.1080/00103624.2020.1865397.*
47. Imran, Amanullah and Abdel Rehman Al-Tawaha.2020a. The Productivity of subsequent wheat enhanced with residual carbon sources and phosphorus under improved irrigation system. *Communication in soil science and plant analysis*, DOI:10.1080/00103624.2020.1763387
48. Imran, Amanullah, Muhammad Arif, Zahir Shah and Abdul Bari. 2020b. Soil application of *Trichoderma* and Peach (*Prunus persica* L.) residues possesses biocontrol potential for weeds and enhances growth and profitability of soybean (Glycine max). *Sarhad Journal of Agriculture* 36 (1) 10-20. DOI | http://dx.doi.org/10.17582/journal.sja/2020/36.1.10.20
49. Baronti S, Vaccari FP, Miglietta F, Calzolari C, Lugato E, Orlandini S, Pini R, Zulian C, Genesio L (2014) Impact of biochar application on plant water relations in *Vitis vinifera* (L.) Euro J Agron 53:38–44
50. Baroowa, B., and N. Gogoi, 2014: Biochemical changes in black gram and green gram genotypes after imposition of drought stress. J. Food Legumes 27, 350–353.
51. Batista, P.S.C., Carvalho, A.J., Portugal, A.F., Bastos, E.A., Cardoso, M.J., Torres, L.G., Júlio, M.M. and de Menezes, C.B., 2019. Selection of sorghum for drought tolerance in a semiarid environment.
52. Batool A, Taj S, Rashid A, Khalid A, Qadeer S, Saleem AR, Ghufran MA (2015) Potential of soil amendments (biochar and gypsum) in increasing water use efficiency of Abelmoschus esculentus L. Moench. Front. Plant Sci 6:1–13

53. Battaglia, M.L.; Lee, C.; Thomason, W. Corn yield components and yield responses to defoliation at different row widths. Agron. J. 2018, 110, 1–16.
54. Battaglia, M.L.; Lee, C.; Thomason, W.; Van Mullekom, J. Effects of corn row width and defoliation timing and intensity on canopy light interception. Crop Sci. 2020, 59, 1718–1731.
55. Berg A, Findell K, Lintner B, Giannini A, Seneviratne SI, van den Hurk B, et al. Land-atmosphere feedbacks amplify aridity increase over land under global warming. Nat Clim Chang. 2016;6:869–74.
56. Berg A, Sheffield J, Milly PCD. Divergent surface and total soil moisture projections under global warming. Geophys Res Lett. 2017;44:236–44.
57. Bernier J (2008) Breeding upland rice for drought resistance. J Sci Food Agric 88(6):927–939
58. Bezerra, B.K.L.; Lima, G.P.P.; dos Reis, A.R.; de Almeida Silva, M.; de Camargo, M.S. Physiological and biochemical impacts of silicon against water deficit in sugarcane. Acta Physiol. Plant. 2019, 41, 1–12.
59. Bhargava S, Sawant K. Drought stress adaptation: metabolic adjustment and regulation of gene expression. Plant Breeding. 2013;132:21–32
60. Bhargavi, B.; Kalpana, K.; Reddy, J.K. Influence of Water Stress on Morphological and Physiological Changes in Andrographis paniculata. Int. J. Pure Appl. Biosci. 2017, 5, 1550–1556
61. Bhatt, R.M.; Rao, N.K.S. Influence of pod load on response of okra to water stress. Indian J. Plant Physiol. 2005, 10, 54–59.
62. Bielach, A.; Hrtyan, M.; Tognetti, V.B. Plants under stress: Involvement of auxin and cytokinin. Int. J. Mol. Sci. 2017, 18, 1427.
63. Birami, B.; Gattmann, M.; Heyer, A.G.; Grote, R.; Arneth, A.; Ruehr, N.K. Heat waves alter carbon allocation and increase mortality of Aleppo pine under dry conditions. Front. For. Glob. Chang. Front. 2018, 1, 8.
64. Borges, A.A.; Jimenez, A.D.; Exposito, R.M.; Sandalio, L.M.; Perez, J.A. Priming crops against biotic and abiotic stresses: MSB as a tool for studying mechanisms. Front. Plant Sci. 2014, 5, 642.
65. Brantley KE, Brye KR, Savin MC, Longer DE (2015) Biochar source and application rate effects on soil water retention determined using wetting curves. Open J Soil Sci 5:1–10
66. Brito, C.; Dinis, L.-T.; Moutinho-Pereira, J.; Correia, C.M. Drought stress effects and olive tree acclimation under a changing climate. Plants 2019, 8, 232.

67. Bruun EW, Petersen CT, Hansen E, Holm JK, Hauggaard-Nielsen H (2014) Biochar amendment to coarse sandy subsoil improves root growth and increases water retention. Soil Use Manage 30:109–118
68. Budak H, Kantar M, YucebilgiliKurtoglu K (2013) Drought tolerance in modern and wild wheat. Sci World J 4:66–75
69. Bukhari, S.A.H.; Peerzada, A.M.; Javed, M.H.; Dawood, M.; Hussain, N.; Ahmad, S. Growth and Development Dynamics in Agronomic Crops Under Environmental Stress. In Agronomic Crops; Springer: Berlin/Heidelberg, Germany, 2019; pp. 83–114
70. Cao, B.; Ma, Q.; Zhao, Q.; Wang, L.; Xu, K. Effects of silicon on absorbed light allocation, antioxidant enzymes and ultrastructure of chloroplasts in tomato leaves under simulated drought stress. Sci. Hortic. 2015, 194, 53–62.
71. Cao, B.L.; Ma, Q.; Xu, K. Silicon restrains drought-induced ROS accumulation by promoting energy dissipation in leaves of tomato. Protoplasma 2020, 257, 537–547.
72. Cardoso VJM (2013) Germinação. In: Kerbauy GB (ed) Fisiologia vegetal. Guanabara Koogan, Rio de Janeiro, pp 384–408
73. Carrijo, D.R.; Lundy, M.E.; Linquist, B.A. Rice yields and water use under alternate wetting and drying irrigation: A meta-analysis. Field Crop Res. 2017, 203, 173–180
74. Chalker-Scott, L. Impact of mulches on landscape plants and the environment—A review. J. Environ. Hortic. 2007, 25, 239–249.
75. Chaves, M.M.; Flexas, J.; Pinheiro, C. Photosynthesis under drought and salt stress: Regulation mechanisms from whole plant to cell. Ann. Bot. 2009, 103, 551–560.
76. Chen, B., Liu, E., Mei, X., Yan, C. and Garré, S., 2018. Modelling soil water dynamic in rain-fed spring maize field with plastic mulching. Agricultural Water Management, 198, pp.19-27.
77. Chen, L.; Fang, Y.; Li, X.; Zeng, K.; Chen, H.; Zhang, H.; Yang, H.; Cao, D.; Hao, Q.; Yuan, S.; et al. Identification of soybean drought-tolerant genotypes and loci correlated with agronomic traits contributes new candidate genes for breeding. Plant. Mol. Biol. 2020, 102, 109–122.
78. Chen, Y.; Saud, S.; Li, X.; Chen, Y.; Zhang, L.; Fahad, S.; Hussain, S.; Sadiq, A. Silicon application increases drought tolerance of kentucky bluegrass by improving plant water relations and morphophysiological functions. Sci. World J. 2014, 2014, 368694.
79. Cho EK, Hong CB (2006) Over-expression of tobacco NtHSP70-1 contributes to drought-stress tolerance in plants. Plant Cell Rep 25(4):349–358

80. Chowdhury, J.; Karim, M.; Khaliq, Q.; Ahmed, A. Effect of drought stress on bio-chemical change and cell membrane stability of soybean genotypes. Bangladesh J. Agric. Res. 2017, 42, 475–485.
81. Christensen, J.H.; Christensen, O.B. A summary of the PRUDENCE model projections of changes in European climate by the end of this century. Clim. Chang. 2007, 81, 7–30.
82. Cohen AC, Bottini R, Pontin M, Berli FJ, Moreno D, Boccanlandro H, Piccoli PN (2015) Azospirillum brasilense ameliorates the response of Arabidopsis thaliana to drought mainly via enhancement of ABA levels. Physiol Plant 153:79–90
83. Collins M, Knutti R, Arblaster J, et al (2013) IPCC WG1AR5 Chapter 12 Long-term climate change: projections, commitments and irreversibility. Clim. Chang. 2013 Phys. Sci. Basis. Contrib. Work. Gr. I to Fifth Assess. Rep. Intergov. Panel Clim. Chang. 1029–1136
84. Cook B, Smerdon J, Seager R, Coats S. Global warming and 21st century drying. Clim Dyn. 2014;43:2607–27. https://doi.org/10.1007/s00382-014-2075-y
85. Cook BI, Ault TR, Smerdon JE. Unprecedented 21st century drought risk in the American Southwest and Central Plains. Sci Adv. 2015;1:e1400082.
86. Corso, D.; Delzon, S.; Lamarque, L.J.; Cochard, H.; Torres-Ruiz, J.M.; King, A.; Brodribb, T. Neither xylem collapse, cavitation, or changing leaf conductance drive stomatal closure in wheat. Plant Cell Env. 2020, 43, 854–865.
87. Cuellar-Ortiz, S.M.; De La Paz Arrieta-Montiel, M.; Acosta-Gallegos, J.; Covarrubias, A.A. Relationship between carbohydrate partitioning and drought resistance in common bean. Plant Cell Environ. 2008, 31, 1399–1409.
88. Dai A. Drought under global warming: a review. Wiley Interdiscip Rev Clim Chang. 2011;2:45–65.
89. Daryanto S, Wang L, Jacinthe P-A, Cordain L, Simopoulos A, Ray D, Mueller N, West P, Foley J, Kadam N et al (2016) Global synthesis of drought effects on maize and wheat production. Hui D, editor. PLoS One 11:e0156362
90. Daryanto, S.; Wang, L.; Jacinthe, P.A. Global synthesis of drought effects on cereal, legume, tuber and root crops production: A review. Agric. Water Manag. 2017, 179, 18–33.
91. De Melo Carvalho MT, de Holanda Nunes Maia A, Madari BE, Bastiaans L, van Oort PAJ, Heinemann AB, Soler da Silva MA, Petter FA, Marimon BH, Meinke H (2014) Biochar increases plant-available

water in a sandy loam soil under an aerobic rice crop system. Solid Earth 5:939–952
92. De Ronde, J.A., Cress, W.A., Krüger, G.H.J., Strasser, R.J. and Van Staden, J., 2004. Photosynthetic response of transgenic soybean plants, containing an Arabidopsis P5CR gene, during heat and drought stress. Journal of plant physiology, 161(11), pp.1211-1224.
93. Deepak, S.B.; Thakur, A.; Singh, S.; Bakshi, M.; Bansal, S. Changes in crop physiology under drought stress: A review. J. Pharmacogn. Phytochem. 2019, 8, 1251–1253.
94. Deka, D.; Singh, A.K.; Singh, A.K. Effect of Drought Stress on Crop Plants with Special Reference to Drought Avoidance and Tolerance Mechanisms: A Review. Int. J. Curr. Microbiol. App. Sci. 2018, 7, 2703–2721.
95. Dela˘c, D.; Grši´c, K.; Nin˘cevi´c, T.; Carovi´c-Stanko, K.; Varga, F.; Grdiša, M. The influence of hydropriming and osmopriming with KNO3 on seed germination of dalmatian pyrethrum (Tanacetum cinerariifolium/Trevir./ Sch. Bip.). Agric. Conspec. Sci. 2018, 83, 205–211.
96. Devincentis, A.J. Scales of Sustainable Agricultural Water Management. Ph.D. Thesis, University of California, Davis, CA, USA, 2020
97. Dixit, S.; Singh, A.; Ma, T.S.C.; Maturan, P.T.; Amante, M.; Kumar, A. Multiple major QTL lead to stable yield performance of rice cultivars across varying drought intensities. BMC Genet. 2014, 15, 16.
98. Dos Reis, S.P.; Marques, D.N.; Lima, A.M.; De Souza, C.R.B. Plant Molecular Adaptations and Strategies Under Drought Stress. In Drought Stress Tolerance in Plants; Springer Science and Business Media LLC: Berlin/Heidelberg, Germany, 2016; Volume 2, pp. 91–122.
99. Egamberdieva D, Reckling M, Wirth S (2017) Biochar-based Bradyrhizobium inoculum improves growth of lupin (*Lupinus angustifolius* L.) under drought stress. Eur J Soil Biol 78:38–42
100. El-Beltagi, H.S., Basit, A., Mohamed, H.I., Ali, I., Ullah, S., Kamel, E.A., Shalaby, T.A., Ramadan, K.M., Alkhateeb, A.A. and Ghazzawy, H.S., 2022. Mulching as a sustainable water and soil saving practice in agriculture: A review. Agronomy, 12(8), p.1881.
101. El-Beltagi, H.S.; Ahmad, I.; Basit, A.; Abd El-Lateef, H.M.; Yasir, M.; Shah, S.T.; Ullah, I.; Mohamed, M.E.M.; Ali, I.; Ali, F.; et al. Effect of azospirillum and azotobacter species on the performance of cherry tomato under different salinity levels. Gesunde Pflanz. 2022, 74, 487–499.

102. El-Beltagi, H.S.; Mohamed, H.I.; Sofy, M.R. Role of ascorbic acid, glutathione and proline applied as singly or in sequence combination in improving chickpea plant through physiological change and antioxidant defense under different levels of irrigation intervals. Molecules 2020, 25, 1702.
103. Elemike, E.E.; Uzoh, I.M.; Onwudiwe, D.C.; Babalola, O.O. The role of nanotechnology in the fortification of plant nutrients and improvement of crop production. Appl. Sci. 2019, 9, 499.
104. El-Ganainy, S.M.; El-Bakery, A.M.; Hafez, H.M.; Ismail, A.M.; El-Abdeen, A.Z.; Ata, A.A.E.; Elraheem, O.A.Y.A.; El Kady, Y.M.Y.; Hamouda, A.F.; El-Beltagi, H.S.; et al. Humic acid-coated Fe3O4 nanoparticles confer resistance to Acremonium wilt disease and improve physiological and morphological attributes of grain Sorghum. Polymers 2022, 14, 3099.
105. Farooq M, Wahid A, Kobayashi N, Fujita D, Basra SMA. Plant drought stress: effects, mechanisms and management. Agron Sustain Dev. 2009;29:185–212.
106. Farooq, A.; Nadeem, M.; Abbas, G.; Shabbir, A.; Khalid, M.S.; Javeed, H.M.R.; Saeed, M.F.; Akram, A.; Younis, A.; Akhtar, G. Cadmium partitioning, physiological and oxidative stress responses in Marigold (*Calendula calypso*) grown on contaminated soil: Implications for phytoremediation. Bull. Environ. Contam. Toxicol. 2020, 105, 270–276.
107. Farooq, M.; Basra, S.M.A.; Wahid, A. Priming of field-sown rice seed enhances germination, seedling establishment, allometry and yield. Plant Growth Regul. 2006, 49, 285–294.
108. Farooq, M.; Wahid, A.; Kobayashi, N.; Fujita, D.; Basra, S. Plant drought stress: Effects, mechanisms and management. In Sustainable Agriculture; Springer: Dordrecht, The Netherlands, 2009; pp. 153–188
109. Foster EJ, Hansen N, Wallenstein M, Cotrufo MF (2016) Biochar and manure amendments impact soil nutrients and microbial enzymatic activities in a semi-arid irrigated maize cropping system. Agri Ecosys Environ 233:404–414
110. Fu Q, Feng S. Responses of terrestrial aridity to global warming. J Geophys Res Atmos. 2014;119:7863–75.
111. Fu, J. J., Cheng, Y. B., Linghu, J. J., Yang, X. H., Kang, L., Zhang, Z. X., et al. (2013). RNA sequencing reveals the complex regulatory network in the maize kernel. Nat. Commun. 4:2832. doi: 10.1038/ncomms3832
112. Gan, Y.; Siddique, K.H.M.; Turner, N.C.; Li, X.G.; Niu, J.Y.; Yang, C.; Liu, L.; Chai, Q. Ridge-Furrow mulching systems-an innovative

technique for boosting crop productivity in semiarid rain-fed environments. Adv. Agron. 2013, 118, 429–476.
113. Gan, Y.T.; Huang, G.B.; Li, L.L.; Liu, J.H.; Hu, Y.G. Unique conservation tillage practices in northwest China. In No-Till Farming Systems, World Association of Soil and Water Conservation; Goddard, T., Zoebisch, M.A., Gan, Y., Ellis, W., Watson, A., Sombatpanit, S., Eds.; Funny Publishing: Bangkok, Thailand, 2008; pp. 429–445.
114. Ge, F.W.; Tao, P.; Zhang, Y.; Wang, J.B. Characterization of AQP gene expressions in Brassica napus during seed germination and in response to abiotic stresses. Biol. Plant. 2014, 58, 274–282.
115. Ghanbari, A.A.; Mousavi, S.H.; Mousapour, A.; Rao, I. Effects of water stress on leaves and seeds of bean (*Phaseolus vulgaris* L.). Turkish J. F. Crop. 2013, 18, 73–77.
116. Gheidary, S.; Akhzari, D.; Pessarakli, M. Effects of salinity, drought, and priming treatments on seed germination and growth parameters of Lathyrus sativus L. J. Plant Nutr. 2017, 40, 1507–1514.
117. Gholami, H.; Farhadi, R.; Rahimi, M.; Zeinalikharaji, A.; Askari, A. Effect of growth hormones on physiology characteristics and essential oil of basil under drought stress condition. J. Am. Sci. 2013, 9, 61–63.
118. Ghooshchi, F., Seilsepour, M. and Jafari, P., 2008. Effects of water stress on yield and some agronomic traits of maize (SC 301). Am. Eurasian J. Agric. Environ. Sci, 4, pp.302-305.
119. Githinji L (2014) Effect of biochar application rate on soil physical and hydraulic properties of a sandy loam. Arch Agron Soil Sci 60:457–470
120. Glick BR (2014) Bacteria with ACC deaminase can promote plant growth and help to feed the world. Microbiol Res 169:30–39
121. Gomes, D.; Agasse, A.; Thiebaud, P.; Delrot, S.; Geros, H.; Chaumont, F. Aquaporins are multifunctional water and solute transporters highly divergent in living organisms. Biochim. Biophys. Acta 2009, 1788, 1213–1228.
122. Goodman, B.A. Utilization of waste straw and husks from rice production, A review. J. Bioresour. Bioprod. 2020, 5, 143–162
123. Gosal SS, Wani SH, Kang MS. Biotechnology and drought tolerance. J Crop Improvement 2009;23:19–54.
124. Gray, S.B.; Brady, S.M. Plant developmental responses to climate Chang. Dev. Biol. 2016, 419, 64–77
125. Guan, S.Y.; Jiao, P.; Jiang, Z.Z.; Qi, Z.; Xia, H.F.; Qu, J.; Ma, Y.Y. Research Progress of MYB Transcription Factors in Plant Abiotic Stress. J. Jilin Agric. Univ. 2019, 41, 253–260.

126. Guan, Y.S.; Serraj, R.; Liu, S.H.; Xu, J.L.; Ali, J.; Wang, W.S.; Venus, E.; Zhu, L.H.; Li, Z.K. Simultaneously improving yield under drought stress and non-stress conditions: A case study of rice (*Oryza sativa* L.). J. Exp. Bot. 2010, 61, 4145–4156
127. Gubis J, Vaňková R, Červená V, Dragúňová M, Hudcovicová M, Lichtnerová H, et al. Transformed tobacco plants with increased tolerance to drought. South Afr J Bot 2007;73: 505–11.
128. Guo C, Ge X, Ma H (2013) The rice OsDIL gene plays a role in drought tolerance at vegetative and reproductive stages. Plant Mol Biol 82(3):239–253
129. Guo, Z.; Yang, W.; Chang, Y.; Ma, X.; Tu, H.; Xiong, F.; Jiang, N.; Feng, H.; Huang, C.; Yang, P.; et al. Genome-Wide Association Studies of Image Traits Reveal Genetic Architecture of Drought Resistance in Rice. Mol. Plant. 2018, 11, 789–805.
130. Gupta, N. K., Gupta, S., and Kumar, A. 2001. Effect of water stress on physiological attributes and their relationship with growth and yield of wheat cultivars at different stages. J. Agron. Crop Sci. 186: 55–62
131. Gururani MA, Upadhyaya CP, Baskar V, Venkatesh J, Nookaraju A, Park SW (2013) Plant growth-promoting rhizobacteria enhance abiotic stress tolerance in Solanum tuberosum through inducing changes in the expression of ROS-scavenging enzymes and improved photosynthetic performance. J Plant Growth Regul 32:245–258
132. Hafez, E.H.; Abou El Hassan, W.H.; Gaafar, I.A.; Seleiman, M.F. Effect of gypsum application and irrigation intervals on clay saline-sodic soil characterization, rice water use efficiency, growth, and yield. J. Agric. Sci. 2015, 7, 208–219.
133. Haider G, Koyro HW, Azam F, Steffens D, Müller C, Kammann C (2015) Biochar but not humic acid product amendment affected maize yields via improving plant-soil moisture relations. Plant Soil 395:141–157
134. Haider G, Steffens D, Moser G, Müller C, Kammann CI (2017) Biochar reduced nitrate leaching and improved soil moisture content without yield improvements in a four-year field study. Agri Ecosys Environ 237:80–94
135. Hanna, S.H.S., Osborne-Lee, I.W., Cesaretti, G.P., Misso, R. and Khalil, M.T., 2016. Ecological agro-ecosystem sustainable development in relationship to other sectors in the economic system, and human ecological footprint and imprint. Agriculture and Agricultural Science Procedia, *8*, pp.17-30.

136. Hayat, S.; Hayat, Q.; Alyemeni, M.N.; Wani, A.S.; Pichtel, J.; Ahmad, A. Role of proline under changing environments: A review. Plant Signal. Behav. 2012, 7, 1456–1466.
137. Helaly, M.N.; El-Hoseiny, H.; El-Sheery, N.I.; Rastogi, A.; Kalaji, H.M. Regulation and physiological role of silicon in alleviating drought stress of mango. Plant Physiol. Biochem. 2017, 118, 31–44.
138. Hirayama, T.; Shinozaki, K. Research on plant abiotic stress responses in the post-genome era: Past, present and future. Plant J. 2010, 61, 1041–1052.
139. Hoang, G.T.; Van Dinh, L.; Nguyen, T.T.; Ta, N.K.; Gathignol, F.; Mai, C.D.; Jouannic, S.; Tran, K.D.; Khuat, T.H.; Do, V.N.; et al. Genome-wide Association Study of a Panel of Vietnamese Rice Landraces Reveals New QTLs for Tolerance to Water Deficit During the Vegetative Phase. Rice 2019, 12, 4.
140. Hou, S.; Zhu, G.; Li, Y.; Li, W.; Fu, J.; Niu, E.; Li, L.; Zhang, D.; Guo, W. Genome-Wide Association Studies Reveal Genetic Variation and Candidate Genes of Drought Stress Related Traits in Cotton (*Gossypium hirsutum* L.). Front. Plant. Sci. 2018, 9, 1276.
141. Hu, L.; Wang, Z.; Huang, B. Effects of cytokinin and potassium on stomatal and photosynthetic recovery of Kentucky bluegrass from drought stress. Crop Sci. 2013, 53, 221–231.
142. Hu, Y.; Schmidhalter, U. Drought and salinity: A comparison of their effects on mineral nutrition of plants. J. Plant Nutr. Soil. Sc. 2005, 168, 541–549.
143. Huang J, Yu H, Guan X, Wang G, Guo R. Accelerated dryland expansion under climate change. Nat Clim Chang. 2016;6:166– 71.
144. https://doi.org/10.1038/nclimate2837.
145. Humphreys MO, Humphreys MW. Breeding for stress resistance: general principles. In: Ashraf M, Harris PJC, editors. Abiotic stresses: plant resistance through breeding and molecular approaches; 2005. p. 19–46.
146. Huseynova, I.M.; Rustamova, S.M. Screening for drought stress tolerance in wheat genotypes using molecular markers. Proc. ANAS (Biol. Sci.) 2010, 65, 132–139.
147. Hussain M, Farooq S, Hasan W, Ul-Allah TM, Farooq M, Nawaz A (2018) Drought stress in sunfower: Physiological efects and its management through breeding and agronomic alternatives. Agric Water Manage 201:152–166
148. Igalavithana AD, Ok YS, Niazi NK, Rizwan M, Al-Wabel MI, Usman AR, Moon DH, Lee SS (2017) Effect of corn residue biochar on the hydraulic properties of sandy loam soil. Sustain 9:1–10

149. Ikram, M.; Ali, N.; Guljan, F.; Khan, N. Endophytic fungal diversity and their interaction with plants for agriculture sustainability under stressful condition. Recent Pat. Food Nutr. Agric. 2019, 10, 1.
150. IPCC (2013) The physical science basis. Contribution of Working Group I to the Fifth Assessment Report of the Intergovernmental Panel on Climate Change 2013. Cambridge University Press, Cambridge
151. IPCC (2014) Mitigation of climate change. Contribution of Working Group III to the Fifth Assessment Report of the Intergovernmental Panel on Climate Change 1454. Ambridge University Press, Cambridge
152. Jabran, K.; Ullah, E.; Hussain, M.; Farooq, M.; Zaman, U.; Yaseen, M.; Chauhan, B.S. Mulching improves water productivity, yield and quality of fine rice under water-saving rice production systems. J. Agron. Crop Sci. 2015, 201, 389–400.
153. Jafar, M.Z.; Farooq, M.; Cheema, M.A.; Afzal, I.; Basra, S.M.A.; Wahid, M.A.; Aziz, T.; Shahid, M. Improving the performance of wheat by seed priming under saline conditions. J Agron. Crop Sci. 2012, 198, 38–45.
154. Jakab, G.; Ton, J.; Flors, V.; Zimmerli, L.; Metraux, J.P.; Mauch, M.B. Enhancing Arabidopsis salt and drought stress tolerance by chemical priming for its abscisic acid responses. Plant Physiol. 2005, 139, 267–274.
155. Jamoussi RJ, Elabbassi MM, Jouira HB, Hanana M, Zoghlami N, Ghorbel A, Mliki A (2014) Physiological responses of transgenic tobacco plants expressing the dehydration-responsive RD22 gene of Vitisvinifera to salt stress. Turk J Bot 38:268–280
156. Jeffery S, Meinders MB, Stoof CR, Bezemer TM, van de Voorde TF, Mommer L, van Groenigen JW (2015) Biochar application does not improve the soil hydrological function of a sandy soil. Geoderma 251:47–54
157. Jeong JS, Kim YS, Baek KH, Jung H, Ha SH, Do Choi Y, Kim JK (2010) Root-specific expression of OsNAC10 improves drought tolerance and grain yield in rice under field drought conditions. Plant Physiol 153(1):185–197
158. Jia, Y.; Li, F.M.; Wang, X.L.; Yang, S.M. Soil water and alfalfa yields as affected by alternating ridges and furrows in rainfall harvest in a semiarid environment. Field Crops Res. 2006, 97, 167–175.
159. Jimenez, M.N.; Pinto, J.R.; Ripoll, M.A.; Sanchez-Miranda, A.; Navarro, F.B. Impact of straw and rock-fragment mulches on soil moisture and early growth of holm oaks in a semiarid area. Catena 2017, 152, 198–206.

160. Kader, M.A.; Senge, M.; Mojid, M.A.; Ito, K. Recent advances in mulching materials and methods for modifying soil environment. Soil Tillage Res. 2017, 168, 155–166.
161. Kader, M.A.; Senge, M.; Mojid, M.A.; Onishi, T.; Ito, K. Effects of plastic-hole mulching on effective rainfall and readily available soil moisture under soybean (*Glycine max*) cultivation. Paddy Water Environ. 2017, 15, 659–668.
162. Kader, M.A.; Singha, A.; Begum, M.A.; Jewel, A.; Khan, F.H.; Khan, N.I. Mulching as water-saving technique in dryland agriculture: Review article. Bull. Natl. Res. Cent. 2019, 43, 147.
163. Kalam, S.; Basu, A.; Ankati, S. Plant Root-Associated Biofilms in Bioremediation. Biofilms Plant Soil Health 2017, 337, 337–355.
164. Kang, J.S.; Singh, H.; Singh, G.; Kang, H.; Kalra, V.P.; Kaur, J. Abiotic Stress and Its Amelioration in Cereals and Pulses: A Review. Int. J. Curr. Microbiol. Appl. Sci. 2017, 6, 1019–1045.
165. Kang, Y., Sakiroglu, M., Krom, N., Stantongeddes, J., Wang, M., Lee, Y., et al. (2015). Genome-wide association of drought-related and biomass traits with HapMap SNPs in Medicago truncatula. Plant Cell Environ. 38, 1997–2011. doi: 10.1111/pce.12520
166. Karim S, Aronsson H, Ericson H, Pirhonen M, Leyman B, Welin B, et al. Improved drought tolerance without undesired side effects in transgenic plants producing trehalose. Plant Mol Biol 2007;64:371–86.
167. Karimpour, M. Effect of Drought Stress on RWC and Chlorophyll Content on Wheat (*Triticum durum* L.) Genotypes. World. Ess. J. 2019, 7, 52–56.
168. Kasim, W.A.; Osman, M.E.; Omar, M.N.; El-Daim, I.A.A.; Bejai, S.; Meijer, J. Control of Drought Stress in Wheat Using Plant-Growth-Promoting Bacteria. J. Plant Growth Regul. 2012, 32, 122–130.
169. Kenneth, O.C.; Nwadibe, E.C.; Kalu, A.U.; Unah, U.V. Plant Growth Promoting Rhizobacteria (PGPR): A Novel Agent for Sustainable Food Production. Am. J. Agric. Biol. Sci. 2019, 14, 35–54.
170. Khalid, K.A. Influence of water stress on growth, essential oil, and chemical composition of herbs (Ocimum sp.). Int. Agrophys. 2006, 20, 289–296.
171. Khan, A.; Pan, X.; Najeeb, U.; Tan, D.K.; Fahad, S.; Zahoor, R.; Luo, H. Coping with drought: Stress and adaptive mechanisms, and management through cultural and molecular alternatives in cotton as vital constituents for plant stress resilience and fitness. Biol. Res. 2018, 51, 47.

172. Khan, M.N.; Zhang, J.; Luo, T.; Liu, J.; Rizwan, M.; Fahad, S.; Xu, Z.; Hu, L. Seed priming with melatonin coping drought stress in rapeseed by regulating reactive oxygen species detoxification: Antioxidant defense system, osmotic adjustment, stomatal traits and chloroplast ultrastructure perseveration. Ind. Crop. Prod. 2019, 140, 111597.
173. Khan, N.; Bano, A. Effects of exogenously applied salicylic acid and putrescine alone and in combination with rhizobacteria on the phytoremediation of heavy metals and chickpea growth in sandy soil. Int. J. Phytoremediat. 2018, 16, 405–414
174. Khan, N.; Bano, A. Rhizobacteria and Abiotic Stress Management. In Plant Growth Promoting Rhizobacteria for Sustainable Stress Management; Springer: Singapore, 2019; pp. 65–80.
175. Khan, N.; Zandi, P.; Ali, S.; Mehmood, A.; Shahid, M.A.; Yang, J. Impact of Salicylic Acid and PGPR on the Drought Tolerance and Phytoremediation Potential of Helianthus annus. Front. Microbiol. 2018, 9, 2507.
176. Kheradmand, M.A.; Fahraji, S.S.; Fatahi, E.; Raoofi, M.M. Effect of water stress on oil yield and some characteristics of Brassica napus. Int. Res. J. Basic Appl. Sci. 2014, 8, 1447–1453.
177. Kim TH (2014) Mechanism of ABA signal transduction: agricultural highlights for improving drought tolerance. J Plant Biol 57(1):1–8
178. Kirigwi, F.M.; Van Ginkel, M.; Brown-Guedira, G.; Gill, B.S.; Paulsen, G.M.; Fritz, A.K. Markers associated with a QTL for grain yield in wheat under drought. Mol. Breed. 2007, 20, 401–413.
179. Kobraei, S., Etminan, A., Mohammadi, R. and Kobraee, S., 2011. Effects of drought stress on yield and yield components of soybean. Annals of Biological Research, 2(5), pp.504-509.
180. Kumar, M.; Kesawat, M.S.; Ali, A.; Lee, S.C.; Gill, S.S.; Kim, H.U. Integration of abscisic acid signaling with other signaling pathways in plant stress responses and development. Plants 2019, 8, 592.
181. Kumawat, K.R.; Sharma, N.K. Effect of Drought Stress on Plants Growth. Popular Kheti 2018, 6, 239–241.
182. Lafitte, H.R.; Yongsheng, G.; Yan, S.; Li, Z.-K. Whole plant responses, key processes, and adaptation to drought stress: The case of rice. J. Exp. Bot. 2006, 58, 169–175.
183. Lang, N.T. and Buu, B.C., 2008. Fine mapping for drought tolerance in rice (*Oryza sativa* L.). Omonrice, 16(2008), pp.9-15.
184. Lata, C.; Prasad, M. Role of DREB in regulation of abiotic stress response in plants. J. Exp. Bot. 2011, 62, 4731–4748.

185. Laxa, M.; Liebthal, M.; Telman, W.; Chibani, K.; Dietz, K.-J. The role of the plant antioxidant system in drought tolerance. Antioxidants 2019, 8, 94.
186. Lee SS, Shah HS, Awad YM, Kumar S, Ok YS (2015) Synergy effects of biochar and polyacrylamide on plants growth and soil erosion control. Environ Earth Sci 74:2463–2473
187. Lee, J.G.; Hwang, H.Y.; Park, M.H.; Lee, C.H.; Kim, P.J. Depletion of soil organic carbon stocks are larger under plastic film mulching for maize. Eur. J. Soil Sci. 2019, 70, 807–818.
188. Levi A, Ovnat L, Paterson AH, Saranga Y. Photosynthesis of cotton near-isogenic lines introgressed with QTLs for productivity and drought related traits. Plant Sci 2009;177: 88–96.
189. Li, C., Wang, C., Wen, X., Qin, X., Liu, Y., Han, J., Li, Y., Liao, Y. and Wu, W., 2017. Ridge–furrow with plastic film mulching practice improves maize productivity and resource use efficiency under the wheat–maize double–cropping system in dry semi–humid areas. Field crops research, 203, pp.201-211.
190. Li, H., Peng, Z., Yang, X., Wang, W., Fu, J., Wang, J., et al. (2013). Genome-wide association study dissects the genetic architecture of oil biosynthesis in maize kernels. Nat. Genet. 45, 43–50. doi: 10.1038/ng.2484
191. Li, Q.; Li, H.; Zhang, L.; Zhang, S.; Chen, Y. Mulching improves yield and water-use efficiency of potato cropping in China, A meta-analysis. Field Crops Res. 2018, 221, 50–60.
192. Li, X.; Topbjerg, H.B.; Jiang, D.; Liu, F. Drought priming at vegetative stage improves the antioxidant capacity and photosynthesis performance of wheat exposed to a short-term low temperature stress at jointing stage. Plant Soil. 2015, 393, 307–318.
193. Li, Y.; Sun, C.; Huang, Z.; Pan, J.; Wang, L.; Fan, X. Mechanisms of progressive water deficit tolerance and growth recovery of Chinese maize foundation genotypes Huangzao 4 and Chang 7-2, which are proposed on the basis of comparison of physiological and transcriptomic responses. Plant Cell Physiol. 2009, 50, 2092–2111.
194. Lin L, Gettelman A, Feng S, Fu Q. Simulated climatology and evolution of aridity in the 21st century. J Geophys Res. 2015;120: 5795–815. https://doi.org/10.1002/2014JD022912.
195. Liu C, Liu F, Ravnskov S, Rubæk GH, Sun Z, Andersen MN (2017) Impact of wood biochar and its interactions with mycorrhizal fungi, phosphorus fertilization and irrigation strategies on potato growth. J Agron Crop Sci 203:131–145

196. Liu, J.G.; Li, Y.B.; Zhang, W.; Sun, Y.Y. The distributing of the residue film and influence on cotton growth under continuous cropping in oasis of Xinjiang. J. Agro-Environ. Sci. 2010, 29, 246–250.
197. Liu, S., Wang, X., Wang, H., Xin, H., Yang, X., Yan, J., et al. (2013). Genomewide analysis of ZmDREB genes and their association with natural variation in drought tolerance at seedling stage of Zea mays L. PLoS Genet. 9:e1003790. doi: 10.1371/journal.pgen.1003790
198. Luo X, Bai X, Sun X, Zhu D, Liu B, Ji W, Liu X (2013) Expression of wild soybean WRKY20 in Arabidopsis enhances drought tolerance and regulates ABA signaling. J Exp Bot 73:60–66
199. Lv SL, Lian LJ, Tao P, Li ZX, Zhang KW, Zhang JR (2009) Overexpression of Thellungiella halophila H+-PPase (TsVP) in cotton enhances drought stress resistance of plants. Planta 229(4):899–910
200. Lyu S, Du G, Liu Z, Zhao L, Lyu D (2016) Effects of biochar on photosystem function and activities of protective enzymes in Pyrus ussuriensis Maxim. under drought stress. Acta Physiol Plant 38:1–10
201. Ma, X.; Feng, F.; Wei, H.; Mei, H.; Xu, K.; Chen, S.; Li, T.; Liang, X.; Liu, H.; Luo, L. Genome-Wide Association Study for Plant Height and Grain Yield in Rice under Contrasting Moisture Regimes. Front. Plant Sci. 2016, 7, 1801.
202. Mafakheri, A., A. Siosemardeh, B. Bahramnejad, P. C. Struik, and Y. Sohrabi, 2010: Effect of drought stress on yield, proline and chlorophyll contents in three chickpea cultivars. Aust. J. Crop Sci. 4, 580–585
203. Maina, F.; Harou, A.; Hamidou, F.; Morris, G.P. Genome-wide association studies identify putative pleiotropic locus mediating drought tolerance in sorghum. Plant Direct. 2022, 6, 413.
204. Majid, S. A., Asghar, R., and Murtaza, G. 2007. Yield stability analysis conferring adaptation of wheat to pre- and post-anthesis drought conditions. Pak. J. Bot. 39: 1623–1637
205. Maleki, A., A. Naderi, R. Naseri, A. Fathi, S. Bahamin, and R. Maleki, 2013: Physiological performance of soybean cultivars under drought stress. Bull. Env. Pharmacol. Life Sci. 2, 38–44.
206. Mansour MMF. Nitrogen containing compounds and adaptation of plants to salinity stress. Biol Plant 2000;43:491–500.
207. Marthandan, V., Geetha, R., Kumutha, K., Renganathan, V.G., Karthikeyan, A. and Ramalingam, J., 2020. Seed priming: a feasible strategy to enhance drought tolerance in crop plants. International Journal of Molecular Sciences, 21(21), p.8258.

208. Martınez, J. P., H. F. L. J. Silva, J. F. Ledent, and M. Pinto, 2007: Effect of drought stress on the osmotic adjustment, cell wall elasticity and cell volume of six cultivars of common beans (*Phaseolus vulgaris* L.). Eur. J. Agron. 26, 30–38.
209. Mathew, I.; Shimelis, H.; Shayanowako, A.; Laing, M.; Chaplot, V. Genome-wide association study of drought tolerance and biomass allocation in wheat. PLoS ONE 2019, 14, e0225383.
210. Mauad, M.; Costa Crusciol, C.A.; Nascente, A.S.; Filho, H.G.; Pereira Lima, G.P. Effects of silicon and drought stress on biochemical characteristics of leaves of upland rice cultivars1. Rev. Cienc. Agron. 2016, 47, 532–539.
211. McMillen, M. The Effect of Mulch Type and Thickness on the Soil Surface Evaporation Rate; California Polytechnic State University: San Luis Obispo, CA, USA, 2013
212. Meena, R.P.; Sendhil, R.; Tripathi, S.; Chander, S.; Chhokar, R.; Sharma, R. Hydro-priming of seed improves the water use efficiency, grain yield and net economic return of wheat under different moisture regimes. SAARC J. Agric. 2013, 11, 149–159.
213. Mei, F.; Chen, B.; Du, L.; Li, S.; Zhu, D.; Chen, N.; Zhang, Y.; Li, F.; Wang, Z.; Cheng, X.; et al. A gain-of-function allele of a DREB transcription factor gene ameliorates drought tolerance in wheat. Plant Cell. 2022, 34, 4472–4494.
214. Mi N., Cai F., Zhang Y., Ji R., Zhang S., Wang Y. (2018): Differential responses of maize yield to drought at vegetative and reproductive stages. Plant Soil Environ., 64: 260-267.
215. Mickan BS, Abbott LK, Stefanova K, Solaiman ZM (2016) Interactions between biochar and mycorrhizal fungi in a water-stressed agricultural soil. Mycorrhiza 26:565–574
216. Miller GAD, Suzuki N, Ciftci-Yilmaz SULTAN, Mittler R (2010) Reactive oxygen species homeostasis and signalling during drought and salinity stresses. Plant Cell Environ 33:453–46
217. Milly PCD, Dunne KA. Potential evapotranspiration and continental drying. Nat Clim Chang. 2016;6:946–9
218. Miranda, M.T.; Da Silva, S.F.; Silveira, N.M.; Pereira, L.; Machado, E.C.; Ribeiro, R.V. Root Osmotic Adjustment and Stomatal Control of Leaf Gas Exchange are Dependent on Citrus Rootstocks Under Water Deficit. J. Plant Growth Regul. 2020, 285, 1–9.
219. Mohamed, H.I.; El-Sayed, A.A.; Rady, M.M.; Caruso, G.; Sekara, A.; Abdelhamid, M.T. Coupling effects of Phosphorus Fertilization Source and rate on Growth and Ion Accumulation of common bean Grown under Salinity stress. PeerJ 2021, 9, 411–463.

220. Mohanty A, Kathuria H, Ferjani A, Sakamoto A, Mohanty P, Murata N, et al. Transgenics of an elite indica rice variety Pusa Basmati 1 harbouring the codA gene are highly tolerant to salt stress. Theor Appl Genet 2002;106:51–7.
221. Mostofa, M.G.; Ghosh, A.; Li, Z.G.; Siddiqui, M.N.; Fujita, M.; Tran, L.S.P. Methylglyoxal–a signaling molecule in plant abiotic stress responses. Free Radic. Biol. Med. 2018, 122, 96–109.
222. Mustafa, H.S.; Mahmood, T.; Ullah, A.; Sharif, A.; Bhatti, A.N.; Nadeem, M.; Ali, R. Role of seed priming to enhance growth and development of crop plants against biotic and abiotic stress. Bull. Biol. All. Sci. 2017, 2, 1–11
223. Mutetwa, M.; Mtaita, T. Effects of mulching and fertilizer sources on growth and yield of onion. J. Glob. Innov. Agric. Soc. Sci. 2014, 2, 102–106.
224. Nawaz, A.; Farooq, M.; Ahmad, R.; Basra, S.M.A.; Lal, R. Seed priming improves stand establishment and productivity of no till wheat grown after direct seeded aerobic and transplanted flooded rice. Eur. J. Agron. 2016, 76, 130–137.
225. Nawaz, J.; Hussain, M.; Jabbar, A.; Nadeem, G.A.; Sajid, M.; Subtain, M.U.; Shabbir, I. Seed priming a technique. Int. J. Agric. Crop Sci. 2013, 6, 1373.
226. Nezhadahmadi, A.; Prodhan, Z.H.; Faruq, G. Drought tolerance in wheat. Sci. World J. 2013, 610721.
227. O'Connell, E. Towards adaptation of water resource Systems to climatic and socio-economic Chang. Water Resour. Manag. 2017, 31, 2965–2984.
228. Ogbonnaya, C.I.; Sarr, B.; Brou, C.; Diouf, O.; Diop, N.N.; Roy-Macauley, H. Selection of cowpea genotypes in hydroponics, pots, and field for drought tolerance. Crop Sci. 2003, 43, 1114–1120
229. Ogundare, S.K.; Babatunde, I.J.; Etukud, O.O. Response of tomato variety (Roma F) yield to different mulching materials and staking in Kabba Kogi State. Niger J. Agric. Stud. 2015, 3, 61–70.
230. Olmo M, Alburquerque JA, Barrón V, del Campillo MC, Gallardo A, Fuentes M, Villar R (2014) Wheat growth and yield responses to biochar addition under Mediterranean climate conditions. Biol Fert Soil 50:1177–1187
231. Osakabe, Y., Yamaguchi-Shinozaki, K., Shinozaki, K., and Tran, L. S. (2014). ABA control of plant macroelement membrane transport systems in response to water deficit and high salinity. New Phytol. 202, 35–49. doi: 10.1111/nph.12613

232. Osuagwu, G.G.E.; Edeoga, H.O.; Osuagwu, A.N. The influence of water stress (drought) on the mineral and vitamin potential of the leaves of Ocimum gratissimum (L). Recent Res. Sci. Technol. 2010, 2, 27–33
233. Pagnani, G.; Galieni, A.; Stagnari, F.; Pellegrini, M.; Del Gallo, M.; Pisante, M. Open field inoculation with PGPR as a strategy to manage fertilization of ancient Triticum genotypes. Biol. Fertil. Soils 2019, 56, 1–14.
234. Pallai, R.; Hynes, R.K.; Verma, B.; Nelson, L.M. Phytohormone production and colonization of canola (*Brassica napus* L.) roots byPseudomonas fluorescens6-8 under gnotobiotic conditions. Can. J. Microbiol. 2012, 58, 170–178.
235. Paneque M, José M, Franco-Navarro JD, Colmenero-Flores JM, Knicker H (2016) Effect of biochar amendment on morphology, productivity and water relations of sunflower plants under non-irrigation conditions. Catena 147:280–287
236. Paparella, S.; Araujo, S.S.; Rossi, G.; Wijaya, S.M.; Carbonera, D.; Balestrazzi, A. Seed priming: State of the art and new perspectives. Plant Cell Rep. 2015, 34, 1281–1293.
237. Park SY, Noh KJ, Yoo JH, Yu JW, Lee BW, Kim JG, Paek NC (2006) Rapid upregulation of Dehyrin3 and Dehydrin4 in response to dehydration is a characteristic of drought-tolerant genotypes in barley. J Plant Biol 49(6):455–462.
238. Pastor, V.; Luna, E.; Mauch, M.B.; Ton, J.; Flors, V. Primed plants do not forget. Environ. Exp. Bot. 2013, 94, 46–56.
239. Patanè, C.; Saita, A.; Sortino, O. Comparative effects of salt and water stress on seed germination and early embryo growth in two cultivars of sweet sorghum. J. Agron. Crop Sci. 2013, 199, 30–37.
240. Peng, H.; Lei, T.; Jiang, Z.; Horton, R. A method for estimating maximum static rainfall retention in pebble mulches used for soil moisture conservation. J. Hydrol. 2016, 537, 346–355
241. Pour-Aboughadareh, A., Mohammadi, R., Etminan, A., Shooshtari, L., Maleki-Tabrizi, N. and Poczai, P., 2020. Effects of drought stress on some agronomic and morpho-physiological traits in durum wheat genotypes. Sustainability, 12(14), p.5610.
242. Prasad, P. V. V., Pisipati, S. R., Momcilovic I., and Ristic, Z. 2011. Independent ´ and combined effects of high temperature and drought stress during grain filling on plant yield and chloroplast EF-Tu expression in spring wheat. J. Agron. Crop Sci. 197: 430–441
243. Pressler Y, Foster EJ, Moore JC, Cotrufo MF (2017) Coupled biochar amendment and limited irrigation strategies do not affect a degraded

soil food web in a maize agroecosystem, compared to the native grassland. GCB Bioenergy. doi:10.1111/gcbb.12429

244. Primo, E.D.; Ruiz, F.; Masciarelli, O.; Giordano, W. Biofilm Formation and Biosurfactant Activity in Plant-Associated Bacteria. In Sustainable Development and Biodiversity; Springer Science and Business Media LLC: Berlin/Heidelberg, Germany, 2015; Volume 12, pp. 337–349.
245. Qayyum, A.; Razzaq, A.; Ahmad, M.; Jenks, M.A. Water stress causes differential effects on germination indices, total soluble sugar and proline content in wheat (*Triticum aestivum* L.) genotypes. Afr. J. Biotechnol. 2011, 10, 14038–14045
246. Qin, F., Shinozaki, K., and Yamaguchi-Shinozaki, K. (2011). Achievements and challenges in understanding plant abiotic stress responses and tolerance. Plant Cell Physiol. 52, 1569–1582. doi: 10.1093/pcp/pcr106
247. Qin, S.; Li, S.; Kang, S ; Du, T., Tong, L.; Ding, R. Can the drip irrigation under film mulch reduce crop evapotranspiration and save water under the sufficient irrigation condition ? Agric. Water Manag. 2016, 177, 128–137.
248. Queiroz, M.S.; Oliveira, C.E.S.; Steiner, F.; Zuffo, A.M.; Zoz, T.; Vendruscolo, E.P.; Menis, V.S.; Mello, B.F.F.R.; Cabral, R.C.; Menis, T.F. Drought stresses on seed germination and early growth of maize and sorghum. J. Agric. Sci. 2019, 11, 310–318.
249. Rabbani J, Emam Y. Yield Response of Maize Hybrids to Drought Stress at Different Growth Stages. Isfahan University of Technology - Journal of Crop Production and Processing 2011; 1 (2) :65-78
250. Rahdari P, Hoseini SM. Drought stress: a review. Intl J Agron Plant Prod. 2012;3:443–6.
251. Rahdari, P.; Hosseini, S.M.; Tavakoli, S. The studying effect of drought stress on germination, proline, sugar, lipid, protein and chlorophyll content in purslane (*Portulaca oleracea* L.) leaves. J. Med. Plants Res. 2012, 6, 1539–1547.
252. Rampino, P.; Pataleo, S.; Gerardi, C.; Mita, G.; Perrotta, C. Drought stress response in wheat: Physiological and molecular analysis of resistant and sensitive genotypes. Plant Cell Environ. 2006, 29, 2143–2152
253. Ramzan, F.; Kim, H.T.; Shim, K.K.; Choi, Y.H.; Younis, A.; Lim, K.B. Genetic diversity and relationship assessment of Lilium lancifolium × Asiatic hybrid ‘Chianti’ progeny by ISSR markers. Eur. J. Hortic. Sci. 2018, 83, 142–150.

254. Rashed, M.A.; Sabry, S.B.S.; Atta, A.H.; Mostafa, A.M. Development of RAPD markers associated drought tolerance in bread wheat (*Triticum aestivum* L.). Egypt. J. Genet. Cytol. 2010, 39, 131–142.
255. Ray, D.K.; West, P.C.; Clark, M.; Gerber, J.S.; Prishchepov, A.V.; Chatterjee, S. Climate change has likely already affected global food production. PLoS ONE 2019, 14, 217148.
256. Raza, A.; Razzaq, A.; Mehmood, S.S.; Zou, X.; Zhang, X.; Lv, Y.; Xu, J. Impact of climate change on crops adaptation and strategies to tackle its outcome: A review. Plants 2019, 8, 34.
257. Rezaei-Chiyaneh, E.; Amirnia, R.; Machiani, M.A.; Javanmard, A.; Maggi, F.; Morshedloo, M.R. Intercropping fennel (*Foeniculum vulgare* L.) with common bean (*Phaseolus vulgaris* L.) as affected by PGPR inoculation: A strategy for improving yield, essential oil and fatty acid composition. Sci. Hortic. 2020, 261, 108951.
258. Rizwan, M.; Ali, S.; Ibrahim, M.; Farid, M.; Adrees, M.; Bharwana, S.A.; Zia-ur-Rehman, M.; Qayyum, M.F.; Abbas, F. Mechanisms of silicon-mediated alleviation of drought and salt stress in plants: A review. Environ. Sci. Pollut. Res. 2015, 22, 15416–15431.
259. Rubin, R.L.; Van Groenigen, K.J.; Hungate, B.A. Plant growth promoting rhizobacteria are more effective under drought: A meta-analysis. Plant Soil 2017, 416, 309–323.
260. Ruehr, N.K.; Grote, R.; Mayr, S.; Arneth, A. Beyond the extreme: Recovery of carbon and water relations in woody plants following heat and drought stress. Tree Physiol. 2019, 39, 1285–1299.
261. Saikia, J.; Sarma, R.K.; Dhandia, R.; Yadav, A.; Bharali, R.; Gupta, V.K.; Saikia, R. Alleviation of drought stress in pulse crops with ACC deaminase producing rhizobacteria isolated from acidic soil of Northeast India. Sci. Rep. 2018, 8, 1–16.
262. Saja-Garbarz, D.; Ostrowska, A.; Kaczanowska, K.; Janowiak, F. Accumulation of silicon and changes in water balance under drought stress in Brassica napus var. napus L. Plants 2021, 10, 280.
263. Salehi-Lisar, S.Y.; Bakhshayeshan-Agdam, H. Drought stress in plants: Causes, consequences, and tolerance. In Drought Stress Tolerance in Plants; Springer: Cham, Switzerland, 2016
264. Samarah, N.H.; Haddad, N.; Alqudah, A.M. Yield potential evaluation in chickpea genotypes under late terminal drought in relation to the length of reproductive stage. Ital. J. Agron. 2009, 3, 111–117.
265. Saranga Y, Menz M, Jiang CX, Wright RJ, Yakir D, Paterson AH. Genomic dissection of genotype × environment interactions conferring adaptation of cotton to arid conditions. Genome Res 2001;11:1988–95.

266. Sarvestani, Z.T.; Pirdashti, H.; Sanavy, S.A.; Balouchi, H. Study of water stress effects in different growth stages on yield and yield components of different rice (*Oryza sativa* L.) cultivars. Pak. J. Biol. Sci. 2008, 11, 1303–1309
267. Scheff J, Frierson DMW. Terrestrial aridity and its response to greenhouse warming across CMIP5 climate models. J Clim. 2015;28:5583–600. https://doi.org/10.1175/JCLI-D-14-00480.1.
268. Seleiman, M.F.; Refay, Y.; Al-Suhaibani, N.; Al-Ashkar, I.; El-Hendawy, S.; Hafez, E.M. Integrative effects of rice-straw biochar and silicon on oil and seed quality, yield and physiological traits of *Helianthus annuus* L. grown under water deficit stress. Agronomy 2019, 9, 637.
269. Selvakumar, G.; Bhatt, R.M.; Upreti, K.K.; Bindu, G.H.; Shweta, K. Citricoccus zhacaiensis B-4 (MTCC 12119) a novel osmotolerant plant growth promoting actinobacterium enhances onion (*Allium cepa* L.) seed germination under osmotic stress conditions. World J. Microbiol. Biotechnol. 2015, 31, 833–839.
270. Serraj R, Sinclair TR. Osmolyte accumulation: can it really increase crop yield under drought conditions? Plant Cell Environ 2002;25:333–41.
271. Setter, T. L., Yan, J., Warburton, M., Ribaut, J. M., Xu, Y., Sawkins, M., et al. (2011). Genetic association mapping identifies single nucleotide polymorphisms in genes that affect abscisic acid levels in maize floral tissues during drought. J. Exp. Bot. 62, 701–716.
272. doi: 10.1093/jxb/erq308
273. Shalaby, T.A.; El-Bialy, S.M.; El-Mahrouk, M.E.; Omara, A.E.; El-Beltagi, H.S.; El-Ramady, H. Acclimatization of in vitro banana seedlings using root-applied bio-nanofertilizer of copper and selenium. Agronomy 2022, 12, 539.
274. Shamsi, K., and Kobraee, S. 2011. Bread wheat production under drought stress conditions. Ann. Biol. Res. 2: 352–358.
275. Shamsi, K., Petrosyan, M., Noor-Mohammadi, G., and Haghparast, R. 2010. The role of water deficit stress and water use efficiency on bread wheat cultivars. J. Appl. Biosci. 35: 2325–2331.
276. Shamsudin, N.A.A.; Swamy, B.P.M.; Ratnam, W.; Cruz, M.T.S.; Sandhu, N.; Raman, A.K.; Kumar, A. Pyramiding of drought yield QTLs into a high quality Malaysian rice cultivar MRQ74 improves yield under reproductive stage drought. Rice 2016, 9, 21
277. Shao, H.B.; Liang, Z.S.; Shao, M.A.; Sun, Q. Dynamic changes of anti-oxidative enzymes of 10 wheat genotypes at soil water deficits. Coll. Surf. B Biointerfaces 2005, 42, 187–195.

278. Sharma, A.; Kumar, V.; Shahzad, B.; Ramakrishnan, M.; Sidhu, G.P.S.; Bali, A.S.; Handa, N.; Kapoor, D.; Yadav, P.; Khanna, K. Photosynthetic response of plants under different abiotic stresses: A review. J. Plant Growth Regul. 2020, 39, 509–531
279. Sharma, A.; Wang, J.; Xu, D.; Tao, S.; Chong, S.; Yan, D.; Li, Z.; Yuan, H.; Zheng, B. Melatonin regulates the functional components of photosynthesis, antioxidant system, gene expression, and metabolic pathways to induce drought resistance in grafted Carya cathayensis plants. Sci. Total Environ. 2020, 713, 136675.
280. Shasany, A.K.; Darokar, M.P.; Dhawan, S.; Gupta, A.K.; Gupta, S.; Shukla, A.K.; Patra, N.K.; Khanuja, S.P. Use of RAPD and AFLP markers to identify interand intraspecific hybrids of Mentha. J. Hered. 2005, 96, 542–549.
281. Shen YG, Du BX, Zhang WK, Zhang JS, Chen SY. AhCMO, regulated by stresses in Atriplex hortensis, can improve drought tolerance in transgenic tobacco. Theor Appl Genet 2002;105:815–21.
282. Shinozaki K, Yamaguchi-Shinozaki K (2007) Gene networks involved in drought stress response and tolerance. J Exp Bot 58(2):221–227
283. Shrestha, R.A.; Turner, N.C.A.; Siddique, K.H.M.A.; Turner, D.W.B.; Speijers, J.C. A water deficit during pod development in lentils reduces flower and pod numbers but not seed size. Aust. J. Agr. Res. 2006, 57, 427–438
284. Singh S, Gupta AK, Kaur N (2012) Infuence of drought and sowing time on protein composition, antinutrients, and mineral contents of wheat. Sci World J 2012:485751
285. Singh, J.; Thakur, J.K. Photosynthesis and abiotic stress in plants. In Biotic and Abiotic Stress Tolerance in Plants; Springer: Singapore, 2018; pp. 27–46.
286. Sobhanian, H.; Pahlavan, S.; Meyfour, A. How does proteomics target plant environmental stresses in a semi-arid area? Mol. Biol. Rep. 2020, 47, 3181–3194.
287. Spindel, J.E.; Dahlberg, J.; Colgan, M.; Hollingsworth, J.; Sievert, J.; Staggenborg, S.H.; Hutmacher, R.; Jansson, C.; Vogel, J.P. Association mapping by aerial drone reveals 213 genetic associations for Sorghum bicolor biomass traits under drought. BMC Genom. 2018, 19, 679.
288. Stephenson N. Actual evapotranspiration and deficit: biologically meaningful correlates of vegetation distribution across spatial scales. J Biogeogr. 1998;25:855–70. https://doi.org/10.1046/j. 1365-2699.1998.00233.x.

289. Stocker T (ed) (2014) Climate change 2013: the physical science basis: working Group I contribution to the Fifth assessment report of the Intergovernmental Panel on Climate Change. Cambridge University Press, Cambridge
290. Sultan, B.; Defrance, D.; Iizumi, T. Evidence of crop production losses in West Africa due to historical global warming in two crop models. Sci. Rep. 2019, 9, 1–15.
291. Sun, M.; Li, Y.; Zheng, J.; Wu, D.; Li, C.; Li, Z.; Zang, Z.; Zhang, Y.; Fang, Q.; Li, W.; et al. A Nuclear Factor Y-B Transcription Factor, GmNFYB17, Regulates Resistance to Drought Stress in Soybean. Int. J. Mol. Sci. 2022, 23, 7242.
292. Sunkar R, Kapoor A, Zhu JK. Post transcriptional induction of two Cu/Zn superoxide dismutase genes in Arabidopsis is mediated by down regulation of miR398 and important for oxidative stress tolerance. Plant Cell 2006;18:2051–65.
293. Swamy, B.P.M.; Shamsudin, N.A.A.; Rahman, S.N.A.; Mauleon, R.; Ratnam, W.; Sta Cruz, M.T.; Kumar, A. Association Mapping of Yield and Yield-related Traits Under Reproductive Stage Drought Stress in Rice (*Oryza sativa* L.). Rice 2017, 10, 21
294. Swamy, M.K.; Akhtar, M.S.; Sinniah, U.R. Response of PGPR and AM Fungi Toward Growth and Secondary Metabolite Production in Medicinal and Aromatic Plants. In Plant, Soil and Microbes; Springer Science and Business Media LLC: Berlin/Heidelberg, Germany, 2016; pp. 145–168
295. Swann, A.L., 2018. Plants and drought in a changing climate. Current Climate Change Reports, 4, pp.192-201.
296. Tabassum, T.; Farooq, M.; Ahmad, R.; Zohaib, A.; Wahid, A.; Shahid, M. Terminal drought and seed priming improves drought tolerance in wheat. Physiol. Mol. Biol. Plants 2018, 24, 845–856.
297. Tang N, Zhang H, Li X, Xiao J, Xiong L (2012) Constitutive activation of transcription factor OsbZIP46 improves drought tolerance in rice. Plant Physiol 158(4):1755–1768
298. Thakur, M.; Kumar, R. Mulching, Boosting crop productivity and improving soil environment in herbal plants. J. Appl. Res. Med. Aromat. Plants 2021, 20, 100–287.
299. Thirunavukkarasu, N., Hossain, F., Arora, K., Sharma, R., Shiriga, K., Mittal, S., et al. (2014). Functional mechanisms of drought tolerance in subtropical maize (*Zea mays* L.) identified using genome-wide association mapping. BMC Genomics 15:1182. doi: 10.1186/1471-2164-15-1182

300. Thorne, S.J.; Hartley, S.E.; Maathuis, F.J.M. The effect of silicon on osmotic and drought stress tolerance in wheat landraces. Plants 2021, 10, 814.
301. Thudi, M., Upadhyaya, H. D., Rathore, A., Gaur, P. M., Krishnamurthy, L., Roorkiwal, M., et al. (2014). Genetic dissection of drought and heat tolerance in chickpea through genome-wide and candidate gene-based association mapping approaches. PLoS ONE 9:e96758. doi: 10.1371/journal.pone.0096758
302. Toscano, S.; Ferrante, A.; Romano, D. Response of Mediterranean ornamental plants to drought stress. Horticulturae 2019, 5, 6
303. Trujillo LE, Sotolongo M, Menendez C, Ochogavia ME, Coll Y, Hernandez I, Hernandez L (2008) SodERF3, a novel sugarcane ethylene responsive factor (ERF), enhances salt and drought tolerance when over expressed in tobacco plants. Plant Cell Physiol 49(4):512–525
304. Ullah A, Akbar A, Luo Q, Khan AH, Manghwar H, Shaban M, Yang X (2019a) Microbiome diversity in cotton rhizosphere under normal and drought conditions. Microb Ecol 77:429–439
305. Ullah A, Heng S, Munis MFH, Fahad S, Yang X (2015a) Phytoremediation of heavy metals assisted by plant growth promoting (PGP) bacteria: a review. Environ Exp Bot 117:28–40
306. Ullah A, Manghwar H, Shaban M, Khan AH, Akbar A, Ali U, Fahad S (2018) Phytohormones enhanced drought tolerance in plants: a coping strategy. Environ Sci Pollut Res 25:33103–33118
307. Ullah A, Mushtaq H, Fahad S, Shah A, Chaudhary HJ (2017) Plant growth promoting potential of bacterial endophytes in novel association with Olea ferruginea and Withania coagulans. Microbiology 86:119–127
308. Ullah A, Sun H, Yang X, Zhang X (2017b) Drought coping strategies in cotton: increased crop per drop. Plant Biotechnol J 15:271–284
309. Uzoma KC, Inoue M, Andry H, Fujimaki H, Zahoor A, Nishihara E (2011) Effect of cow manure biochar on maize productivity under sandy soil condition. Soil Use Manage 27:205–212
310. Vaccari FP, Maienza A, Miglietta F, Baronti S, Di Lonardo S, Giagnoni L, Lagomarsino A, Pozzi A, Pusceddu E, Ranieri R, Valboa G, Genesio L (2015) Biochar stimulates plant growth but not fruit yield of processing tomato in a fertile soil. Agri Ecosys Environ 207:163–170
311. Vacheron J, Desbrosses G, Boufaud ML, Touraine B, Moënne-Loccoz Y, Muller D, Prigent-Combaret C (2013) Plant growth-promoting rhizobacteria and root system functioning. Front Plant Sci 4:356

312. Valliyodan B, Nguyen HT (2006) Understanding regulatory networks and engineering for enhanced drought tolerance in plants. Curr Opin Plant Biol 9(2):189–195
313. Vandenbroucke K, Metzlaff M (2013) Abiotic stress tolerant crops: genes, pathways and bottlenecks. In Sustainable Food Production. Springer, New York, pp 1–17
314. Vander, W.; Postaire, C.; Tournaire-Roux, O.C.; Boursiac, Y.; Maurel, C. Expression and inhibition of aquaporins in germinating Arabidopsis seeds. Plant Cell Physiol. 2006, 47, 1241–1250.
315. Varshney, R.K., Graner, A., Sorrells, M.E. 2005. Genic microsatellite markers in plants: features and applications. Trends Biotechnol., 23: 48–55.
316. Verma, V.; Ravindran, P.; Kumar, P.P. Plant hormone-mediated regulation of stress responses. BMC Plant Biol. 2016, 16, 86.
317. Verslues, P. E., Lasky, J. R., Juenger, T. E., Liu, T. W., and Kumar, M. N. (2014). Genome-wide association mapping combined with reverse genetics identifies new effectors of low water potential-induced proline accumulation in Arabidopsis. Plant Physiol. 164, 144–159. doi: 10.1104/pp.113. 224014
318. Verslues, P.E., Agarwal, M., Katiyar-Agarwal, S., Zhu, J. and Zhu, J.K., 2006. Methods and concepts in quantifying resistance to drought, salt and freezing, abiotic stresses that affect plant water status. The Plant Journal, 45(4), pp.523-539.
319. Vinh NT, Paterson AH. Genome mapping and its implication for stress resistance in plants. In: Ashraf M, Harris PJC, editors. Abiotic stresses: plant resistance through breeding and molecular approaches; 2005.
320. Wallace, J. G., Zhang, X., Beyene, Y., Semagn, K., Olsen, M., Prasanna, B. M., et al. (2016). Genome-wide association for plant height and flowering time across 15 tropical maize populations under managed drought stress and wellwatered conditions in sub-saharan africa. Crop Sci. 56, 2365–2378. doi: 10.2135/ cropsci2015.10.0632
321. Wang YJ, Hao YJ, Zhang ZG, Chen T, Zhang JS, Chen SY. Isolation of trehalose-6- phosphate phosphatase gene from tobacco and its functional analysis in yeast cells. J Plant Physiol 2005;162:215–23
322. Wang, H. and Qin, F., 2017. Genome-wide association study reveals natural variations contributing to drought resistance in crops. Frontiers in plant science, 8, p.1110.
323. Wang, X.; Vignjevic, M.; Liu, F.; Jacobsen, S.; Jiang, D.; Wollenweber, B. Drought priming at vegetative growth stages improves tolerance to drought and heat stresses occurring during grain filling in spring wheat. Plant Growth Regul. 2015, 75, 677–687.

324. Wang, X.; Wang, H.; Liu, S.; Ferjani, A.; Li, J.; Yan, J.; Yang, X.; Qin, F. Genetic variation in ZmVPP1 contributes to drought tolerance in maize seedlings. Nat. Genet. 2016, 48, 1233–1241.
325. Wang, Y.P.; Li, X.G.; Fu, T.T.; Wang, L.; Turner, N.C.; Siddique, K.H.M.; Li, F.M. Multi-site assessment of the effects of plastic-film mulch on the soil organic carbon balance in semiarid areas of China. Agric. For. Meteorol. 2016, 228, 42–51.
326. Warner, K.; Afifi, T. Where the rain falls: Evidence from 8 countries on how vulnerable households use migration to manage the risk of rainfall variability and food insecurity. Clim. Dev. 2014, 6, 1–17.
327. Wehner, G. G., Balko, C. C., Enders, M. M., Humbeck, K. K., and Ordon, F. F. (2015). Identification of genomic regions involved in tolerance to drought stress and drought stress induced leaf senescence in juvenile barley. BMC Plant Biol. 15:125. doi: 10.1186/s12870-015-0524-3
328. Wu R, Garg A. Engineering rice plants with trehalose-producing genes improves tolerance to drought, salt, and low temperature. ISB News Report; 2003.
329. Wu, X.; Feng, H.; Wu, D.; Yan, S.; Zhang, P.; Wang, W.; Zhang, J.; Ye, J.; Dai, G.; Fan, Y.; et al. Using high-throughput multiple optical phenotyping to decipher the genetic architecture of maize drought tolerance. Genome Biol. 2021, 22, 185.
330. Xiang, D.-B.; Peng, L.-X.; Zhao, J.-L.; Zou, L.; Zhao, G.; Song, C. Effect of drought stress on yield, chlorophyll contents and photosynthesis in tartary buckwheat (*Fagopyrum tataricum*). J. Food Agric. Environ. 2013, 11, 1358–1363.
331. Xiao B., Chen X, Xiang CB, Tang N, Zhang QF, Xiong LZ (2009) Evaluation of seven function-known candidate genes for their effects on improving drought resistance of transgenic rice under field conditions. Mol Plant 2(1):73–83
332. Xiao Q, Zhu LX, Shen YF, Li SQ (2016) Sensitivity of soil water retention and availability to biochar addition in rainfed semi-arid farmland during a three-year field experiment. Field Crops Res 196:284–293
333. Xiao, Y.M.; Luo, R.P.; Hayes, J.; Kim, S.N.; Ding, M.; Liao, A.K. Structure basis for directional R-loop formation and substrate handover mechanisms in type I CRISPR-Cas system cryo-EM structures of type I CRISPR-Cas system resolve the mechanisms governing the PAM-dependent R-loop formation, Cas3 recruitment, and subst. Cell 2017, 170, 48–60.e11.

334. Xiong L, Schumaker KS, Zhu JK (2002) Cell signaling during cold, drought, and salt stress. Plant Cell Online 14(1):165–183
335. Xue, Y., Warburton, M. L., Sawkins, M., Zhang, X., Setter, T., Xu, Y., et al. (2013). Genome-wide association analysis for nine agronomic traits in maize under well-watered and water-stressed conditions. Theor. Appl. Genet. 126, 2587–2596. doi: 10.1007/s00122-013-2158-x
336. Yamada M, Morishita H, Urano K, Shiozaki N, Yamaguchi-Shinozaki K, Shinozaki K, et al. Effects of free proline accumulation in petunias under drought stress. J Exp Bot 2005;56:1975–81.
337. Yan M (2015) Seed priming stimulate germination and early seedling growth of Chinese cabbage under drought stress. S Afr J Bot 99:88–92
338. Yang, H.; Huntingford, C.; Wiltshire, A.; Sitch, S.; Mercado, L. Compensatory climate effects link trends in global runoff to rising atmospheric CO2 concentration. Environ. Res. Lett. 2019, 14, 124075.
339. Yang, N.; Sun, Z.; Feng, L.; Zheng, M.; Chi, D. Plastic film mulching for water- efficient agricultural applications and degradable films materials development research. Mater. Manuf. Process. 2015, 30, 143–154.
340. Yin, J.; Gentine, P.; Zhou, S.; Sullivan, S.C.; Wang, R.; Zhang, Y.; Guo, S. Large increase in global storm runoff extremes driven by climate and anthropogenic changes. Nat. Commun. 2018, 9, 1–10.
341. Yu, Y.Y.; Turner, N.C.; Gong, Y.H.; Li, F.M.; Fang, C.; Ge, L.J.; Ye, J.S. Benefits and limitations to straw- and plastic-film mulch on maize yield and water use efficiency, a meta-analysis across hydrothermal gradients. Eur. J. Agron. 2018, 99, 138–147.
342. Zandalinas, S.I., Balfagón, D., Arbona, V. and Gómez-Cadenas, A., 2017. Modulation of antioxidant defense system is associated with combined drought and heat stress tolerance in citrus. Frontiers in Plant Science, 8, p.953.
343. Zandalinas, S.I.; Fritschi, F.B.; Mittler, R. Signal transduction networks during stress combination. J. Exp. Bot. 2020, 71, 1734–1741.
344. Zargar, S.M.; Mahajan, R.; Bhat, J.A.; Nazir, M.; Deshmukh, R. Role of silicon in plant stress tolerance: Opportunities to achieve a sustainable cropping system. 3 Biotech. 2019, 9, 1–16.
345. Zhang GH, Su Q, An LJ, Wu S. Characterization and expression of a vacuolar Na+/H+ antiporter gene from the monocot halophyte Aeluropus littoralis. Plant Physiol Biochem 2008;46:117–26.
346. Zhang Z, Quick MK, Kanelakis KC, Gijzen M, Krishna P (2003) Characterization of a plant homolog of hop, a cochaperone of hsp90. Plant Physiol 131(2):525–535

347. Zhang, Y.; Shi, Y.; Gong, H.; Zhao, H.; Li, H.; Hu, Y.; Wang, Y. Beneficial effects of silicon on photosynthesis of tomato seedlings under water stress. J. Integr. Agric. 2018, 17, 2151–2159.
348. Zhao T, Dai A. The magnitude and causes of global drought changes in the 21st century under a low-moderate emissions scenario. J Clim. 2015;28:4490–512. https://doi.org/10.1175/JCLI-D-14- 00363.1.
349. Zhao, H.; Wang, R.Y.; Ma, B.L.; Xiong, Y.C.; Qiang, S.C.; Wang, C.L. Ridge-furrow with full plastic film mulching improves water use efficiency and tuber yields of potato in a semiarid rain-fed ecosystem. Field Crops Res. 2014, 161, 137–148.
350. Zhao, W., Liu, L., Shen, Q., Yang, J., Han, X., Tian, F. and Wu, J., 2020. Effects of water stress on photosynthesis, yield, and water use efficiency in winter wheat. Water, 12(8), p.2127.
351. Zheng X, Chen B, Lu G, Han B (2009) Overexpression of a NAC transcription factor enhances rice drought and salt tolerance. Biochem Biophy Res C 379(4):985–989
352. Zheng, X.; Qiao, L.; Liu, Y.; Wei, N.; Zhao, J.; Wu, B.; Yang, B.; Wang, J.; Zheng, J. Genome-Wide Association Study of Grain Number in Common Wheat from Shanxi Under Different Water Regimes. Front. Plant Sci. 2022, 12, 806295.
353. Zhou, G.A.; Chang, R.Z.; Qiu, L.J. Overexpression of soybean ubiquitin-conjugating enzyme gene GmUBC2 confers enhanced drought and salt tolerance through modulating abiotic stress-responsive gene expression in Arabidopsis. Plant Mol. Biol. 2010, 72, 357–369.
354. Zhou, J.B.; Wang, C.Y.; Zhang, H.; Dong, F.; Zheng, X.F.; Gale, W.; Li, S.X. Effect of water saving management practices and nitrogen fertilizer rate on crop yield and water use efficiency in a winter wheat-summer maize cropping system. Field Crops Res. 2011, 122, 157–163.
355. Zhou, L.M.; Li, F.M.; Jin, S.L.; Song, Y. How two ridges and the furrow mulched with plastic film affect soil water, soil temperature and yield of maize on the semiarid Loess Plateau of China. Field Crops Res. 2009, 113, 41–47.
356. Zhu, M.-L.; Wu, X.-Q.; Wang, Y.-H.; Dai, Y. Role of Biofilm Formation by Bacillus pumilus HR10 in Biocontrol against Pine Seedling Damping-Off Disease Caused by Rhizoctonia solani. Forest 2020, 11, 652.
357. Zhu, Y.; Gong, H. Beneficial effects of silicon on salt and drought tolerance in plants. Agron. Sustain. Dev. 2014, 34, 455–472.

358. Zlatev, Z.; Lidon, F.C. An overview on drought induced changes in plant growth, water relations and photosynthesis. Emir. J. Food Agric. 2012, 24, 57–72.
359. Zoghi, Z.; Hosseini, S.M.; Kouchaksaraei, M.T.; Kooch, Y.; Guidi, L. The effect of biochar amendment on the growth, morphology and physiology of Quercuscastaneifolia seedlings under water-deficit stress. Eur. J. For. Res. 2019, 138, 967–979.
360. Zribi, W.; Aragüés, R.; Medina, E.; Faci, J.M. Efficiency of inorganic and organic mulching materials for soil evaporation control. Soil Tillage Res. 2015, 148, 40–45.

CHAPTER SEVEN

APPROCHES FOR SUSTINABLE CROP PRODUCTION AND DRY LAND MANAGEMENT

IMRAN[1,2,3]

[1]DEPARTMENT OF AGRONOMY, THE UNIVERSITY OF AGRICULTURE, PESHAWAR

[2]AGRICULTURE OFFICER (EXTENSION WING), OFFICE OF THE DISTRICT DIRECTOR AGRICULTURE EXTENSION, DISTRICT KURRAM, MERGED DISTRICTS KHYBER PAKHTUNKHWA, PAKISTAN

[3]RHIZOSPHERE LABORATORY, UNIVERSITY OF CUKUROVA, FACULTY OF AGRICULTURE, DEPARTMENT OF SOIL SCIENCE AND PLANT NUTRITION, ADANA, TURKEY

Abstract

Global climate crisis and natural disasters (such as floods, droughts, rise in sea levels and forest fires), decreases our water resources day by day. We need to store water and improve our water resources in the soil and increase the efficiency of water use as sustainable dryland management. Drought stress has vital limiting factors in agriculture and mycorrhizal colonization increase plant drought tolerance. In this context, mycorrhizal fungus can be used to reduce plant water stress, as well as organic wastes such as biochar, which have a large surface area and are stable in the soil. There are many studies showing that mycorrhiza increases plant resistance to water stress. However, the water uptake and transport mechanism of mycorrhiza to plant tissues is still a subject of research. In addition, limited irrigation technique can be used in arid and semi-arid regions. Cover crops can also reduce losses by covering the soil surface in agricultural production, increasing soil carbon and water storage. Biochar has been applied to the soil as an amendment to improve soil water holding capacity due to its different porous nature. In order to better understand the water retention mechanisms of biochar, it has vital to examine the behavior of water with biochar in soil. Moreover, the CSA technique, which is a new technique, should be increased and expanded.

Keywords: Climate change, biochar, mycorrhizea, soil health, water conservation, dry land

Introduction

Due to the increasing world population, which is expected to increase by over 9 billion by 2050. Global warming and anthropological reasons, soil, water and environmental pollution is increasing day by day. In addition to reclaim the degraded soil, intensive agricultural production also puts great pressure on the amount of water available per capita. As known, just 3% of the world's water resources are fresh water (Ungureanu et al., 2020). Almost 70% of these resources are used in agriculture. It has been determined that intensive inputs applied to the soil (such as excessive fertilization, irrigation and soil tillage) cause deterioration in soil quality and productivity in a short time (Boretti and Rosa, 2019). It has been reported that the management of the inputs applied to the soil causes an increase in the release of greenhouse gases like CO_2, CH_4 and N_2O, etc. (Lal, 2019), as well as the increase in temperature due to these gases increases the drought.

While the fresh water per capita in the world is 1.34 m3, this amount is expected to decrease gradually due to the increasing world population, unplanned urbanization and climate changes (irregular precipitation, drought, etc.) (Ortaş, 2022). Drought caused by water scarcity raises the possibility that some countries (especially some African and Middle Eastern countries) may give up on producing agricultural products. (Anonymous, 2014). This will seriously affect the social and economic growth of countries. Some agro-ecological techniques can be used to overcome or to reduce drought. The first of these techniques is the use of organic wastes with a large surface area such as biochar. Biochar is a biological coal produced at temperature of 250 ^{0}C and above (Lehmann et al., 2015). In addition to increasing plant growth and yield (Bai et al., 2022), biochar is now used as a soil recoveries in agriculture such as; soil nutrient cycle (Sigua et al., 2016) , Cation Exchange Capacity (CEC) (Zornoza et al., 2016), porosity, bulk density, water holding capacity (Githinji, 2014) and carbon stability (Tag et al., 2016). Studies on the effect of biochar on water holding capacity, which is one of its properties, have been increasing recently (Streubel et al., 2011). While many articles report that biochar were increase on carbon sequestration potential and nutrient trapping, there have been only a few studies about the effect of biochar on water holding capacity (Yu et al., 2013).It can be used to increase water use efficiency and water management in organic fertilizers such as biochar, as well as biological fertilizers can be used like mycorrhiza.

Mycorrhizal symbiosis occurs in nearly 80-90% of terrestrial plants in the world (Bast et al., 2014). Symbiosis between mycorrhizal fungi and terrestrial plants; While the mycorrhizal fungus provides nutrients (such as P, Zn and Cu elements with low mobility in the soil) and water to the plants, the plants provide photosynthesis products as an energy source to the mycorrhizal fungi (Solaiman, 2014). In addition, deficit irrigation, cover crops and water recycling techniques, which have increased in importance recently, can also be used. Deficit irrigation is an irrigation technique that aims at maximum efficiency with minimum water by hitting the plant under water stress at certain periods or throughout its development (Subhan et al., 2021). While water saving is achieved with limited irrigation, it has been reported to increase the efficiency in many studies (Romero et al., 2022; Tong et al., 2022; Yu et al., 2020).

Cover crops are another effective soil protection practice to decrease water erosion and water runoff (Baumhardt and Blanco-Canqui, 2014). Cover crop diversification strategies can change soil water dynamics through changes in, loss pathways, transpiration and evaporation over the course of the plant rotation (Renwick et al., 2019). Recently, there is a new approach to save environment, soil and water as Climate-smart agriculture (CSA). CSA is an approach to support agricultural systems worldwide, concurrently addressing three challenge areas: increased adaptation to climate change, mitigation of climate change, and ensuring global food security – through innovative policies, practices, and financing (Torquebiau et al., 2018).

In this context, we should use water as effectively in agriculture and increase water use efficiency with different new techniques, approaches, fertilizations and agricultural policies, which are decreasing because of climate changes and a source of life. So, the aim of the study is to explain the techniques of managing our water resources, which are decreasing due to climate changes in agriculture. The aim of this chapter is that to research sustainable and new agricultural techniques against drought.

Water Scarcity Induces Water Stress

Usually water stress (drought) is a natural hazard that can be well-defined as the deficiency of water over an extended period of time causing problems to some activities, groups, and other environmental sectors (Mishra and Singh, 2010). It can be mostly classified to four types, such as (Fao, 2008); I) Meteorological, II) Hydrological, III) Socio-economic and IV) Agricultural Drought. These types of drought are related to each other; though, our focus would be fixated on agricultural drought as it is considered as one of the most important subjects in most of the countries in terms of social stability, food security and economic (Hazaymeh and Hassan, 2016). For agriculture, drought may be defined as the deficiency of adequate moisture needed for normal plant growth to complete its life cycle (Ouda et al., 2020; Zhu, 2002). Water stress affects every aspect of plant growth, morphology, anatomy, physiology, and biochemistry. Decreasing water availability under water scarcity commonly results in limited nutrients uptake of plant. Another important impact of water scarcity is that the decrease of nutrient acquisition by the root and its carriage to shoots (Ouda et al., 2020). At present, almost 80% of the world's population is open to high levels of threat to water security (Bunn, 2016).

Among different abiotic stresses factors, water stress (drought) is considered the most disastrous stressor with its serious impact on the plant growth and yield (Manna et al., 2021). Drought can be chronic in climatic regions with low water availability or random and unpredictable because of changes in weather conditions throughout the period of plant growth (De Oliveira et al., 2014). Drought is frequently accompanied by comparatively high temperatures, which encourage evapotranspiration and affects photosynthetic, therefore intensifying the effects of drought and additional decreasing crop yields. Plant responses to soil water deficit have been extensively examined at developmental, physiological and molecular stages and the complex nature of growth regulation under drought stress conditions has been highlighted (De Oliveira et al., 2014; Harb et al., 2010; Hirayama and Shinozaki, 2010). Severe water stress can result in the arrest of photosynthesis activity, disturbance of metabolism and finally the death of plant (Tarafdar et al., 2022).

It is seen in fig 1. Results of the global drought risk assessment for agricultural systems (rain fed and irrigated) at the pixel level. And, it is seen in fig 2. The degree of fragmentation, water stress as the proportion of withdrawal to availability, developed area and relative proportion of nonnative fish species of Europe.

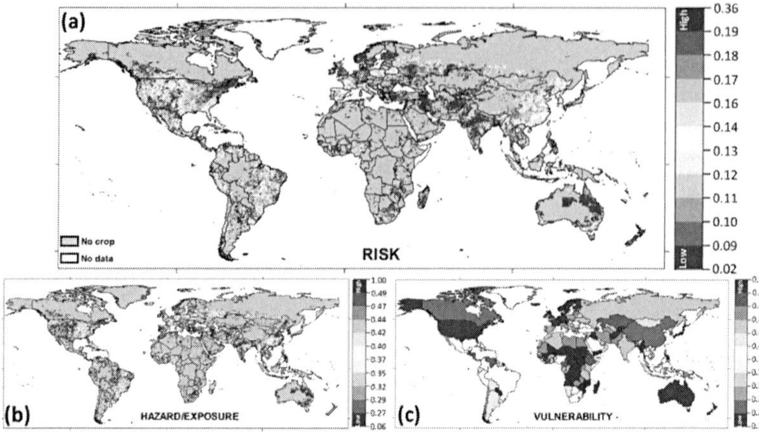

Figure 1. Risk of drought (a), hazard and exposure (b) and vulnerability (c) for irrigated agricultural systems. The legends were defined by assigning the median of the value distribution to the yellow color in the center, the 90th percentile to the deepest red color and the 10th percentile to the deepest blue color and by determining the class ranges of the other colors by linear interpolation. Risk was directly calculated by multiplying hazard and exposure by vulnerability (Meza et al., 2020).

(A)

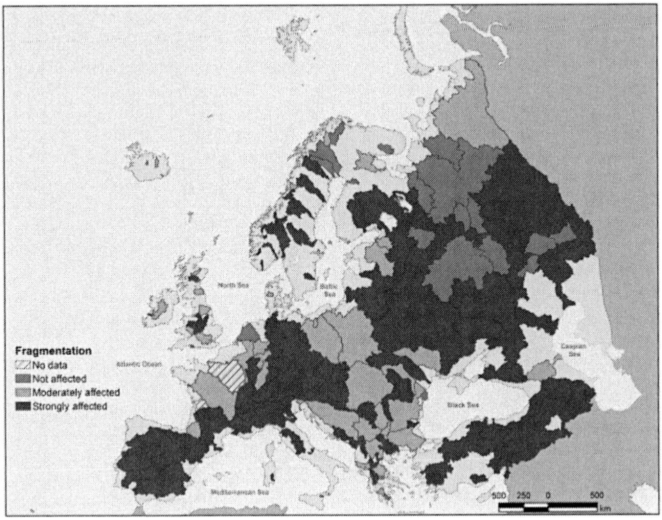
Fragmentation
No data
Not affected
Moderately affected
Strongly affected
500 250 0 500
km

(B)

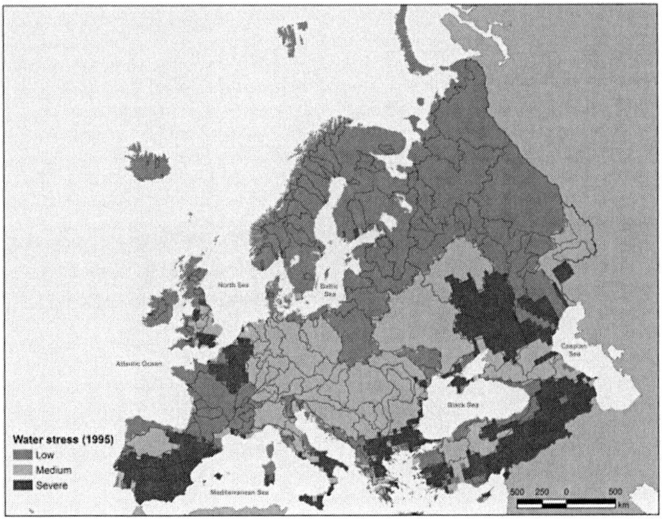
Water stress (1995)
Low
Medium
Severe
500 250 0 500
km

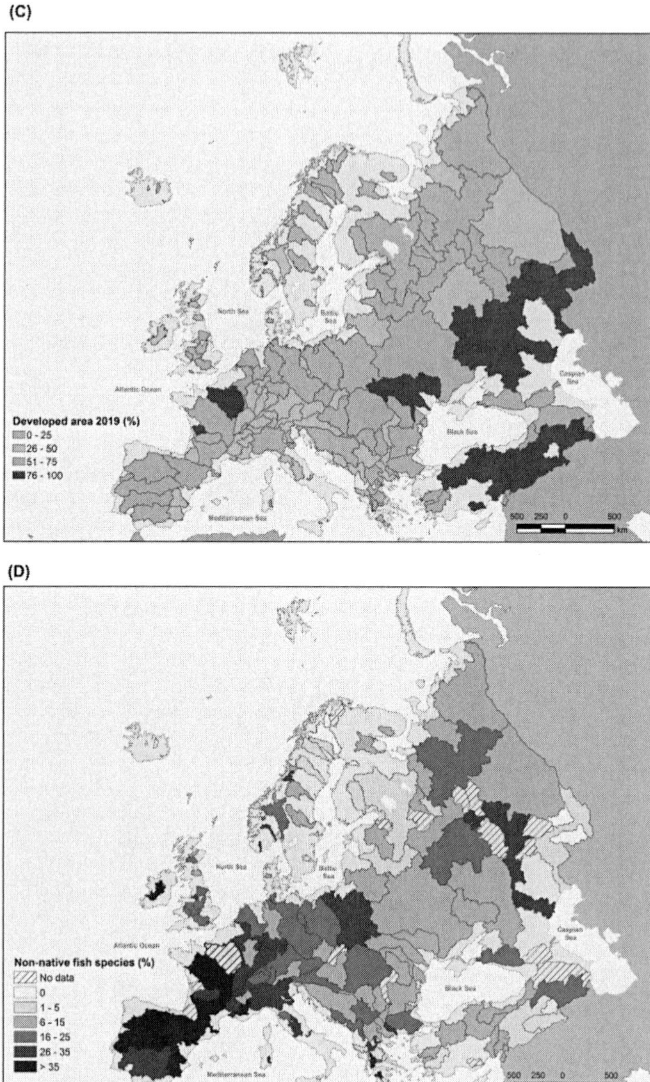

Figure 2. (A) Degree of fragmentation, (B) water stress as the proportion of withdrawal to availability, (C) developed area (urban, arable, and pasture in % of the catchment), (D) relative proportion (%) of nonnative fish species (Tockner et al., 2022)

Biochar to Store Water

Water is one of the vital restrictive factors in agricultural production (Aznar-Sanchez et al., 2019). There is an urgent need to develop agricultural methods that balance water supply and demand while at the same time improve resilience to climate variability (Fischer et al., 2019a). One of these agricultural methods is biochar, which has been studied a lot recently and is still not entirely understood. Biomass feedstock, pyrolysis temperature and pyrolysis time were important for biochar physio-chemical properties (Ghodake et al., 2021). Because of its high specific surface area (Graber et al., 2012) and the hydrocarbon group (like –COO– (–COOH) ve –O– (–OH)) it contains (Yuan et al., 2011), biochar can hold water molecules. So there were so many study about biochar increase plant production and water holding capacity under low water condition (Awan et al., 2021; Chen et al., 2021; Fischer et al., 2019a; Gupta and Su, 2022; Yu et al., 2013). Therefore, it is very important to quantify the availability of water and identify innovations modifiable both productive (like transpiration) and non-productive (like leakage, evaporation and runoff) (Fischer et al., 2019b; Vico and Porporato, 2015). As shown fig 3, the soil water balance relation with temporal changes in the quantity of available soil water (w), with water inputs to the soil through irrigation (QI) and rainfall (P), and water losses from the soil through transpiration and evaporation combined as evapotranspiration (ET), leakage (L) and runoff (QR):

$$\frac{dw}{dt} = P + QI - ET - L - QR \quad (1)$$

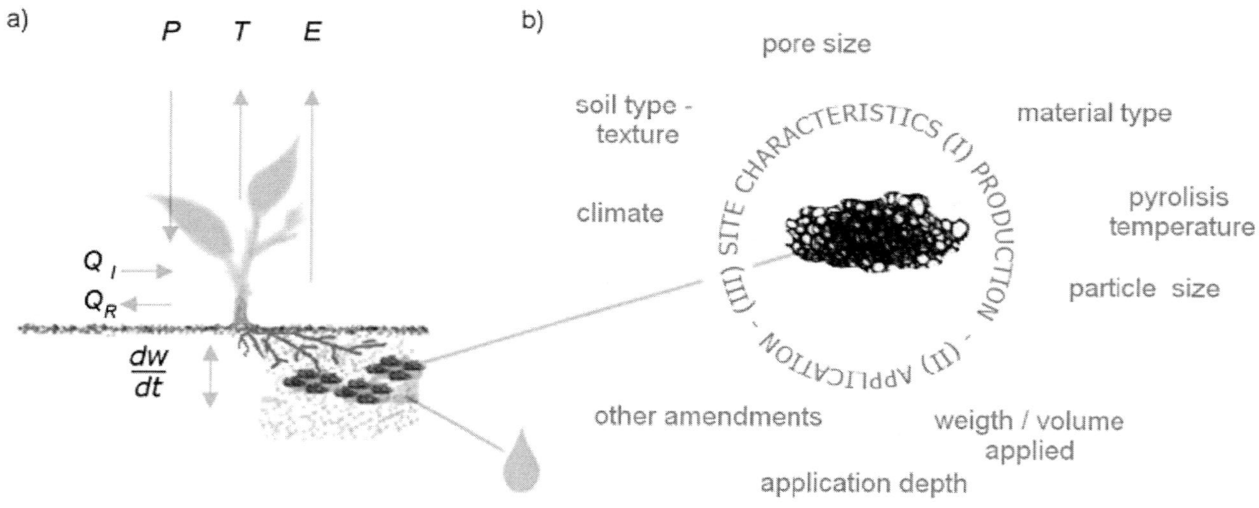

Figure 3. (a) Scheme of the soil – plant – atmosphere system showing water fluxes and soil storage where biochar is added. Capital letters indicate rainfall (P), transpiration (T), soil evaporation (E), irrigation (Qi), runoff (QR) and available soil water (w). (b) Different factors affecting the functioning of biochar which originate from (I) production, (II) biochar application, i.e., user decisions and (III) site characteristics.

In addition, Biochar has been applied to the soil as an amendment to improve soil water holding capacity due to its different porous nature. In order to better understand the water retention mechanisms of biochar, it has vital to examine the behavior of water with biochar in soil (Bikbulatova et al., 2018). The main factors that impact the behavior of water in porous materials are both the chemical (functional groups) and physical (porosity) structures of material (Bikbulatova et al., 2018). For example, Marshall et al. (2019) has been reported that the biochar produced at the appropriate pyrolysis temperature and feedstock increases the water holding capacity by 23% compared to the clay-type soil. This is due to the porous and high surface area of the biochar. Fischer et al. (2019a), in their work, reported that the maximum noted volumetric soil water contents values were 40%, 43% and 35%, and the minimum water content values were 30%, 25% and 23% in the BC1, BC2, and control treatment, in respectively. In general, the soil matric potential in the biochar treatments was 0.002 MPa higher than control (without biochar) treatment.

Figure 4. Effect of Biochar application on maize growth under deficit irrigation condition.

Because biochar decrease bulk density, it can increase soil water infiltration. And water can move easily in soil. Biochar was amend to soil for the purpose of improving soil hydraulic conductivity (Novak et al., 2016). Yu et al. (2013) were reported that biochar applications at different rates (between 0% and 10%) increase the water holding capacity linearly, but KARADERE reported that biochar application higher than 2% in their study under field conditions may adversely affect plant root growth.

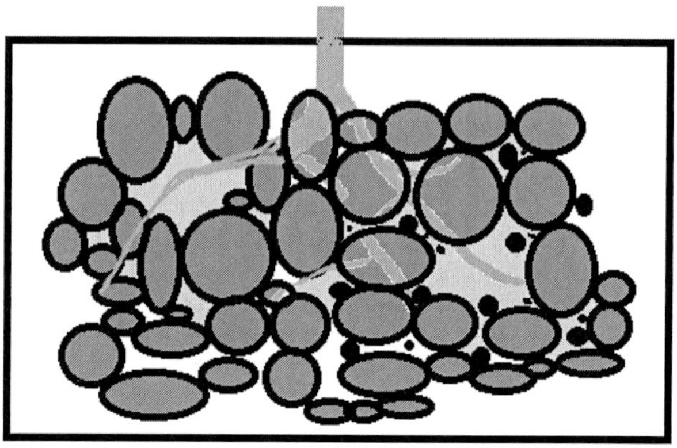

Effect of Mycorrhizal Fungi on Water Deficit Tolerance

Mycorrhizas have amongst the most well-known and antique symbioses on word. These plant–fungus associations improve nutrition and water relations of plants (Johnson and Jansa, 2017). There are so many benefit of mycorrhiza like as; improve carbon storage (Jevon and Lang, 2022), maintaining plant production and reducing pathogens (da Silva Campos, 2020) and enhancing abiotic stress tolerance (Malhi et al., 2021).

Drought stress has vital limiting factors in agriculture and mycorrhizal colonization increase plant drought tolerance. There are many studies showing that mycorrhiza increases plant resistance to water stress. However, the water uptake and transport mechanism of mycorrhiza to plant tissues is still a subject of research.

Physiological mechanisms of AM fungi involved in drought tolerance are studied. Although few studies have been conducted on molecular levels, little work has been done around regulatory genes, functional genes, and transcription factors, which is not enough to decipher the molecular mechanisms of mycorrhizal enhancement of plant drought tolerance. In short, AM fungi modulate various functional and regulatory genes involved in enhancing drought tolerance of host plants (Wang et al., 2023).

AMF contribute to the alleviation of numerous abiotic (like drought and heavy metal) and biotic (like disease) plant stresses. Especially, the symbiotic relationship between AMF and the roots of plants aids plants to deal with drought stress (Leventis et al., 2021). This has been shown for a number of fungal species and host plants (Moradtalab et al., 2019). The capability of AMF-colonized plants to tolerate drought effectively is based on both direct and indirect mechanisms. Water uptake and transport by AMF mycelium, in addition to nutritional and physiological benefits induced by mycorrhization have been shown (Leventis et al., 2021; Smith and Smith, 2011). Because of their small diameter (between 2 and 20 μm), their high surface-volume ratio and their capacity for rapid exponential growth, fungal hyphae have access to soil pores inaccessible to plant roots and root hairs, resulting in more efficient water extraction by mycorrhizal than by non-mycorrhizal plants. Furthermore AMF change root morphology, mainly by promoting lateral root production (Gutjahr and Paszkowski, 2013). Physiological effects include modifications of foliar water relationship characteristics (Wu et al., 2007), adjustments in root-to-shoot signaling and hormone production, biosynthesis of secondary metabolites, as strigolactones that act as cues in establishing and maintaining AMF symbiosis with the host plant (Leventis et al., 2021; Stassen et al., 2021) and regulation of antioxidative enzymes (Rivero et al., 2018).

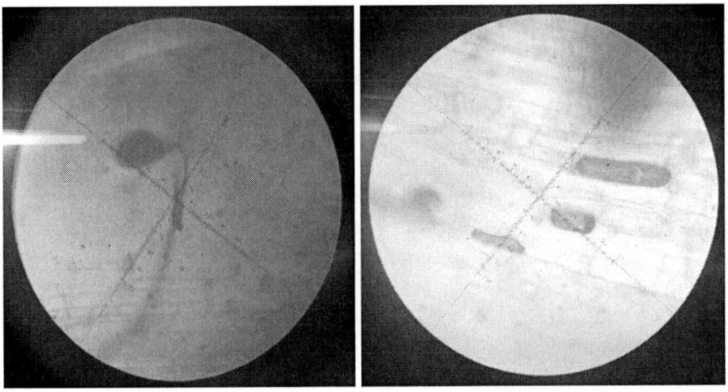

Figure 5. A Picture from Mycorrhizal Infection

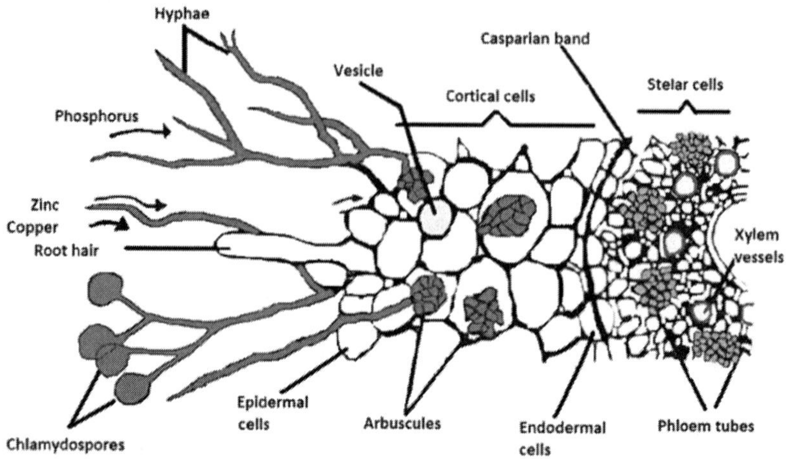

Figure 6. Diagrammatic representation of a cross-section of a vesicular-arbuscular mycorrhiza (VAM) root (Maiti and Ghosh, 2020).

Cover Crop as a source of Water conservation

An agricultural production with high inputs and exploitation of the land. If the soil surface is bare, we will have problems with erosion and water storage, we must do grazing and agriculture in a planned and sufficient way. Cover crops are another effective soil conservation practice to reduce runoff and water erosion. Cover crops are not a new practice, but there is renewed interest in using cover crops to conserve soil and improve soil productivity (Baumhardt and Blanco-Canqui, 2014). Also, these crops provide many other ecosystem services including improvement in soil fertility, biological diversity, soil properties and water quality. Cover crops perform a dual function because they protect the soil surface and improve soil properties.

On the other hand, cover crops can contend with the trees for nutrients and water, particularly in new orchards or where cover crops grow along tree rows which host detrimental organisms and other pests, such as rodents (Fogliatto et al., 2020). Moreover, the bare soil surface can lose water with the wind, but thanks to the cover crops, the soil can store water. Furthermore, it may be an option for water loss in windbreaks.

Cereal cover crops are extremely effective at reducing NO3− leaching and water-dependent emissions of N_2O compared to legumes (Basche et al., 2014). Findings on the net effect of crop rotation on N_2O emissions are mixed, with increases in N_2O emissions reported for addition of soybean to continuous corn but reductions in global warming potential and total greenhouse gas intensity (yield-scaled global warming potential) reported for diversification of small grain systems with legumes, differences that were attributed to soil-plant N cycling shifts (Renwick et al., 2019).

Deficit Irrigation

Reduce Deficit Irrigation (RDI) was developed in the 1980s as a strategy to decrease growth of vigorous trees and to save water (Chalmers et al., 1981). Deficit irrigation (DI) strategies are becoming popular in the areas where water supply is inadequate under irregular climate situations (Asrey et al., 2018; Fereres and Soriano, 2007), thereby increasing water use efficiency (WUE) (Sidhu et al., 2021). Furthermore, the DI strategies offer great chances for saving water without cooperating production. Some researchers have reported water savings from 43% to 65% under RDI strategy with a minor reduction in yield, but with advanced quality of product (Asrey et al., 2018).

The product grown with this method is subjected to water stress at certain levels at certain periods or throughout the entire development, and ensures effective use of water (Bekele and Tilahun, 2007).. Expectation in regular restricted irrigation; Instead of losses in irrigation conditions in field capacity, it is to take the best possible yield from the plant under water stress and save the water resources for the next crop production (Eck et al., 1987). Especially, numerous crops have different sensitivities to drought stress at various stages of development, and the DI program must be designed to manage the stress that yield drop is minimized. Information regarding the crop response to DI is essential to achieve such objectives when water is limited (Sidhu et al., 2021).

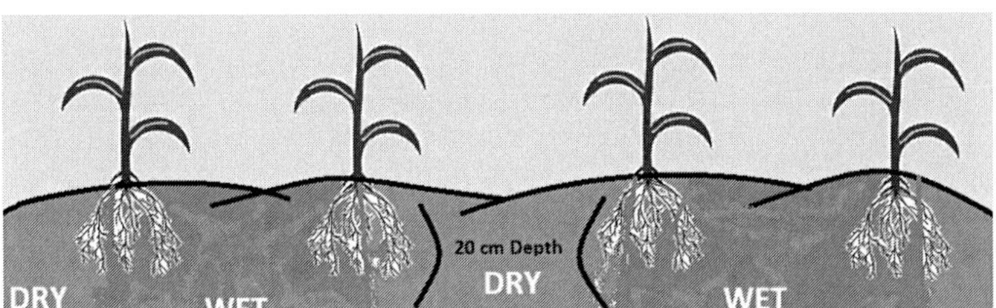

Climate Smart Agriculture (CSA)

Climate-smart agriculture (CSA) is an approach to support agricultural systems worldwide, concurrently addressing three challenge areas: increased adaptation to climate change, mitigation of climate change, and ensuring global food security – through innovative policies, practices, and financing (Torquebiau et al., 2018). CSA is a very new topic and developing countries are insufficient in this regard. It can help that carbon-smart, water-smart, energy-smart, and knowledge-smart technologies and practices meaningfully contributed directly or indirectly to advance productivity, improve resilience and decrease Green Gas Ememission (Imran et al., 2019). World Bank suggested CSA (Worldbank, 2021) with great money to reduce GHG Emission and food security.

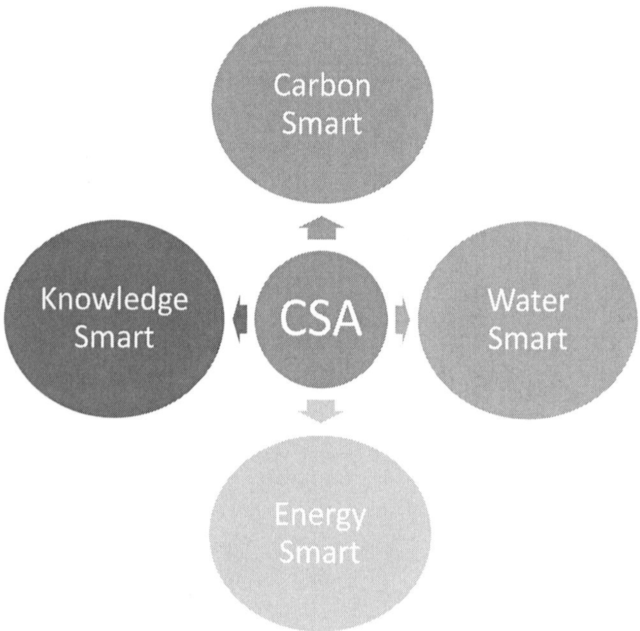

Figure 7. CSA can help that carbon-smart, water-smart, energy-smart, and knowledge-smart technologies

Low-cost sustainable agriculture practices such as conservation agriculture, agro-ecology, ecosystem-based management, small-scale irrigation, agroforestry, soil/water conservation, and grazing land management have been implemented for decades (Chandra et al., 2018).

Conclusion

As it is known, the global climate crisis, as well as natural disasters (such as floods, floods, rise in sea levels, earthquakes, fires and drought), decreases our water resources day by day. We need to store our water resources in the soil and increase the efficiency of water use as sustainable dryland management. In this context, mycorrhizal fungus can be used to reduce plant water stress, as well as organic wastes such as biochar, which have a large surface area and are stable in the soil. In addition, limited irrigation technique can be used in arid and semi-arid regions. Cover crops can also reduce losses by covering the soil surface in agricultural production, increasing soil carbon and water storage. Moreover, the CSA technique, which is a new technique, should be increased and expanded.

References

1. Anonim, 2014. https://www.bloomberght.com/analiz/haber/1601593-analiz-dunyanin-basi-kuraklik-tehlikesi-ile-belada.
   Erişim Tarihi: 25.01.2023
2. Aguégué, M. R., Ahoyo Adjovi, N. R., Agbodjato, N. A., Noumavo, P. A., Assogba, S., Salami, H., Salako, V. K., Ramón, R., Baba-Moussa, F., and Adjanohoun, A. (2021). Efficacy of native strains of arbuscular mycorrhizal fungi on maize productivity on ferralitic soil in Benin. *Agricultural Research*, 1-15.
3. Akpınar, Ç., Demirbaş, A., and Ortaş, İ. (2020). The Effects of Different Mycorrhiza Species, Compost and Rock Phosphate Applications on the Growth of Sorghum Plant. *Turkish Journal of Agriculture-Food Science and Technology* **8**, 2583-2589.
4. Amir, H., Gensous, S., Cavaloc, Y., and Wantiez, L. (2021). Phosphorus Fertilization of an Ultramafic Soil Reduced Effects of Arbuscular Mycorrhizal Fungi but not Mycorrhizal Colonization. *Journal of soil science and plant nutrition*, 1-11.
5. Anli, M., Symanczik, S., El Abbassi, A., Ait-El-Mokhtar, M., Boutasknit, A., Ben-Laouane, R., Toubali, S., Baslam, M., Mäder, P., and Hafidi, M. (2021). Use of arbuscular mycorrhizal fungus Rhizoglomus irregulare and compost to improve growth and physiological responses of Phoenix dactylifera 'Boufgouss'. *Plant*

Biosystems-An International Journal Dealing with all Aspects of Plant Biology **155**, 763-771.

6. Arruda, B., Herrera, W. F. B., Rojas-García, J. C., Turner, C., and Pavinato, P. S. (2021). Cover crop species and mycorrhizal colonization on soil phosphorus dynamics. *Rhizosphere* **19**, 100396.
7. Asrey, R., Kumar, S., and Meena, N. K. (2018). Influence of water quality on postharvest fruit and vegetable quality. *In* "Preharvest modulation of postharvest fruit and vegetable quality", pp. 169-187. Elsevier.
8. Atteya, A. K., Sami, R., Al-Mushhin, A. A., Ismail, K. A., and Genaidy, E. A. (2021). Response of seeds, oil yield and fatty acids percentage of Jojoba Shrub strain EAI to mycorrhizal fungi and moringa leaves extract. *Horticulturae* **7**, 395.
9. Awan, S., Ippolito, J. A., Ullman, J., Ansari, K., Cui, L., and Siyal, A. (2021). Biochars reduce irrigation water sodium adsorption ratio. *Biochar* **3**, 77-87.
10. Aznar-Sanchez, J. A., Piquer-Rodriguez, M., Velasco-Munoz, J. F., and Manzano-Agugliaro, F. (2019). Worldwide research trends on sustainable land use in agriculture. *Land use policy* **87**, 104069.
11. Bai, S. H., Omidvar, N., Gallart, M., Kämper, W., Tahmasbian, I., Farrar, M. B., Singh, K., Zhou, G., Muqadass, B., and Xu, C.-Y. (2022). Combined effects of biochar and fertilizer applications on yield: A review and meta-analysis. *Science of The Total Environment* **808**, 152073.
12. Basche, A. D., Miguez, F. E., Kaspar, T. C., and Castellano, M. J. (2014). Do cover crops increase or decrease nitrous oxide emissions? A meta-analysis. *Journal of Soil and Water Conservation* **69**, 471-482.
13. Bast, A., Wilcke, W., Graf, F., Lüscher, P., and Gärtner, H. (2014). The use of mycorrhiza for eco-engineering measures in steep alpine environments: effects on soil aggregate formation and fine-root development. *Earth surface processes and landforms* **39**, 1753-1763.
14. Bastami, A., Amirnia, R., Sayyed, R., and Enshasy, H. A. E. (2021). The Effect of Mycorrhizal Fungi and Organic Fertilizers on Quantitative and Qualitative Traits of Two Important Satureja Species. *Agronomy* **11**, 1285.
15. Baumhardt, R. L., and Blanco-Canqui, H. (2014). Soil: Conservation Practices. *In* "Encyclopedia of Agriculture and Food Systems" (N. K. Van Alfen, ed.), pp. 153-165. Academic Press, Oxford.
16. Bekele, S., and Tilahun, K. (2007). Regulated deficit irrigation scheduling of onion in a semiarid region of Ethiopia. *Agricultural water management* **89**, 148-152.

17. Bencherif, K., Laruelle, F., Dalpé, Y., and Lounès-Hadj Sahraoui, A. (2021). Inoculum Sources Modulate Mycorrhizal Inoculation Effect on Tamarix articulata Development and Its Associated Rhizosphere Microbiota. *Plants* **10**, 2716.
18. Bhantana, P., Rana, M. S., Sun, X.-c., Moussa, M. G., Saleem, M. H., Syaifudin, M., Shah, A., Poudel, A., Pun, A. B., and Bhat, M. A. (2021). Arbuscular mycorrhizal fungi and its major role in plant growth, zinc nutrition, phosphorous regulation and phytoremediation. *Symbiosis* **84**, 19-37.
19. Bharali, A., Baruah, K. K., Bhattacharyya, P., and Gorh, D. (2017). Integrated nutrient management in wheat grown in a northeast India soil: impacts on soil organic carbon fractions in relation to grain yield. *Soil and Tillage Research* **168**, 81-91.
20. Bikbulatova, S., Tahmasebi, A., Zhang, Z., Rish, S. K., and Yu, J. (2018). Understanding water retention behavior and mechanism in bio-char. *Fuel Processing Technology* **169**, 101-111.
21. Boretti, A., and Rosa, L. (2019). Reassessing the projections of the world water development report. *NPJ Clean Water* **2**, 15.
22. Bunn, S. E. (2016). Grand challenge for the future of freshwater ecosystems. Vol. 4, pp. 21. Frontiers Media SA.
23. Castañeda-Gómez, L., Powell, J. R., Ellsworth, D. S., Pendall, E., and Carrillo, Y. (2021). The influence of roots on mycorrhizal fungi, saprotrophic microbes and carbon dynamics in a low-phosphorus Eucalyptus forest under elevated CO2. *Functional Ecology* **35**, 2056-2071.
24. Çekiç, F. Ö., Ünyayar, S., and Ortaş, İ. (2012). Effects of arbuscular mycorrhizal inoculation on biochemical parameters in Capsicum annuum grown under long term salt stress. *Turkish Journal of Botany* **36**, 63-72.
25. Celik, I., Ortas, I., and Kilic, S. (2004). Effects of compost, mycorrhiza, manure and fertilizer on some physical properties of a Chromoxerert soil. *Soil and Tillage Research* **78**, 59-67.
26. Chalmers, D., Mitchell, P., and Van Heek, L. (1981). Control of Peach Tree Growth and Productivity by Regulated Water Supply, Tree Density, and Summer Pruning1. *Journal of the American Society for Horticultural Science* **106**, 307-312.
27. Chandra, A., McNamara, K. E., and Dargusch, P. (2018). Climate-smart agriculture: perspectives and framings. *Climate Policy* **18**, 526-541.
28. Chen, X., Yang, S., Ding, J., Jiang, Z., and Sun, X. (2021). Effects of biochar addition on rice growth and yield under water-saving irrigation. *Water* **13**, 209.

29. Cheng, H.-Q., Giri, B., Wu, Q.-S., Zou, Y.-N., and Kuča, K. (2021). Arbuscular mycorrhizal fungi mitigate drought stress in citrus by modulating root microenvironment. *Archives of Agronomy and Soil Science*, 1-12.
30. Chu, Q., Zhang, L., Zhou, J., Lixing, Y., Fanjun, C., Fusuo, Z., Gu, F., and Zed, R. (2020). Soil plant-available phosphorus levels and maize genotypes determine the phosphorus acquisition efficiency and contribution of mycorrhizal pathway. *Plant and soil* **449**, 357-371.
31. Churchman, G. J., Singh, M., Schapel, A., Sarkar, B., and Bolan, N. (2020). CLAY MINERALS AS THE KEY TO THE SEQUESTRATION OF CARBON IN SOILS. *Clays and Clay Minerals* **68**, 135-143.
32. da Silva Campos, M. A. (2020). Bioprotection by arbuscular mycorrhizal fungi in plants infected with Meloidogyne nematodes: A sustainable alternative. *Crop Protection* **135**, 105203.
33. De Oliveira, A. C., Marini, N., and Farias, D. (2014). Climate change: New breeding pressures and goals.
34. Demirbaş, A., Kaya, Z., Akpınar, Ç., and Ortaş, İ. (2019). The Effects of Applications of Fertigation and Mycorrhiza on Yield and Nutrient Uptake of Pepper Plant (Capsicum annum L.) under Field Conditions. *Turkish Journal of Agriculture-Food Science and Technology* **7**, 152-161.
35. Deveautour, C., Donn, S., Bennett, A. E., Power, S., and Powell, J. R. (2021). Variability of arbuscular mycorrhizal fungal communities within the root systems of individual plants is high and influenced by host species and root phosphorus. *Pedobiologia* **84**, 150691.
36. Eck, H., Mathers, A., and Musick, J. (1987). Plant water stress at various growth stages and growth and yield of soybeans. *Field Crops Research* **17**, 1-16.
37. Edwards, C. A. (1998). The use of earthworms in the breakdown and management of organic wastes. In: Edwards, C.A. (Ed.), Earth-worm Ecology. . *CRC Press, Boca Raton, FL, pp.* , 327±354.
38. Edwards, C. A., and Burrows, I. (1988). The potential of earthworm compost as plant growth media. In: Edwards, C.A., Neuhauser, E.F. (Eds.), Earthworms in Waste and Environmental Management *SPB Academic Publishers, The Netherlands,pp.* , 21-32.
39. Edwards, C. A., Burrows, I., Fletcher, K. E., and Jones, B. A. (1985). The use of earthworms for composting farm wastes. In: Gasser, J.K.R. (Ed.), Com-posting Agricultural and Other Wastes *Elsevier, London and New York, pp*, 229-241.

40. Fao, N. (2008). A review of drought occurrence and monitoring and planning activities in the Near East Region. *FAO, Rome.*
41. Fereres, E., and Soriano, M. A. (2007). Deficit irrigation for reducing agricultural water use. *Journal of experimental botany* **58**, 147-159.
42. Fischer, B. M., Manzoni, S., Morillas, L., Garcia, M., Johnson, M. S., and Lyon, S. W. (2019a). Improving agricultural water use efficiency with biochar–A synthesis of biochar effects on water storage and fluxes across scales. *Science of the Total Environment* **657**, 853-862.
43. Fischer, B. M. C., Manzoni, S., Morillas, L., Garcia, M., Johnson, M. S., and Lyon, S. W. (2019b). Improving agricultural water use efficiency with biochar – A synthesis of biochar effects on water storage and fluxes across scales. *Science of The Total Environment* **657**, 853-862.
44. Fogliatto, S., Ferrero, A., and Vidotto, F. (2020). Chapter Six - Current and future scenarios of glyphosate use in Europe: Are there alternatives? *In* "Advances in Agronomy" (D. L. Sparks, ed.), Vol. 163, pp. 219-278. Academic Press.
45. Gattinger, A., Muller, A., Haeni, M., Skinner, C., Fliessbach, A., Buchmann, N., Mader, P., Stolze, M., Smith, P., Scialabba Nel, H., and Niggli, U. (2013). Reply to Leifeld et al.: Enhanced top soil carbon stocks under organic farming is not equated with climate change mitigation. *Proc Natl Acad Sci U S A* **110**, E985.
46. Ghodake, G. S., Shinde, S. K., Kadam, A. A., Saratale, R. G., Saratale, G. D., Kumar, M., Palem, R. R., Al-Shwaiman, H. A., Elgorban, A. M., Syed, A., and Kim, D.-Y. (2021). Review on biomass feedstocks, pyrolysis mechanism and physicochemical properties of biochar: State-of-the-art framework to speed up vision of circular bioeconomy. *Journal of Cleaner Production* **297**, 126645.
47. Githinji, L. (2014). Effect of biochar application rate on soil physical and hydraulic properties of a sandy loam. *Archives of Agronomy and Soil Science* **60**, 457-470.
48. Golchin, A., Oades, J. M., Skjemstad, J. O., and Clarke, P. (1994). Soil-structure and carbon cycling. *Australian Journal of Soil Research* **32**, 1043-1068.
49. Graber, E., Tsechansky, L., Gerstl, Z., and Lew, B. (2012). High surface area biochar negatively impacts herbicide efficacy. *Plant and Soil* **353**, 95-106.
50. Gujre, N., Soni, A., Rangan, L., Tsang, D. C., and Mitra, S. (2021). Sustainable improvement of soil health utilizing biochar and arbuscular mycorrhizal fungi: A review. *Environmental pollution* **268**, 115549.

51. Gull, M., Hafeez, F., Saleem, M., and Malik, K. (2004). Phosphorus uptake and growth promotion of chickpea by co-inoculation of mineral phosphate solubilising bacteria and a mixed rhizobial culture. *Australian Journal of Experimental Agriculture* **44**, 623-628.
52. Gupta, S., and Su, R. (2022). An efficient differential evolution with fitness-based dynamic mutation strategy and control parameters. *Knowledge-Based Systems* **251**, 109280.
53. Gupta, S., Thokchom, S. D., and Kapoor, R. (2021). Arbuscular mycorrhiza improves photosynthesis and restores alteration in sugar metabolism in Triticum aestivum L. grown in arsenic contaminated soil. *Frontiers in plant science* **12**, 334.
54. Gutjahr, C., and Paszkowski, U. (2013). Multiple control levels of root system remodeling in arbuscular mycorrhizal symbiosis. *Frontiers in plant science* **4**, 204.
55. Harb, A., Krishnan, A., Ambavaram, M. M., and Pereira, A. (2010). Molecular and physiological analysis of drought stress in Arabidopsis reveals early responses leading to acclimation in plant growth. *Plant physiology* **154**, 1254-1271.
56. Hazaymeh, K., and Hassan, Q. K. (2016). Remote sensing of agricultural drought monitoring: A state of art review. *AIMS Environmental Science* **3**, 604-630.
57. He, Y., Yang, R., Lei, G., Li, M., Li, T., Zhan, F., and Li, Y. (2021). Arbuscular mycorrhizal fungus–induced decrease in phosphorus loss due to leaching in red soils under simulated heavy rainfall. *Journal of Soils and Sediments* **21**, 881-889.
58. Hirayama, T., and Shinozaki, K. (2010). Research on plant abiotic stress responses in the post-genome era: past, present and future. *The plant journal* **61**, 1041-1052.
59. Hou, L., Zhang, X., Feng, G., Li, Z., Zhang, Y., and Cao, N. (2021). Arbuscular mycorrhizal enhancement of phosphorus uptake and yields of maize under high planting density in the black soil region of China. *Scientific reports* **11**, 1-11.
60. Hussain, S., Sharif, M., and Ahmad, W. (2021). Selection of efficient phosphorus solubilizing bacteria strains and mycorrhizea for enhanced cereal growth, root microbe status and N and P uptake in alkaline calcareous soil. *Soil Science and Plant Nutrition*, 1-10.
61. Imran, M. A., Ali, A., Ashfaq, M., Hassan, S., Culas, R., and Ma, C. (2019). Impact of climate smart agriculture (CSA) through sustainable irrigation management on Resource use efficiency: A sustainable production alternative for cotton. *Land Use Policy* **88**, 104113.

62. Jabborova, D., Annapurna, K., Paul, S., Kumar, S., Saad, H. A., Desouky, S., Ibrahim, M. F., and Elkelish, A. (2021). Beneficial features of biochar and arbuscular mycorrhiza for improving spinach plant growth, root morphological traits, physiological properties, and soil enzymatic activities. *Journal of Fungi* **7**, 571.
63. Jansa, J., Mozafar, A., Kuhn, G., Anken, T., Ruh, R., Sanders, I., and Frossard, E. (2003). Soil tillage affects the community structure of mycorrhizal fungi in maize roots. *Ecological Applications* **13**, 1164-1176.
64. Jastrow, J. D., Miller, R. M., and Lussenhop, J. (1998). Contributions of interacting biological mechanisms to soil aggregate stabilization in restored prairie. *Soil Biology & Biochemistry* **30**, 905-916.
65. Jevon, F. V., and Lang, A. K. (2022). Tree biomass allocation differs by mycorrhizal association. *Ecology* **103**, e3688.
66. Jiang, F., Zhang, L., Zhou, J., George, T. S., and Feng, G. (2021). Arbuscular mycorrhizal fungi enhance mineralisation of organic phosphorus by carrying bacteria along their extraradical hyphae. *New phytologist* **230**, 304-315.
67. Johnson, N. C., and Jansa, J. (2017). Mycorrhizas: at the interface of biological, soil, and earth sciences. *In* "Mycorrhizal Mediation of Soil", pp. 1-6. Elsevier.
68. Khresat, S., Al-Bakri, J., and Al-Tahhan, R. (2008). Impacts of land use/cover change on soil properties in the Mediterranean region of northwestern Jordan. *Land Degradation & Development* **19**, 397-407.
69. Kuyper, T. W., Wang, X., and Muchane, M. N. (2021). The interplay between roots and arbuscular mycorrhizal fungi influencing water and nutrient acquisition and use efficiency. *The Root Systems in Sustainable Agricultural Intensification*, 193-220.
70. Lal, R. (2010). Managing Soils and Ecosystems for Mitigating Anthropogenic Carbon Emissions and Advancing Global Food Security. *Bioscience* **60**, 708-721.
71. Lal, R. (2019). Carbon Cycling in Global Drylands. *Current Climate Change Reports* **5**, 221-232.
72. Larionova, A. A., Kurganova, I. N., de Gerenyu, V. O. L., Zolotareva, B. N., Yevdokimov, I. V., and Kudeyarov, V. N. (2010). Carbon dioxide emissions from agrogray soils under climate changes. *Eurasian Soil Science* **43**, 168-176.
73. Lehmann, J., Kuzyakov, Y., Pan, G., and Ok, Y. S. (2015). Biochars and the plant-soil interface. Springer.
74. Leventis, G., Tsiknia, M., Feka, M., Ladikou, E. V., Papadakis, I. E., Chatzipavlidis, I., Papadopoulou, K., and Ehaliotis, C. (2021).

Arbuscular mycorrhizal fungi enhance growth of tomato under normal and drought conditions, via different water regulation mechanisms. *Rhizosphere* **19**, 100394.
75. Liu, M., Shen, Y., Li, Q., Xiao, W., and Song, X. (2021a). Arbuscular mycorrhizal fungal colonization and soil pH induced by nitrogen and phosphorus additions affects leaf C: N: P stoichiometry in chinese fir (Cunninghamia lanceolata) forests. *Plant and soil* **461**, 421-440.
76. Liu, Y., Zhang, G., Luo, X., Hou, E., Zheng, M., Zhang, L., He, X., Shen, W., and Wen, D. (2021b). Mycorrhizal fungi and phosphatase involvement in rhizosphere phosphorus transformations improves plant nutrition during subtropical forest succession. *Soil Biology and Biochemistry* **153**, 108099.
77. Lourdes, G.-C. M., Stéphane, D., and Maryline, C.-S. (2021). Impact of increasing chromium (VI) concentrations on growth, phosphorus and chromium uptake of maize plants associated to the mycorrhizal fungus Rhizophagus irregularis MUCL 41833. *Heliyon* **7**, e05891.
78. Ma, X., Li, X., and Ludewig, U. (2021). Arbuscular mycorrhizal colonization outcompetes root hairs in maize under low phosphorus availability. *Annals of Botany* **127**, 155-166.
79. Maiti, S. K., and Ghosh, D. (2020). Plant–soil interactions as a restoration tool. *In* "Climate change and soil interactions", pp. 689-730. Elsevier.
80. Maitra, P., Zheng, Y., Wang, Y.-L., Mandal, D., Lü, P.-P., Gao, C., Babalola, B. J., Ji, N.-N., Li, X.-C., and Guo, L.-D. (2021). Phosphorus fertilization rather than nitrogen fertilization, growing season and plant successional stage structures arbuscular mycorrhizal fungal community in a subtropical forest. *Biology and Fertility of Soils* **57**, 685-697.
81. Malhi, G. S., Kaur, M., Kaushik, P., Alyemeni, M. N., Alsahli, A. A., and Ahmad, P. (2021). Arbuscular mycorrhiza in combating abiotic stresses in vegetables: An eco-friendly approach. *Saudi Journal of Biological Sciences* **28**, 1465-1476.
82. Manna, M., Thakur, T., Chirom, O., Mandlik, R., Deshmukh, R., and Salvi, P. (2021). Transcription factors as key molecular target to strengthen the drought stress tolerance in plants. *Physiologia Plantarum* **172**, 847-868.
83. Marshall, J., Muhlack, R., Morton, B. J., Dunnigan, L., Chittleborough, D., and Kwong, C. W. (2019). Pyrolysis temperature effects on biochar–Water interactions and application for improved water holding capacity in vineyard soils. *Soil Systems* **3**, 27.

84. Martínez-Viveros, O., Jorquera, M., Crowley, D., Gajardo, G., and Mora, M. (2010). Mechanisms and practical considerations involved in plant growth promotion by rhizobacteria. *Journal of soil science and plant nutrition* **10**, 293-319.
85. Metwally, R. A., Soliman, S. A., Latef, A. A. H. A., and Abdelhameed, R. E. (2021). The Individual and interactive role of arbuscular mycorrhizal fungi and Trichoderma viride on growth, protein content, amino acids fractionation, and phosphatases enzyme activities of onion plants amended with fish waste. *Ecotoxicology and Environmental Safety* **214**, 112072.
86. Meza, I., Siebert, S., Döll, P., Kusche, J., Herbert, C., Eyshi Rezaei, E., Nouri, H., Gerdener, H., Popat, E., and Frischen, J. (2020). Global-scale drought risk assessment for agricultural systems. *Natural Hazards and Earth System Sciences* **20**, 695-712.
87. Mishra, A. K., and Singh, V. P. (2010). A review of drought concepts. *Journal of hydrology* **391**, 202-216.
88. Moradtalab, N., Hajiboland, R., Aliasgharzad, N., Hartmann, T. E., and Neumann, G. (2019). Silicon and the association with an arbuscular-mycorrhizal fungus (Rhizophagus clarus) mitigate the adverse effects of drought stress on strawberry. *Agronomy* **9**, 41.
89. Nouri, E., Surve, R., Bapaume, L., Stumpe, M., Chen, M., Zhang, Y., Ruyter-Spira, C., Bouwmeester, H., Glauser, G., and Bruisson, S. (2021). Phosphate Suppression of Arbuscular Mycorrhizal Symbiosis Involves Gibberellic Acid Signalling. *Plant and Cell Physiology*.
90. Novak, J., Sigua, G., Watts, D., Cantrell, K., Shumaker, P., Szogi, A., Johnson, M. G., and Spokas, K. (2016). Biochars impact on water infiltration and water quality through a compacted subsoil layer. *Chemosphere* **142**, 160-167.
91. Orozco, F. H., Cegarra, J., Trujillo, L. M., and Roig, A. (1996). Vermicomposting of coffee pulp using the earthwormEisenia fetida: effects on C and N contents and the availability of nutrients *Biol Fertil Soils* **22** 162-166.
92. Ortas, I. (2019). Under filed conditions, mycorrhizal inoculum effectiveness depends on plant species and phosphorus nutrition. *Journal of plant nutrition* **42**, 2349-2362.
93. Ortas, I. (2020). "Mycorrhizas in fruit nutrition: Important breakthroughs."
94. Ortaş, İ. (2022). "Türkiyenin Gıda Güvencesi Güvende mi?," Akademisyen Kitabevi.

95. Ortas, I., Akpinar, C., and Lal, R. (2013). Long-term impacts of organic and inorganic fertilizers on carbon sequestration in aggregates of an Entisol in Mediterranean Turkey. *Soil Science* **178**, 12-23.
96. Ortas, I., Bulutekin, B., and Öztürk, F. (2019). RHIZOSPHERE AND NON-RHIZOSPHERE SOIL STRUCTURE DEVELOPMENT UNDER PHOSPHORUS FERTILIZATION APPLICATION. *In* "Фундаментальные концепции физики почв: развитие, современные приложения и перспективы", pp. 369-379.
97. Ortas, I., and Bykova, A. (2020). Effects of long-term phosphorus fertilizer applications on soil carbon and CO(2)flux. *Communications in Soil Science and Plant Analysis* **51**, 2270-2279.
98. Ortaş, I., Lal, R., and Kapur, S. (2017a). Carbon Sequestration and Mycorrhizae in Turkish Soils. *In* "Carbon Management, Technologies, and Trends in Mediterranean Ecosystems", pp. 139-149. Springer.
99. Ortaş, I., and Rafique, M. (2017). The mechanisms of nutrient uptake by arbuscular mycorrhizae. *In* "Mycorrhiza-Nutrient Uptake, Biocontrol, Ecorestoration", pp. 1-19. Springer.
100. Ortaş, I., Rafique, M., and Ahmed, İ. A. (2017b). Application of arbuscular mycorrhizal fungi into agriculture. *In* "Arbuscular mycorrhizas and stress tolerance of plants", pp. 305-327. Springer.
101. Ortas, İ., Rafique, M., Akpinar, C., and Kacar, Y. A. (2017). Growth media and mycorrhizal species effect on acclimatization and nutrient uptake of banana plantlets. *Scientia Horticulturae* **217**, 55-60.
102. Ortas, I., Rafique, M., and Çekiç, F. (2021). Do Mycorrhizal Fungi Enable Plants to Cope with Abiotic Stresses by Overcoming the Detrimental Effects of Salinity and Improving Drought Tolerance? *In* "Symbiotic Soil Microorganisms", pp. 391-428. Springer.
103. Ortaş, I., Rafique, M., and Iqbal, M. T. (2019). Mycorrhizae Resource Allocation in Root Development and Root Morphology. *In* "Plant Microbe Interface", pp. 1-26. Springer.
104. Ortaş, I., Razzaghi, S., and Rafique, M. (2016). Arbuscular mycorrhizae: Effect of rhizosphere and relation with carbon nutrition. *In* "Plant-Microbe Interaction: An Approach to Sustainable Agriculture" (D. Choudhary, A. Varma and N. Tuteja, eds.), pp. 125-152. Springer, Singapore.
105. Ortas, I., Sari, N., Akpinar, Ç., and Yetisir, H. (2011). Screening mycorrhiza species for plant growth, P and Zn uptake in pepper seedling grown under greenhouse conditions. *Scientia Horticulturae* **128**, 92-98.

106. Ortas, I., and Yucel, C. (2020). Do mycorrhizae influence cover crop biomass production? *Acta Agriculturae Scandinavica Section B-Soil and Plant Science* **70**, 657-666.
107. Ouda, S., Zohry, A. E.-H., Noreldin, T., Ouda, S., and Zohry, A. E.-H. (2020). Water scarcity leads to food insecurity. *Deficit Irrigation: A Remedy for Water Scarcity*, 1-13.
108. Palacios, Y. M., Gleadow, R., Davidson, C., Gan, W., and Winfrey, B. (2021). Do mycorrhizae increase plant growth and pollutant removal in stormwater biofilters? *Water Research*, 117381.
109. Poveda, J., Hermosa, R., Monte, E., and Nicolás, C. (2019). Trichoderma harzianum favours the access of arbuscular mycorrhizal fungi to non-host Brassicaceae roots and increases plant productivity. *Scientific reports* **9**, 1-11.
110. Qu, L., Wang, M., and Biere, A. (2021). Interactive Effects of Mycorrhizae, Soil Phosphorus, and Light on Growth and Induction and Priming of Defense in Plantago lanceolata. *Frontiers in plant science* **12**.
111. Quan, L., Zhang, J., Wei, Q., Wang, Y., Qin, C., Hu, F., Chen, Y., Shen, Z., and Xia, Y. (2021). Promotion of zinc tolerance, acquisition and translocation of phosphorus in Mimosa pudica L. mediated by arbuscular mycorrhizal fungi. *Bulletin of Environmental Contamination and Toxicology* **106**, 507-515.
112. Rafique, M., Ortas, I., Rizwan, M., Sultan, T., Chaudhary, H. J., Işik, M., and Aydin, O. (2019). Effects of Rhizophagus clarus and biochar on growth, photosynthesis, nutrients, and cadmium (Cd) concentration of maize (Zea mays) grown in Cd-spiked soil. *Environmental Science and Pollution Research* **26**, 20689-20700.
113. Renwick, L. L. R., Bowles, T. M., Deen, W., and Gaudin, A. C. M. (2019). Chapter 4 - Potential of Increased Temporal Crop Diversity to Improve Resource Use Efficiencies: Exploiting Water and Nitrogen Linkages. *In* "Agroecosystem Diversity" (G. Lemaire, P. C. D. F. Carvalho, S. Kronberg and S. Recous, eds.), pp. 55-73. Academic Press.
114. Riaz, M., Kamran, M., Fang, Y., Wang, Q., Cao, H., Yang, G., Deng, L., Wang, Y., Zhou, Y., and Anastopoulos, I. (2021). Arbuscular mycorrhizal fungi-induced mitigation of heavy metal phytotoxicity in metal contaminated soils: A critical review. *Journal of hazardous materials* **402**, 123919.
115. Rivero, J., Álvarez, D., Flors, V., Azcón-Aguilar, C., and Pozo, M. J. (2018). Root metabolic plasticity underlies functional diversity in

mycorrhiza-enhanced stress tolerance in tomato. *New phytologist* **220**, 1322-1336.
116. Roba, T. B. (2018). Review on: The Effect of Mixing Organic and Inorganic Fertilizer on Productivity and Soil Fertility. *Open Access Library Journal* **5**, 1.
117. Rodrigues, M. Â., Piroli, L. B., Forcelini, D., Raimundo, S., da Silva Domingues, L., Cassol, L. C., Correia, C. M., and Arrobas, M. (2021). Use of commercial mycorrhizal fungi in stress-free growing conditions of potted olive cuttings. *Scientia Horticulturae* **275**, 109712.
118. Romero, P., Navarro, J. M., and Ordaz, P. B. (2022). Towards a sustainable viticulture: The combination of deficit irrigation strategies and agroecological practices in Mediterranean vineyards. A review and update. *Agricultural Water Management* **259**, 107216.
119. Roy-Bolduc, A., and Hijri, M. (2011). The use of mycorrhizae to enhance phosphorus uptake: a way out the phosphorus crisis. *J. Biofertil. Biopestici* **2**, 1-5.
120. Ryan, J., Masri, S., Ibrikci, H., Singh, M., Pala, M., and Harris, H. C. (2008). Implications of cereal-based crop rotations, nitrogen fertilization, and stubble grazing on soil organic matter in a Mediterranean-type environment. *Turkish Journal of Agriculture and Forestry* **32**, 289-297.
121. Saboor, A., Ali, M. A., Hussain, S., El Enshasy, H. A., Hussain, S., Ahmed, N., Gafur, A., Sayyed, R., Fahad, S., and Danish, S. (2021). Zinc nutrition and arbuscular mycorrhizal symbiosis effects on maize (Zea mays L.) growth and productivity. *Saudi journal of biological sciences* **28**, 6339-6351.
122. Sagar, A., Rathore, P., Ramteke, P. W., Ramakrishna, W., Reddy, M. S., and Pecoraro, L. (2021). Plant Growth Promoting Rhizobacteria, Arbuscular Mycorrhizal Fungi and Their Synergistic Interactions to Counteract the Negative Effects of Saline Soil on Agriculture: Key Macromolecules and Mechanisms. *Microorganisms* **9**, 1491.
123. Sidhu, R. K., Kumar, R., Rana, P. S., and Jat, M. L. (2021). Chapter Five - Automation in drip irrigation for enhancing water use efficiency in cereal systems of South Asia: Status and prospects. *In* "Advances in Agronomy" (D. L. Sparks, ed.), Vol. 167, pp. 247-300. Academic Press.
124. Sigua, G., Novak, J., Watts, D., Johnson, M., and Spokas, K. (2016). Efficacies of designer biochars in improving biomass and nutrient uptake of winter wheat grown in a hard setting subsoil layer. *Chemosphere* **142**, 176-183.

125. Singh, M., Sarkar, B., Biswas, B., Churchman, J., and Bolan, N. S. (2016). Adsorption-desorption behavior of dissolved organic carbon by soil clay fractions of varying mineralogy. *Geoderma* **280**, 47-56.
126. Smith, S. E., and Smith, F. A. (2011). Roles of arbuscular mycorrhizas in plant nutrition and growth: new paradigms from cellular to ecosystem scales. *Annual review of plant biology* **62**, 227-250.
127. Solaiman, Z. M. (2014). Contribution of arbuscular mycorrhizal fungi to soil carbon sequestration. *In* "Mycorrhizal fungi: Use in sustainable agriculture and land restoration", pp. 287-296. Springer.
128. Sorensen, J., Larsen, J., and Jakobsen, I. (2008). Pre-inoculation with arbuscular mycorrhizal fungi increases early nutrient concentration and growth of field-grown leeks under high productivity conditions. *Plant and soil* **307**, 135-147.
129. Stassen, M. J., Hsu, S.-H., Pieterse, C. M., and Stringlis, I. A. (2021). Coumarin communication along the microbiome–root–shoot axis. *Trends in Plant Science* **26**, 169-183.
130. Streubel, J., Collins, H., Garcia-Perez, M., Tarara, J., Granatstein, D., and Kruger, C. (2011). Influence of contrasting biochar types on five soils at increasing rates of application. *Soil Science Society of America Journal* **75**, 1402-1413.
131. Subhan, F., Malik, A., Ul Haq, Z., and Khalil, T. M. (2021). Effect of deficit irrigation under different furrow irrigation techniques on cauliflower yield and water productivity in Mardan, Pakistan. *Sarhad Journal of Agriculture* **37**.
132. Tag, A. T., Duman, G., Ucar, S., and Yanik, J. (2016). Effects of feedstock type and pyrolysis temperature on potential applications of biochar. *Journal of analytical and applied pyrolysis* **120**, 200-206.
133. Tarafdar, M., Bahadur, V., Rana, S., and Singh, R. K. (2022). A Review: Abiotic Stress on Transpiration, Stomatal Diffusive Resistance and Photosynthetic Rate. *Pharma Innov. J* **11**, 1632-1635.
134. Tockner, K., Tonolla, D., Bremerich, V., Jähnig, S. C., Robinson, C. T., and Zarfl, C. (2022). Introduction to European rivers. *In* "Rivers of Europe", pp. 1-26. Elsevier.
135. Tong, X., Wu, P., Liu, X., Zhang, L., Zhou, W., and Wang, Z. (2022). A global meta-analysis of fruit tree yield and water use efficiency under deficit irrigation. *Agricultural Water Management* **260**, 107321.
136. Torquebiau, E., Rosenzweig, C., Chatrchyan, A. M., Andrieu, N., and Khosla, R. (2018). Identifying Climate-smart agriculture research needs.
137. Tran, B. T., Cavagnaro, T. R., Jewell, N., Brien, C., Berger, B., and Watts-Williams, S. J. (2021). High-throughput phenotyping reveals

growth of Medicago truncatula is positively affected by arbuscular mycorrhizal fungi even at high soil phosphorus availability. *Plants, People, Planet* **3**, 600-613.
138. Tuomisto, H. L., Hodge, I., Riordan, P., and Macdonald, D. W. J. J. o. e. m. (2012). Does organic farming reduce environmental impacts?–A meta-analysis of European research. **112**, 309-320.
139. Turan, V. (2021). Arbuscular mycorrhizal fungi and pistachio husk biochar combination reduces Ni distribution in mungbean plant and improves plant antioxidants and soil enzymes. *Physiologia Plantarum* **173**, 418-429.
140. Ungureanu, N., Vlăduț, V., and Voicu, G. (2020). Water scarcity and wastewater reuse in crop irrigation. *Sustainability* **12**, 9055.
141. van't Padje, A., Bonfante, P., Ciampi, L. T., and Kiers, E. T. (2021a). Quantifying nutrient trade in the arbuscular mycorrhizal symbiosis under extreme weather events using quantum-dot tagged phosphorus. *Frontiers in Ecology and Evolution* **9**, 153.
142. van't Padje, A., Galvez, L. O., Klein, M., Hink, M. A., Postma, M., Shimizu, T., and Kiers, E. T. (2021b). Temporal tracking of quantum-dot apatite across in vitro mycorrhizal networks shows how host demand can influence fungal nutrient transfer strategies. *The ISME journal* **15**, 435-449.
143. Vico, G., and Porporato, A. (2015). Ecohydrology of agroecosystems: quantitative approaches towards sustainable irrigation. *Bulletin of mathematical biology* **77**, 298-318.
144. Vosnjak, M., Likar, M., and Osterc, G. (2021). The Effect of Mycorrhizal Inoculum and Phosphorus Treatment on Growth and Flowering of Ajania (Ajania pacifica (Nakai) Bremer et Humphries) Plant. *Horticulturae* **7**, 178.
145. Wang, F., and Feng, G. (2021). Arbuscular mycorrhizal fungi interactions in the rhizosphere. *In* "Rhizosphere Biology: Interactions Between Microbes and Plants", pp. 217-235. Springer.
146. Wang, F., Li, K., and Shi, Z. (2021a). Phosphorus fertilization and mycorrhizal colonization change silver nanoparticle impacts on maize. *Ecotoxicology* **30**, 118-129.
147. Wang, G., Wang, L., Ma, F., Yang, D., and You, Y. (2021b). Earthworm and arbuscular mycorrhiza interactions: Strategies to motivate antioxidant responses and improve soil functionality. *Environmental pollution* **272**, 115980.
148. Wang, X. Y., He, T. H., Gen, S. Y., Zhang, X. Q., Wang, X., Jiang, D., Li, C. Y., Li, C. S., Wang, J. L., Zhang, W. Y., and Li, C. D. (2020).

Soil properties and agricultural practices shape microbial communities in flooded and rainfed croplands. *Applied Soil Ecology* **147**.

149. Wang, Y., Zou, Y.-N., Shu, B., and Wu, Q.-S. (2023). Deciphering molecular mechanisms regarding enhanced drought tolerance in plants by arbuscular mycorrhizal fungi. *Scientia Horticulturae* **308**, 111591.
150. Wu, Q.-S., Zou, Y.-N., Xia, R.-X., and Wang, M.-Y. (2007). Five Glomus species affect water relations of Citrus tangerine during drought stress. *Bot Stud* **48**, 147-154.
151. Wu, Y.-H., Wang, H., Liu, M., Li, B., Chen, X., Ma, Y.-T., and Yan, Z.-Y. (2021). Effects of native arbuscular mycorrhizae isolated on root biomass and secondary metabolites of Salvia miltiorrhiza Bge. *Frontiers in plant science* **12**, 66.
152. Yadav, R., Ror, P., Rathore, P., Kumar, S., and Ramakrishna, W. (2021). Bacillus subtilis CP4, isolated from native soil in combination with arbuscular mycorrhizal fungi promotes biofortification, yield and metabolite production in wheat under field conditions. *Journal of Applied Microbiology* **131**, 339-359.
153. Yazici, M. A., Asif, M., Tutus, Y., Ortas, I., Ozturk, L., Lambers, H., and Cakmak, I. (2021). Reduced root mycorrhizal colonization as affected by phosphorus fertilization is responsible for high cadmium accumulation in wheat. *Plant and soil*, 1-17.
154. Yu, L., Zhao, X., Gao, X., and Siddique, K. H. (2020). Improving/maintaining water-use efficiency and yield of wheat by deficit irrigation: A global meta-analysis. *Agricultural Water Management* **228**, 105906.
155. Yu, O.-Y., Raichle, B., and Sink, S. (2013). Impact of biochar on the water holding capacity of loamy sand soil. *International Journal of Energy and Environmental Engineering* **4**, 1-9.
156. Yuan, J.-H., Xu, R.-K., and Zhang, H. (2011). The forms of alkalis in the biochar produced from crop residues at different temperatures. *Bioresource Technology* **102**, 3488-3497.
157. Zare, M., Ganjeali, A., and Lahouti, M. (2021). Rosmarinic and caffeic acids contents in Basil (Ocimum basilicum L.) are altered by different levels of phosphorus and mycorrhiza inoculation under drought stress. *Acta Physiologiae Plantarum* **43**, 1-10.
158. Zhang, L., Chu, Q., Zhou, J., Rengel, Z., and Feng, G. (2021). Soil phosphorus availability determines the preference for direct or mycorrhizal phosphorus uptake pathway in maize. *Geoderma* **403**, 115261.
159. Zhu, J.-K. (2002). Salt and drought stress signal transduction in plants. *Annual review of plant biology* **53**, 247-273.

160. Zornoza, R., Moreno-Barriga, F., Acosta, J., Muñoz, M., and Faz, A. (2016). Stability, nutrient availability and hydrophobicity of biochars derived from manure, crop residues, and municipal solid waste for their use as soil amendments. *Chemosphere* **144**, 122-130.

CHAPTER EIGHT

LONG TERM SOIL FERTILITY: NANO-BLACK CARBON IS AN ORGANIC TOOL

IMRAN[1,2,3]

[1]DEPARTMENT OF AGRONOMY, THE UNIVERSITY OF AGRICULTURE, PESHAWAR

[2]AGRICULTURE OFFICER (EXTENSION WING), OFFICE OF THE DISTRICT DIRECTOR AGRICULTURE EXTENSION, DISTRICT KURRAM, MERGED DISTRICTS KHYBER PAKHTUNKHWA, PAKISTAN

[3]RHIZOSPHERE LABORATORY, UNIVERSITY OF CUKUROVA, FACULTY OF AGRICULTURE, DEPARTMENT OF SOIL SCIENCE AND PLANT NUTRITION, ADANA, TURKEY

Abstract

Organic biomass thermal degradation in limited oxygen forms a solid coproduct known as Nano-black carbon (biochar). There is a growing interest worldwide in Nano-black carbon application in agriculture and farm produce. The main reasons for Nano-black carbon application in agricultural land and cultural practices might be due to (a) pyrolysis can be a source of renewable biofuels. (ii) Many biomass waste materials can be treated by pyrolysis and thus converted into fuel sources. (iii) the long-term sequence of carbon dioxide that can be introduced into the atmosphere by adding Nano-black carbon to the soil;) Nano-black carbon amendment of soil aids in improving soil fertility and crop productivity. To meet global food security, crop production faces several real challenges, including soil fertility and organic matter depletion, climate change, unmanaged rainfall patterns, and pathogens. Abiotic pressures such as heat and cold, drought, salinity, water storage, and heavy metal pressure are emerging as important factors in reducing crop production globally. Nano-black carbon is a high carbonate modification created by the absence of oxygen or limited supply of a wide variety of feedstocks. Nano-black carbon typically has a high level of area, capacity for cation exchange, reactive functional groups, and highly reactive carbon. Due to these characteristics, Nano-black carbon has been widely associated with multiple abiotic stress reductions to enhance crop growth, development, production, soil fertility, and soil quality.

Keywords: biochar, soil fertility, nutrients status, availability, soil replenishment

1 Introduction

The adverse impact of environmental factors on a living organism in specific circumstances known as abiotic stress *(Imran et al. 2020; Rolf et al. 2004).* The non-living variable must influence the environment beyond its normal range of variation to adversely affect the population performance or individual physiology of the organism in a significant way *(*Akhtar et al. 2015b; *Ji-Ping; et al. 2007;* Khan et al. 2016; Zhan et al. 2011*).* Abiotic stress is essentially unavoidable. Abiotic stress effects on plants are especially dependent, if not solely dependent, on environmental factors, so it is particularly constraining. Abiotic stress is the most harmful factor concerning the growth and productivity of crops worldwide. *Mittler (2006)* stated that abiotic stressors are most harmful when they occur together, in combinations of abiotic stress factors. *These findings were concerning those of Roelofs et al. (2008) who revealed that salinity and drought stress combine reduced the grain yield of maize.*

Abiotic stress comes in many forms (Amonette and Joseph. 2009; Armor et al., 2005; Arshad et al., 2008). The most common of the stressors are the easiest to identify, but there are many other, less recognizable abiotic stress factors that affect environments constantly *(*Asad et al., 2015; Ayaz, et al., 2000*).* The most basic Abiotic stressors include High winds, Extreme temperatures, Drought, Flood, and other natural disasters, such as tornadoes and wildfires (Babar et al. 2016; *Wang et al., 2007*). Lesser-known stressors generally occur on a smaller scale. They include poor edaphic conditions like rock content and pH levels, high radiation, compaction, contamination, and other, highly specific conditions like rapid rehydration during seed germination (Bruinsma, 1963; *Brussaard et al., 2007*). Abiotic stress, as a natural part of every ecosystem, will affect the plant in a variety of ways (Bohnert and Jensen. 1996; Bos, et al., 2005). Although these effects may be either beneficial or detrimental, the location of the area is crucial in determining the extent of the impact that abiotic stress will have (Khan et al., 2015; Akhtar et al., 2015a; Imran 2018; Imran and Khan, 2015; Imran et al., 2015). The higher the latitude of the area affected, the greater the impact of abiotic stress will be on that area (Capell, and Doerffling. 1993; Chimenti, et al., 2006). So boreal forest is at the mercy of whatever abiotic stress factors may come along, while tropical zones are much less susceptible to such stressors *(Cramer et al., 2011;* Babar et al. 2016; Baber et al. 2015*).*

A plant's first line of defense against abiotic stress is in its roots (Imran et al., 2021a). If the soil holding the plant is healthy and biologically diverse, the plant will have a higher chance of surviving stressful conditions. (Conde, 2011; Cramer, et al., 2011; Imran 2018; Imran and Khan, 2017; Imran and Khan, 2015; Duan, et al., 2007; Farahbakhsh, 2012*).* The plant responses to stress are dependent on the tissue or organ affected by the stress. For example, transcriptional responses to stress are tissue or cell-specific in roots and are quite different depending on the stress involved *(Maestre et al., 2007).* Various publications report a generally positive effect of Nano-black carbon soil amendment on field crops under stress conditions. Major et al., (2010) reported that a single application of 20 t ha^{-1} Nano-black carbon to a Colombian savanna soil resulted in an increase in maize yield by 28 to 140% as compared with the unamended control under the saline stress condition (Badridze, et al., 2009; Bates et al., 1973). Nano-black carbon acts synergistically to improve crop performance under stress conditions. Specific mechanisms underlying the contribution of Nano-black carbon to plant response are poorly understood (Flowers, 2004; Gasco, et al., 2012; *Goncalves et al., 2001; Gonzalez et al., 2018*). Regional conditions including climate, soil chemistry, and soil condition all influence Nano-black carbon agronomic benefits (Imran et al., 2015; Imran et al., 2017). Evidence is mounting that Nano-black carbon in soil has significant effects on soil microorganisms (Lehmann et al., 2011; Imran et al., 2016). In the majority of the studies assessed that microbial biomass increased in Nano-black carbon amended soils (Imran et al., 2018) which act as a biocontrol agent and mitigate abiotic stress induce damages (Heans, 1984; Imran 2018; Imran and Khan, 2015).

2 Salinity stress mitigation with nano-black carbon

Nano-black carbon application in agriculture possesses the potential to tolerate abiotic stress (Khan, et al., 2010; Imran et al., 2019) such as high salinity is the disruption of the Na^+/K^+ ratio in the cytoplasm of the plant cell (Jackson, 1962; Jagatheeswari and Ranganathan, 2012). High concentrations of Na+, for example, can decrease the capacity for the plant to take up water and also alter enzyme and transporter functions (Joseph and Jini, 2011; Kamara, et al., 2014). Soil salinization, the accumulation of water-soluble salts to levels that negatively impact plant production, is a global phenomenon affecting approximately 831 million hectares of land. (Martinez and Manzur 2005; Kammann, et al., 2011; Lashari et al., 2013; Lehmann and Joseph. 2009). More specifically, the phenomenon threatens 19.5% of the world's irrigated agricultural land and 2.1% of the world's non-irrigated (dry-land) agricultural lands *(Lei et al., 2011; Maestre,* et al., 2007; Martinez and Manzur 2005; Imran et al., 2017*)*. High soil salinity content can be harmful to plants because water-soluble salts can alter osmotic potential gradients and consequently inhibit many cellular functions (*Raghothama, K. G 1999).* For example, high soil salinity content can inhibit the process of photosynthesis by limiting a plant's water uptake; high levels of water-soluble salts in the soil can decrease the osmotic potential of the soil and consequently decrease the difference in water potential between the soil and the plant's roots, thereby limiting electron flow from H_2O to P680 in Photosystem II's reaction center *(Rubio et al.,2001;* Lehmann, et al., 2011; *Lei,* et al., 2011; Lu. And Vonshak, 2002*)*.

The enzyme activation in plants with nano-black carbon application has been proposed to increase tolerance to abiotic stresses in crop plants which may increase a stress-inducing chemical in the plant so that the plant begins preparing defense mechanisms (Paz et al., 2014; Peng, et al., 2008; Ping et al., 2007; Prado et al., 2000; Raghothama, 1999). Thus, when abiotic stress occurs, the plant has already prepared defense mechanisms that can be activated faster and increase tolerance (Congming and Vonshak. 2002). Over generations, many plants have mutated and built different mechanisms to counter salinity effects. *(Pant et al., 2015; Rajalakshmi et al., 2015; Rashid et al., 2003; Imran, 2018; Imran, 2015; Iqbal et al., 2017).* A good combatant of salinity in plants is the Nano-black carbon amendment. Nano-black carbon is known for regulating plant growth and development and dealing with stress conditions (Palm, and Volkenburgh, 2012; Palta et al., 2006; Pant et al., 2015). With Nano-black carbon application under stress conditions, plants adapt very differently from one another, even from a plant living in the same area (Imran et al., 2021b). When a group of different plant species was prompted by a variety of different stress signals, such as drought or cold, each plant responded uniquely. Hardly any of the responses were similar, even though the plants had become accustomed to the same home environment *(Neto et al., 2004; Rolf et al., 2004; Rubio et al., 2001; Savvides et al., 2015; Shahida et al., 2016; Shamsi and Kobraee. 2013; Shamsuddin et al., 2015; Shazma et al., 2016).*

3 Plant wilting stress mitigation with Nano-black carbon amendments

Plant wilting occurs when water is scarce (Imran et al., 2021c). Drought stress is one of the main causes of crop loss within the agricultural world. This is due to water's necessity in so many fundamental processes in plant growth. It has become especially important in recent years to find a way to combat drought stress (Sidari, et al., 2008; Karim and Imran. 2019; Tank and Saraf. 2010; Tiwari et al., 2011; Tombesi, et al., 2018; Ulfat et al., 2016). A decrease in precipitation and a subsequent increase in drought is extremely likely in the future due to an increase in global warming. *(Zargar, et al., 2017;* Van et al., 2001; *Wang* et al., 2007; Wang and Xu. 2013; Wang and Jin. 2007; Wolfe 2007*).* Plants have come up with many mechanisms and adaptations to try and deal with drought stress (Fig. 1). One of the leading ways that plants combat drought stress is by increasing water infiltration rate and soil porosity (Xu et al., 2008; Yeo et al., 1990; Younis et al., 2015; Zakaria et al., 2012; Zargar et al., 2017). Nano-black carbon application can enhance soil porosity and increasing water infiltration in various types of soil. It has been reported that Nano-black carbon application regulating hormone secretion that maintains stomatal opening and closing (Imran et al., 2021d). The hormone which is regulating by Nano-black carbon application is abscisic acid (ABA). Synthesis of ABA causes the ABA to bind to receptors. This binding then affects the opening of ion channels thereby decreasing turgor pressure in the stomata and causing them to close. Recent studies showed that how ABA levels increased in drought-stressed plants (Van et al., 2001; Wang et al., 2007; Wang and Xu. 2013; Wang and Jin. 2007). They showed that when plants were placed in a stressful situation they produced more ABA to try and conserve any water they had in their leaves *(Goncalves et al., 200; Rolf* et al., 2004; *Rubio* et al., 2001; *Savvides* et al., 2015; Shahida et al., 2016; Shamsi and Kobraee. 2013*).* Nano-black carbon application can regulate another extremely important factor dealing with drought stress and uptake of water is aquaporins (AQPs). AQPs are integral membrane proteins that makeup channels. These channel's main job is the transport of water and other necessary solutes. AQPs are both transcriptionally and post-transcriptionally regulated by many different factors such as ABA, GA3, pH, and $Ca2+$ and the specific levels of AQPs in certain parts of the plant, such as roots or leaves, helps to draw as much water into the plant as possible *(Rubio et al., 2001; Savvides et al., 2015; Shahida et al., 2016; Shamsi and Kobraee. 2013; Shamsuddin et al., 2015).*

4 Relationship of soil microbiota with nano-black carbon

Evidence is mounting that Nano-black carbon in soil has significant effects on soil micro-organisms, as reviewed recently by Lehmann et al. (2011) and Imran et al. (2020a). In the majority of the studies assessed in that review, microbial biomass increased in Nano-black carbon-amended soils (Palm, and Volkenburgh, 2012; Palta et al., 2006; Pant et al., 2015; Paz et al., 2014; Peng, et al., 2008; Ping et al., 2007; Prado et al., 2000; Raghothama, 1999). Nano-black carbon addition also caused significant changes in microbial community composition and enzyme activities in both bulk soil and the rhizosphere (Lehmann, et al., 2011; Lei, et al., 2011; Lu. And Vonshak, 2002). For instance, the Nano-black carbon amendment was generally characterized by an increase in the relative abundance of members of the Actinobacteria and Bacteriodetes phyla (Je-sus et al., 2009; O'Neill et al., 2009; Khodadad et al., 2011; Kolton et al., 2011). While little is understood regarding the mechanisms by which Nano-black carbon affects the microbial abundance and community structure, it is well-known that soil micro-organisms can have a tremendous impact on plant productivity (Imran et al., 2020b). Therefore, Nano-black carbon-induced changes in soil micro-organisms may certainly play a role in 'The Nano-black carbon Effect' (Graber et al., 2010; Lehmann, et al., 2011; Lei, et al., 2011; Lu. And Vonshak, 2002).

5 Nutrients stress mitigation with nano-black carbon

In the agriculture sector, very little attention has been given to Nano-black carbon utilization (Imran, 2021), because crop response and soil health benefits have yet to be quantified, and because the mechanisms by which it improves soil fertility are poorly understood (Bohnert and Jensen. 1996; Bos, et al., 2005; Bruinsma, 1963; Brussaard et al., 2007; Capell, and Doerffling. 1993; Chimenti, et al., 2006). The positive effects of Nano-black carbon on crop productivity under conditions of extensive agriculture are frequently attributed to direct effects of Nano-black carbon (Imran et al., 2021; Ilyas et al., 2020) supplied nutrients and several other indirect effects, including increased water and nutrient retention, improvements in soil pH, increased soil cation exchange capacity, effects on P and S transformations and turnover, neutralization of phytotoxic compounds in the soil, improved soil physical properties (Imran, 2021), promotion of mycorrhizal fungi, and alteration of soil microbial populations and functions (Izhar et al., 2019; Imran and Khan, 2015; Imran et al., 2015; Imran et al., 2014; Jackson, 1962; Jagatheeswari and Ranganathan. 2012; Joseph and Jini, 2011; Kamara, et al., 2014). Yet, the Nano-black carbon effect is also evident under conditions of intensive production where many of these parameters are not limited (Gasco, et al., 2012; Goncalves et al., 2001). Nano-black carbon addition to soil alters microbial populations in the rhizosphere, albeit via mechanisms not yet understood, and may cause a shift towards beneficial microorganism populations that promote plant growth and resistance to biotic stresses. In addition to some scant evidence for Nano-black carbon-induced plant protection against soil-borne diseases, the induction of systemic resistance towards several foliar pathogens in three crop systems has been demonstrated (Khan, et al., 2010; Imran et al., 2019; Imran, Amanullah. 2018; Imran et al., 201; Imran et al., 2015; Lashari et al., 2013; Lehmann and Joseph. 2009). There are indications that Nano-black carbon induces responses along with both systemic acquired resistance (SAR) and induced systemic resistance (ISR) pathways, resulting in a broad spectrum controlling capacity in the canopy (Lei, et al., 2011; Lu. And Vonshak, 2002). This review examines the effects of Nano-black carbon soil amendment on the different soil-plant-microbe interactions that may have a role in plant health. Improvement of plant responses to disease can be one of the benefits gained from applying Nano-black carbon to soil (Paz et al., 2014; Peng, et al., 2008; Ping et al., 2007; Prado et al., 2000; Raghothama, 1999; Rajalakshmi et al., 2015; Rashid et al., 2003; Imran, 2018; Imran, 2015; Iqbal et al., 2017; Raza 2012; Rengasamy, 2002; Roelofs et al., 2008).

6 Plants response to nano-black carbon amendments

Under field and greenhouse conditions, Nano-black carbon has a positive effect on plant growth and development. Early studies reported that charcoal added to soil increased the yield of moong, soybean, and pea (Paz et al., 2014; Peng, et al., 2008; *Ping* et al., 2007; Prado et al., 2000; *Raghothama, 1999;* Rajalakshmi et al., 2015; Rashid et al., 2003; Imran, 2018; Imran, 2015; Iqbal et al., 2017; Raza 2012; Rengasamy, 2002; *Roelofs* et al., 2008; Iswaran et al., 1980) and of soybean (Kishimoto and Sugiura, 1985). Shoot and root biomass of birch and pine were greater in charcoal amended soil (Rashid et al., 2003; Imran, 2018; Imran, 2015; Iqbal et al., 2017). Similarly, five years following the soil application of charcoal, biomass production of sugi trees (Cryptomeria japonica) was substantially increased (Sidari, et al., 2008; Karim and Imran. 2019). A single application of 20 t ha^{-1}Nano-black carbon to a Colombian savanna soil resulted in an increase in maize yield by 28 to 140% as compared with the unamended control in the 2nd to 4th years after application (Peng, et al., 2008; Ping et al., 2007; Prado et al., 2000). With the addition of Nano-black carbon (at 90 g kg^{-1}) to a tropical, low-fertility Ferralsol, the proportion of N fixed by bean plants (Phaseolus vulgaris) increased from 50% (without Nano-black carbon) to 72%, and biomass production and bean yield were significantly improved (Van et al., 2001; Wang et al., 2007; Wang and Xu. 2013; Wang and Jin. 2007; Wolfe 2007; Xu et al., 2008; Yeo et al., 1990; Younis et al., 2015). On the same type of soil, total N recovery in soil, crop residues, and grains were significantly higher on compost (16.5%), charcoal (18.1%), and charcoal plus compost treatments (17.4%) in comparison to mineral-fertilized plots (10.9%) (Steiner et al., 2008b). Nano-black carbon soil application resulted in higher upland rice (Oryza sativa) grain yields at sites in northern Laos with low P availability and improved the response to N and NP chemical fertilizer treatments (Munns et al., 2006; Munns, 2002; Munns, and Tester, 2008; Nazarli et al., 2011; Neto et al., 2004). Large vol-ume applications of Nano-black carbon (30 and 60 t ha^{-1}) in the Mediterranean basin increased durum wheat (Triticum durum) biomass and yield by up to 30%, an effect which was sustained for two consecutive seasons (Nazarli et al., 2011; Neto et al., 2004). Overall, these results demonstrate the potential of Nano-black carbon application to improve plant productivity. The means by which Nano-black carbon improves crop response can be attributed to direct effects via bio-char-supplied nutrients (Munns et al., 2006; Munns, 2002; Munns, and Tester, 2008), and to several other indirect effects, including: increased nutrient retention (Rajalakshmi et al., 2015; Rashid et al., 2003; Imran, 2018; Imran, 2015; Iqbal et al., 2017; Raza 2012; Rengasamy, 2002; Roelofs et al., 2008); improvements in soil pH (Yama-to et al., 2006; Steiner et al., 2007; Novak et al., 2009); increased soil cation exchange capacity (Rashid et al., 2003; Imran, 2018; Imran, 2015; Iqbal et al., 2017; Raza 2012; Rengasamy, 2002; Roelofs et al., 2008); effects on P and S transformations and turnover (Imran, 2015; Iqbal et al., 2017; Raza 2012; Rengasamy, 2002; Roelofs et al., 2008); neutralization of phytotoxic compounds in the soil (Rajalakshmi et al., 2015; Raza 2012; Rengasamy, 2002; Roelofs et al., 2008); improved soil physical properties including water retention (Rajalakshmi et al., 2015; Rashid et al., 2003; Imran, 2018; Imran, 2015; Iqbal et al., 2017; Raza 2012; Rengasamy, 2002; Roelofs et al., 2008); promotion of mycorrhizal fungi (Rajalakshmi et al., 2015; Rashid et al., 2003; Imran, 2018; Imran, 2015; Iqbal et al., 2017; Raza 2012; Rengasamy, 2002; Roelofs et al., 2008); and alteration of soil microbial populations and functions (Rubio et al., 2001; Savvides et al., 2015; Shahida et al., 2016; Shamsi and Kobraee. 2013; Shamsuddin et al., 2015). Many of these effects are interrelated and may act synergistically to improve crop performance (*Rubio* et al., 2001; *Savvides* et al., 2015; Shahida et al., 2016; Shamsi and Kobraee. 2013; Shamsuddin et al., 2015). Specific mechanisms underlying the contribution of Nano-black carbon to plant response are poorly understood. Regional conditions including climate, soil chemistry, and soil condition all influence Nano-black carbon agronomic benefits. Besides, dissimilar biomass feedstocks and pyrolysis conditions create Nano-black carbon's with different physical and chemical properties (Savvides et al., 2015; Shahida et al., 2016; Shamsi and Kobraee. 2013; Shamsuddin et al., 2015), resulting in different impacts on plant response (Shahida et al., 2016; Shamsi and Kobraee. 2013; Shamsuddin et al., 2015). Given that the Nano-black carbon-soil-plant-water-environment is highly complex, it is difficult to isolate those factors which play an instrumental role in the 'Nano-black carbon Effect'. To reduce the number of potential factors involved, Savvides et al. (2015) tested whether Nano-black carbon addition could impact plant growth when nutritional and soil physical aspects of Nano-black carbon amendment were eliminated. This was achieved by examining the impact of a nutrient-poor, wood-derived Nano-black carbon on tomato.

7 Nano-black carbon application and Phosphate starvation in plants

Many researchers reported that with no P fertilization, plant growth and development significantly reduced and affect the plant dry biomass and productivity which ultimately leading to food insurgency (Imran et al., 2020a; Imran et al., .2020b; Imran et al., 2019a; Ali et al., 2019). It has been reported that P is an important plant metabolism component and making 0.2% of the plant dry weight (Karim and Imran. 2019; Imran et al., 2018; Imran 2018a). Therefore P fertilization is needed as it is an important major plant nutrient after N. In limited P application or no P fertilization, plants cannot grow as it is a key component of nucleic acids, phospholipids, and ATP, and without these components, the plant cannot survive (Imran 2018b; Imran. 2018c; Imran 2018d). Phosphorus is involved in controlling and regulating several enzymatic functions and reactions in the metabolic pathway and thus regulating plant growth and development and ultimately plant productivity and yield (Imran and Amanullah. 2018; Imran. 2018e; Imran and Khan 2017). Other researchers reported that P is very rarely available in soil or only available in the soil out of the rhizosphere which cannot support the plant for high growth rate and development (Imran et al., 2020c; Imran et al., 2017a). Phosphorus application is necessary for agriculture systems to ensure crop productivity and food security (Imran et al., 2017b; Imran 2017; Imran et al., 2015a). Limited P supply in bulk soil affect P uptake by plants as it is only moving by the process of diffusion and do not uptake or very low amount is uptake by the crop plants which is an alarming threat for low productivity (Imran, et al., 2015b; Imran et al., 2019b). Phosphorus (P) is an essential macronutrient required for plant growth and development, but most of the world's soil is limited in this important plant nutrient (Palta et al., 2006; Pant et al., 2015; Paz et al., 2014; Peng, et al., 2008; Ping et al., 2007; Prado et al., 2000). Plants can utilize P mainly in the form of soluble inorganic phosphate (Pi) but are subjected to abiotic stress of P-limitation when there are not sufficient soluble PO_4 available in the soil. Phosphorus forms insoluble complexes with Ca and Mg in alkaline soils and Al and Fe in acidic soils that makes it unavailable for plant roots (Imran, 2018; Imran, 2015; Iqbal et al., 2017; Raza 2012; Rengasamy, 2002; Roelofs et al., 2008; Rolf et al., 2004; Rubio et al., 2001; Savvides et al., 2015). When there is limited bioavailable P in the soil, plants show extensive abiotic stress phenotype such as short primary roots and more lateral roots and root hairs to make more surface available for Pi absorption, exudation of organic acids, and phosphatase to release Pi from complex P containing molecules and make it available for growing plants organs (Imran, 2018; Imran, 2015; Iqbal et al., 2017; Raza 2012; Rengasamy, 2002; *Roelofs* et al., 2008; *Rolf* et al., 2004; *Rubio* et al., 2001; *Savvides* et al., 2015). It has been shown that Nano-black carbon application is a master regulator of P-starvation response in plants *(Gonzalez et al., 2018)*. Nano-black carbon amended soil has been shown to regulate extensive secondary metabolites during phosphorus limitation stress *(Bos, et al., 2005; Bruinsma, 1963; Brussaard et al., 2007; Capell, and Doerffling. 1993)*.

8 Conclusion

Nano-black carbon is an activated carbon soil conditioner that can alleviate the negative impacts of salinity, drought, heat and cold, water-logging, and heavy metal stresses. Drought and salt stress negatively affect soil fertility and plant growth. The application of Nano-black carbon ameliorates the negative effects of drought and salt stress on plants. The Nano-black carbon application increased the plant growth, biomass, and yield under either drought or salt stress and also increased photosynthesis, nutrient uptake, and modified gas exchange characteristics in drought and salt-stressed plants. Under drought stress, Nano-black carbon increased the water holding capacity of the soil and improved the physical and biological properties of soils. Under salt stress, Nano-black carbon decreased Na^+ uptake, while increased K^+ uptake by plants. Nano-black carbon-mediated increase in salt tolerance of plants is primarily associated with improvement in soil properties, thus increasing plant water status, reduction of Na^+ uptakc, increasing uptake of minerals, and regulation of stomatal conductance and phytohormones. This chapter highlights both the potential of Nano-black carbon in alleviating abiotic stresses in plants and the prospect of the role of Nano-black carbon under abiotic stresses in plants.

Figure 1. Biochar chamber (pyrolyser) and various type of biochar prepared on various temperature (Cukurova University Turkey).

Summary

Organic biomass thermal degradation in limited oxygen forms a solid coproduct known as Nano-black carbon (biochar). There is a growing interest worldwide in Nano-black carbon application in agriculture and farm produce. The main reasons for Nano-black carbon application in agricultural land and cultural practices might be due to (a) pyrolysis can be a source of renewable biofuels. (ii) Many biomass waste materials can be treated by pyrolysis and thus converted into fuel sources. (iii) the long-term sequence of carbon dioxide that can be introduced into the atmosphere by adding Nano-black carbon to the soil;) Nano-black carbon amendment of soil aids in improving soil fertility and crop productivity. To meet global food security, crop production faces several real challenges, including soil fertility and organic matter depletion, climate change, unmanaged rainfall patterns, and pathogens. Abiotic pressures such as heat and cold, drought, salinity, water storage, and heavy metal pressure are emerging as important factors in reducing crop production globally. Nano-black carbon is a high carbonate modification created by the absence of oxygen or limited supply of a wide variety of feedstocks. Nano-black carbon typically has a high level of area, capacity for cation exchange, reactive functional groups, and highly reactive carbon. Due to these characteristics, Nano-black carbon has been widely associated with multiple abiotic stress reductions to enhance crop growth, development, production, soil fertility, and soil quality.

References

1. Imran, Zada H, Naveed S, Khattak I And Ahmad S. 2016. Variable Rates of Phosphorous Application Influenced Phenological Traits Of Green Gram (*Vignaradiata* L.) J Agri Res 2016, 1(3).
2. Imran, A.A. Khan, I.U. Khan and S. Naveed. 2016. Weeds density and late sown maize productivity influenced by compost application and seed rates under temperate environment. Pak. J. Weed Sci. Res. 22(1): 169-181.
3. A. Khan, M. N. Khan, I. uallah, S. Shah, I. R. Arshad, I. M. A. Zeb, and Imran. 2015. Effect of potash application on growth, yield and yield components of spring maize hybrids. Pure and Applied Biology Vol. 4, Issue 2, 195-203.
4. Akhtar, S. S., M. N. Andersen, and F. Liu. 2015a. Nano-black carbon mitigates salinity stress in potato. *Journal of Agronomy and Crop Sciences* 201:321–400.
5. Akhtar, S. S., M. N. Andersen, and F. Liu. 2015b. Residual effects of Nano-black carbon on improving growth, physiology and yield of wheat under salt stress. *Agricultural Water Management* 158:61–8.
6. Imran, A.A. Khan, Inamullah and Luqman. 2015. Weeding stages and their effect on yield and yield components of rice in upper Swat, Pakistan. Pak. J. Weed Sci. Res. 21(4): 555-563.
7. Imran, Abdul Bari, Roshan Ali, Naeem Ahmad, Zulfiqar Ahmad, Muhammad Idrees Khattak, Akhtar Ali, Fayaz Ahmad, Inayatullah Khan and Shahida Naveed. 2017. Traditional Rice Farming Accelerate CH4 & N2O Emissions Functioning as A Stronger Contributors of Climate Change. Int J Environ Sci Nat Res. 2017; DOI: 10.19080/ARTOAJ.2017.09.555765
8. Imran, Amanullah, A. Bari and R. Ali. 2018. Peach sources, phosphorous and beneficial microbes enhance productivity of soybean. Soy. Res. 16(2): 39-48.
9. Akhtar, S. S., M. N. Andersen, M. Naveed, Z. A. Zahir, and F. Liu. 2015. Interactive effect of Nano-black carbon and plant growth-promoting bacterial endophytes on ameliorating salinity stress in maize. *Functional Plant Biology* 42:770–81.
10. Amir Zaman Khan, Imran, Asim Muhammad, Aiman Khalil, Hasina Gul, Habib Akbar and Said Wahab.2016. Impact of fertilizer priming

on seed germination behavior and vigor of maize. Pure and Applied Biology. Vol. 5, Issue 4, pp744-751. http://dx.doi.org/10.19045/bspab.2016.50093
11. Amonette, J. E., and S. Joseph. 2009. Characteristics of Nano-black carbon: Microchemical properties. In *Nano-black carbon for environmental management: Science and technology*, eds. J. Lehmann and S. Joseph, 33–52. Earthscan, London: Sterling.
12. Armor, N. B., K. B. Hamed, A. Debez, and C. Grignonand Crabbedly. 2005. Physiological and antioxidant responses of perennial halophyte *Crithmum maritimum* to salinity. *Plant Science*4:889–99.
13. Arshad, M., B. Shaharoona, and T. Mahmood. 2008. Inoculation with *Pseudomonas* spp. containing ACC-deaminase partially eliminates the effects of drought stress on growth, yield, and ripening of pea (*Pisum sativum* L.). *Pedosphere* 18:611–20.
14. Asad Ali Khan, Imran* Fawad Ali Shah Inamullah, Laiq Zada Muhammad Naeem Muhammad Nouman Khan.2015. Phenological Traits of Rice as Influenced by Seedling Age and Number of Seedling per Hill under Temperate Region. J. Bio, Agric and H. care ISSN 2224-3208 (Paper) ISSN 2225-093X (Online).
15. Ayaz, F. A., A. Kadioglu, and T. R. Turgut. 2000. Water stress effects on the content of low molecular weight carbohydrates and phenolic acids in *Cienanthe setosa*. *Candian Journal of Plant Sciences* 80:373–8.
16. Babar Iqbal, Bashir Ahmad, Inam Ullah, Imran, Asad Ali Khan, Shazma Anwar, Asad Ali, Khurrum Shahzadand Shehryar Khan. 2016. Effect of phosphorus, sulphur and different irrigation levels on phenological traits of Triticale. Pure and Applied Biology. Vol. 5, Issue 2, pp 303-310. http://dx.doi.org/10.19045/bspab.2016.50039
17. Babar Iqbal, Mohammad Tariq Jan, Zar Muhammad, Asad Ali Khan, Shazma Anwar, Imran and Khurrum Shahzad. 2016. Phenological traits of Maize influenced by integrated management of compost and fertilizer Nitrogen. J. Pure & Applied biology. 5(1): 58-63.
18. Babar Iqbal[1], Muhammad Tariq Jan[1], Inamullah[1], Imran*[1], Asad Ali Khan*[1], Zar Muhammad[2], Shazma Anwar[1], Shehryar Khan[1], Shamsuddin Baqa[1], Amjad Usman[3] .2015. Integrated Management Of Compost Type And Fertilizer-N In Maize. J. Pure & Applied biology Vol. 4, Issue 4.
19. Baber Iqbal, M.T. Jan, Imran, Zar M, A.A. Khan, S. Anwar and K. Shahzad 2015. Growth and phenology of maize as affected by integrated management of compost and fertilizers Nitrogen. J. Pure & Applied biology Vol. 4, Issue 4.

20. Badridze, G., A. Weidner, F. Asch, and A. Borner. 2009. Variation in salt tolerance within Georgian wheat germplasm collection. *Genetic Resources and Crop Evolution* 56:1125–30.
21. Bates, L. S., R. P. Waldren, and I. D. Teare. 1973. Rapid determination of free proline for water stress studies. *Plant and Soil* 39:205–7.
22. Bohnert, H. J., and R. G. Jensen. 1996. Strategies for engineering water stress tolerance in plants. *Trends in Biotechnology* 14:89–97.
23. Imran, Shahida Naveed, Asad Ali Khan And Inayat Khattak. 2015. Impact Of Phosphorus Levels And Seed Rates On Growth And Yield Of Late Sown Maize On High Elevation In Swat, Pakistan. Pakistan J. Agric. Res. Vol. 28 No. 4:406-413
24. Imran. 2018. Ecological Environmental Variability Influence Growth and Yield Potential of Rice under Northern Climatic Scenario. Russian Agricultural Sciences, 2018, Vol. 44, No. 1, pp. 18–24.
25. Bos, C., B. Juillet, H. Fouillet, L. Turlan, S. Dare, C. Luengo, R. Nto unda, R. Benamouzig, N. Gausseres, V. Tome, and C. Gaudichon. 2005. Postprandial metabolic utilization of wheat protein in humans. *The American Journal of Clinical Nutrition* 81:87–94.
26. Bruinsma, J. 1963. The quantitative analysis of chlorophyll a and b in plant extract. *Photochemistry and Photobiology* 2:241–9.
27. *Brussaard, Lijbert; de Ruiter, Peter C.; Brown, George G. (2007). "Soil biodiversity for agricultural sustainability". Agriculture, Ecosystems and Environment.* ***121*** *(3): 233–244.* doi:10.1016/j.agee.2006.12.013.
28. Capell, B., and K. Doerffling. 1993. Genotype specific difference in chilling tolerance of maize in relation to chilling induce changes in water status and abscisic acid accumulation. *Physiologia Plantarum* 88:638–46.
29. Chimenti, C. A., M. Marcantonio, and J. A. Hall. 2006. Divergent selection for osmotic adjustment results in improved drought tolerance in maize (*Zea mays* L.) in both early growth and flowering phases. *Field Crops Research* 95:305–15.
30. *Conde, Artur (2011).* "Membrane Transport, Sensing and Signaling in Plant Adaptation to Environmental Stress" (PDF). *Plant & Cell Physiology.* ***52*** *(9): 1583–1602.* doi:10.1093/pcp/pcr107. PMID 21828102 – *via Google Scholar.*
31. *Cramer, Grant R; Urano, Kaoru; Delrot, Serge; Pezzotti, Mario; Shinozaki, Kazuo (2011-11-17).* "Effects of abiotic stress on plants: a systems biology perspective". *BMC Plant Biology.* ***11****:163.* doi:10.1186/1471-2229-11-163. ISSN 1471-2229. PMC 3252258. PMID 22094046.

32. Duan, D. Y., W. Q. Li, X. J. Liu, H. P. Ouyang, and P. An. 2007. Seed germination and seedling growth of *Suaeda salsa* under salt stress. *Annales Botanici Fennici* 44:161–9.
33. Dubois, M., K. Gilles, J. K. Hammiltron, P. A. Robers, and F. Smith. 1951. A colorimetric method for the determination of sugars. *Nature* 168:167–8.
34. Farahbakhsh, H. 2012. Germination and seedling growth in un-primed and primed seeds of fennel as affected by reduced water potential induced by NaCl. *International Research Journal of Applied and Basic Sciences* 3:737–44.
35. Imran 2018. Phosphorous fertilization influenced weeds attributes and phenological characteristics Of mungbean cultivars (vigna radiata l.) Russian Agricultural Sciences, 2018, Vol. 44, No. 3, pp. 229–234.
36. Imran and Khan, A. A. (2017). Canola Yield And Quality Enhanced With Sulphur Fertilization Russ. Agricult. Sci. 43: 113 10.3103/S1068367417020100
37. Flowers, T. J. 2004.Improving crop salt tolerance. *Journal of Experimental Botany* 55:307–19.
38. Gasco, G., Paz-Ferreiro, and A. Mendez. 2012. Thermal analysis of soil amended with sewage sludge and Nano-black carbon from sewage sludge pyrolysis. *Journal of Thermal Analysis and Calorimetry* 108:769–75.
39. Giannopolitis, C. N., and S. K. Ries. 1977. Superoxide dismutase occurance in higher plants. *Plant Physiology* 59:304–14.
40. *Goncalves-Alvim, Silmary J.; Fernandez, G. Wilson (2001). "Biodiversity of galling insects: historical, community and habitat effects in four neotropical savannas". Biodiversity and Conservation.* ***10****: 79–98.*
doi:10.1023/a:1016602213305
41. Imran, A. A. Khan, I. Ullah, H. zada, F. Ahmad, S.T. Shah, A. Usman and I. Ullah. 2015. Yield and yield attributes of rapeseed cultivars as influence by sulfur level under swat valley conditions. J. Pure & Applied biology Vol. 4, Issue 3.
42. *Gonzalez-Villagra, Jorge; Rodrigues-Salvador, Acácio; Nunes-Nesi, Adriano; Cohen, Jerry D.; Reyes-Díaz, Marjorie M. (2018). "Age-related mechanism and its relationship with secondary metabolism and abscisic acid in Aristotelia chilensis plants subjected to drought stress". Plant Physiology and Biochemistry.* ***124****: 136–145.*
doi:10.1016/j.plaphy.2018.01.010. ISSN 0981-9428. PMID 29360623.

43. Imran, I. Hussain, I. khattak, A. Rehman[2], F. Ahmad and S.T. shah .2015. *Roots nodulation, yield and yield contributing parameters of mungbean cultivars as influenced by different phosphorous level in swat-Pakistan.* J. Pure & Applied biology Vol. 4, Issue 4.
44. Heans, D. L. 1984. Determination of total organic carbon in soils by an improved chromic acid digestion and spectrophotometric procedure. *Communications in Soil Science and Plant Analysis* 15:1191–213.
45. Imran, I. Khattak, I. Hussain, A. Rehman, S. Anwar[1], F. Ahmad, A. A. Khan[1] and H. Zada. 2015. Growth and Yield of Maize Hybrids as effected by different Sowing Dates in Swat Pakistan. J. Pure & Applied biology Vol. 4, Issue 4.
46. Imran, Amanullah, A.A Khan, Abdul bari..... 2019. Production statistics and modern technology of maize cultivation in Khyber Pakhtunkhwa Pakistan. Plant sci. Arc.2 (2) 1-12.
47. Imran, Amanullah. 2018. Review. Global impact of climate change on water, soil resources and threat towards food security: evidence from Pakistan. Adv Plants Agric Res. 2018;8(5):350–355. DOI: 10.15406/apar.2018.08.00349
48. Imran, Asad Ali Khan, Inamullah Inam & Fayaz Ahmad. (2016). Yield and yield attributes of Mungbean (*Vigna radiata* L.) cultivars as affected by phosphorous levels under different tillage systems, *Cogent Food & Agriculture* (2016), 2: 1151982.
49. Imran. 2018. Organic Matter Amendments Improve Soil Health, Productivity and Profitability of Maize and Soybean. Ann Rev Resear. 1(3): 555564.
50. Iqbal A, Amanullah, Ali A, Iqbal M, Ikramullah and Imran. 2017. Integrated use of phosphorus and organic matter improve fodder yield of moth bean (Vigna aconitifolia (Jacq.) under irrigated and dryland conditions of Pakistan. Journal of AgriSearch 4(1): 10-15
51. ISTA. 2005. *International rules for seed testing*. Zurich, Switzerland: International Seed Testing Association.
52. Izhar Ali, Asad Ali Khan, Imran, inamullah, Aman Khan .... 2019. Humic acids and Nitrogen levels optimizing productivity of green gram (vigna radiate L.). Russian Agriculture Sciences, 45(1), 43-47.
53. Jackson, M. 1962. *Soil chemical analysis*, 496. London: Constable and Company Limited.
54. Jagatheeswari, D., and P. Ranganathan. 2012. Influence of Mercuric Chloride on seeds germination, seedling growth and biochemical analysis of green gram (*Vigna radiata* (L.) Wilczek. Var. Vamban-

3). *International Journal of Pharmaceutical and Biological Archives* 3:291–5.

55. Joseph B., and D. Jini. 2011. Development of salt stress-tolerant plants by gene manipulation of antioxidant enzymes. *Asian Journal of Agricultural Research* 5:17–27.
56. Kamara, A., A. Kamara, M. M. Mansaray, and P. A. Sawyerr. 2014. Effects of Nano-black carbon derived from maize stover and rice straw on the germination of their seeds. *American Journal of Agriculture and Forestry* 2:246–9.
57. Kammann, C. I., S. Linsel, J. W. Gobling, and H. Koyro. 2011. Influence of Nano-black carbon ondrought tolerance of *Chenopodium quinoa* willd and on soil-plant relation. *Plant and Soil* 345:195–210.
58. Khan, M. A., E. Islam, M. U. Shirazi, S. Mumtaz, S. M. Mujtaba, M. A. Khan, A. Shereen, M. Y. Ashrafand, and G. M. Kaleri. 2010. Physiological responses of various wheat genotypes to salinity. *Pakistan Journal of Botany* 42:3497–505.
59. Lashari M.S., Y. Liu, W. Li, J. Pan, G. Pan, J. Zheng, G. Zheng, X. Z hang, and X. Yu. 2013. Effect of amendment of Nano-black carbon manure compost in conjunction with pyroligneous solution on soil quality and wheat yield of a salt stressed cropeland from central China Great Plain. *Food Crop Research* 144:113–8.
60. Lehmann J., and S. Joseph. 2009. Nano-black carbon for environmental management: An introduction. In *Nano-black carbon for environmental management: Science and technology*, eds. J. Lehmann and S. Joseph, 1–12. London: Earthscan.
61. Lehmann, J., M. C. Rillig, J. Thies, C. A. Masiello, W. C. Hockaday, and D. Crowley. 2011. Nano-black carbon effects on soil biota. *Soil Biology and Biochemistry* 43:1812–36.
62. *Lei, Gang; Shen, Ming; Li, Zhi-Gang; Zhang, Bo; Duan, Kai-Xuan; Wang, Ning; Cao, Yang-Rong; Zhang, Wan-Ke; Ma, Biao (2011). "EIN2 regulates salt stress response and interacts with a MA3 domain-containing protein ECIP1 in Arabidopsis". Plant, Cell & Environment. **34** (10):1678–1692.* doi:10.1111/j.13653040.2011.02363.x. ISSN 1365-3040.
63. Lu. Congming, A. Vonshak. (2002). Effects of salinity stress on photosystem II function in cyanobacterial Spirulina platensis cells. Physiol. Plant 114 405-413.
64. *Maestre, Fernando T.; Cortina, Jordi; Bautista, Susana (2007). "Mechanisms underlying the interaction between Pinus halepensis and the native late-successional shrub Pistacia lentiscus in a semi-arid plantation". Ecography. **27** (6):776–786.*

doi:10.1111/j.0906-7590.2004.03990.x.
65. Martinez-Beltran J, Manzur CL. (2005). Overview of salinity problems in the world and FAO strategies to address the problem. Proceedings of the international salinity forum, Riverside, California, April 2005, 311–313.
66. Mazhar, R., N. Ilyas, M. Saeed, F. Bibi, and N. Batool. 2016. Biocontrol and salinity tolerance potential of *Azospirillum lipoferum* and its inoculation effect in wheat crop. *International Journal of Agriculture and Biology* 3:494–500.
67. Ministry of Food, Agriculture and Livestock, 2007. *Agricultural statistics of Pakistan*. Islamabad: Ministry of Food, Agriculture and Livestock, Government of Pakistan.
68. Miri, Y., and S. A. Mirjalili. 2013. Effect of salinity stress on seed germination and some physiological traits in primary stages of growth in purple coneflower (*Echinacea purpurea*). *International Journal of Agronomy and Plant Production* 4:142–6.
69. *Mittler, Ron (2006). "Abiotic stress, the field environment and stress combination". Trends in Plant Science.* ***11*** *(1): 15–19.* doi:10.1016/j.tplants.2005.11.002. PMID 16359910.
70. Muhammad Islam[1], Shazma Anwar[1], Saqib Bashir[1], Wajid Ali Khattak[1], Imran, Murad Ali[2] and Nadar Khan .2015. Growth and Yield Components of Wheat Varieties As Affected By Dual Purpose Practices. J. Pure & Applied biology.
71. Munns, R. R., A. James, and A. Lauchli. 2006. Approaches to increasing the salt tolerance of wheat and other cereals. *Journal of Experimental Botany* 57:1025–43.
72. Munns, R. 2002. Comparative physiology of salt and water stress. *Plant Cell Environment*25:239–50.
73. Munns, R., and M. Tester. 2008. Mechanisms of salinity tolerance. *Annual Review of Plant Biology* 59:651–81.
74. Nazarli, H., F. Faraji, and M. R. Zardashti. 2011. Effect of drought stress and polymer on osmotic adjusment and photosynthetic pigments of sunflower. *Carcetriari Agronomica in Maldova* 1:45.
75. *Neto, Azevedo; De, André Dias; Prisco, José Tarquinio; Enéas-Filho, Joaquim; Lacerda, Claudivan Feitosa de; Silva, José Vieira; Costa, Paulo Henrique Alves da; Gomes-Filho, Enéas (2004-04-01). "Effects of salt stress on plant growth, stomatal response and solute accumulation of different maize genotypes". Brazilian Journal of Plant Physiology.* ***16****(1): 31 38.* doi:10.1590/S1677-04202004000100005. ISSN 1677-0420.

76. Olsen, S. R., and L. E. Sommers. 1982. Phosphorous. In *Methods ofsoil analysis: Chemical and microbiological properties*, ed. D. Bunton, 403–27. 2nd ed. Madison, WI: Soil Science Society of America.
77. *Palm, Brady; Van Volkenburgh (2012). "Serpentine tolerance in Mimuslus guttatus does not rely on exclusion of magnesium". Functional Plant Biology.*
78. Palta, Jiwan P. and Farag, Karim. September 2006. "Methohasds for enhancing plant health, protecting plants from biotic and abiotic stress related injuries and enhancing the recovery of plants injured as a result of such stresses." United States Patent 7101828.
79. *Pant, Bikram Datt; Burgos, Asdrubal; Pant, Pooja; Cuadros-Inostroza, Alvaro; Willmitzer, Lothar; Scheible, Wolf-Rüdiger (2015).* "The transcription factor PHR1 regulates lipid remodeling and triacylglycerol accumulation in Arabidopsis thaliana during phosphorus starvation". *Journal of Experimental Botany.* ***66*** *(7): 1907–1918.* doi:10.1093/jxb/eru535. ISSN 0022-0957. PMC 4378627. PMID 25680792.
80. *Pant, Bikram-Datt; Pant, Pooja; Erban, Alexander; Huhman, David; Kopka, Joachim; Scheible, Wolf-Rüdiger (2015). "Identification of primary and secondary metabolites with phosphorus status-dependent abundance in Arabidopsis, and of the transcription factor PHR1 as a major regulator of metabolic changes during phosphorus limitation". Plant, Cell & Environment.* ***38*** *(1): 172–187.* doi:10.1111/pce.12378. ISSN 1365-3040. PMID 24894834.
81. Paz-Ferreiro, J., S. Fu, A. Mendez, and G. Gasco. 2014. Interactive effects of Nano-black carbon and the earhworm *Pontoscolex corethrurus* on plant productivity and soil enzymes activities, *Journal of Soils Sediments*, 14:483–94.
82. Peng, Y., Z. Gao, Y. Gao, G. Liu, L. Shengand D. Wang. 2008. Ecophysiological characteristics of alfalfa seedlings in response to various mixed salt-alkaline stresses. *Journal of Integrative Plant Biology* 50:29–39.
83. *Ping Ji, Geo K. Ron M. (2007). "Understanding Abiotic Stress Tolerance Mechanisms: Recent Studies on Stress Response in Rice". Journal of Integrative Plant Biology.* ***49*** *(6): 742–750.* doi:10.1111/j.1744-7909.2007.00495.x.
84. Prado, F. E., C. Boero, M. Gallarodo, and A. J. Gonzalez. 2000. Effect of NaCl on germination, growth and soluble sugar content in *Chenopodium quinoa* wil seeds. *Botanical Bulletin - Academia Sinica Taipei* 14:27–34.

85. *Raghothama, K. G. (1999). "Phosphate Acquisition". Annual Review of Plant Physiology and Plant Molecular Biology. **50** (1): 665–693.* doi:10.1146/annurev.arplant.50.1.665. PMID 15012223.
86. Rajalakshmi, A. G., S. K. Kumar, C. D. Bharathi, R. Karthika, K. D. Visalakshi, R. Meera, and S. Mohanapriya. 2015. Effect of Nano-black carbon in seed germination: *In-vitro* study. *International Journal of Biosciences and Nanosciences* 2:132–6.
87. Rashid, A., Q. Saleem, A. Nazir, and H. S. Kazım. 2003. Yield potential and stability of nine wheat varieties under water stress conditions. *International Journal of Agriculture and Biology* 5:7–9.
88. Raza, A. 2012. Grain and field annual 2012 gain report. Global Agricultural Information Network. *USDA Foreign Agrilcultural services* 1–3.
89. Rengasamy, P. 2002. Transient salinity and subsoil constraints to dryland farming in Australian sodic soils: An overview. *Australian Journal of Experimental Agriculture* 42:351–61.
90. *Roelofs, D, Jiwan P and kwangy I.. (2008). "Functional ecological genomics to demonstrate general and specific responses to abiotic stress". Functional Ecology. **22**: 8–18.* doi:10.1111/j.1365-2435.2007.01312.x.
91. *Rolf D, Vinebrooke K., Mittler J. (2004). "Impacts of multiple stressors on biodiversity and ecosystem functioning: the role of species co-tolerance". OIKOS. **104** (3): 451–457.* doi:10.1111/j.0030-1299.2004.13255.x.
92. *Rubio, Vicente; Linhares, Francisco; Solano, Roberto; Martín, Ana C.; Iglesias, Joaquín; Leyva, Antonio; Paz-Ares, Javier (2001).* "A conserved MYB transcription factor involved in phosphate starvation signaling both in vascular plants and in unicellular algae". *Genes & Development. **15** (16): 2122–2133.* doi:10.1101/gad.204401. ISSN 0890-9369. PMC 312755. PMID 11511543.
93. *Savvides, Andreas (2015).* "Chemical Priming of Plants Against Multiple Abiotic Stresses: Mission Possible?". *Trends in Plant Science. **21** (4): 329 340.* doi:10.1016/j.tplants.2015.11.003. hdl:10754/596020.
94. Scholander, P. F., H. T. Hammel, E. D. Bradstreet, and E. A. Hemminogsen. 1965. Sap pressure in vascular plants: Negative hydrostatic pressure can be measured in plants. *Plant Sciences* z48:339–46.
95. Shahida Naveed, Muhammad Ibrar, Inayat Khattak, Imtiaz Khan, Imran, Muhammad Isfhaq Khan and Haroon Khan. 2016.

Anthelmintic, Antilice, Insecticidal, Cytotoxic and Phytotoxic Potential Of Ethanolic Extracts Of Two Wild Medicinal Plants *Iphiona Grantioides* And *Plucheaarguta.* J. Woulfenia (23) 11: 13-25

96. Shamsi, K., and S. Kobraee. 2013. Biochemical and physiological responses of three wheat cultivars (*Triticum aestivum* L.) to salinity stress. *Annals of Biological Research* 4:180–5.
97. Shamsuddin Baqa[1], Amir Zaman Khan[1], Inamullah[1], Imran*[1], Asad Ali Khan*[1], Shazma Anwar[1], Babar Iqbal[1], Shehryar Khan[1], Amjad Usman[2]. 2015. Influence Of Farm Yard Manure And Phosphorus Application On Yield And Yield Components Of Wheat. J. Pure & Applied biology Vol. 4, Issue 4.
98. Shazma Anwar, Israeel, Babar Iqbal, Asad Ali Khan, Imran, Wajid Ali Shah, Mohammad Islam, Wajid Ali Khattak, Ikramullah, Wasim Akram and Wasim Abbas. 2016. Nitrogen and phosphorus fertilization of improved varieties for enhancing phenological traits of wheat. Pure and Applied Biology. Vol. 5, Issue 3, pp511-519. http://dx.doi.org/10.19045/bspab.2016.50065
99. Sidari, M., C. Mallamaci, and A. Muscolo. 2008. Drought, salinity and heat differently affect seed germination of Pinus pinea. *Journal of Forest Research* 13:326–330.
100. Syed Arshad Karim and Imran. 2019. Reduction in substrate moisture content reduce final yield of wheat. Biomed J. Sci & Tec Res. 1-5.
101. Tank, N, and M. Saraf. 2010. Salinity-resistant plant growth promoting rhizobacteria ameliorates sodium chloride stress on tomato plants. *Journal of Plant Interaction* 5:51–8.
102. Tiwari S., P. Singh, R. Tiwari, K. K. Meena, M. Yandigeri, D. P. Singh, and D. K. Arora. 2011. Salt-tolerant rhizobacteria-mediated induced tolerance in wheat (*Triticum aestivum*) and chemical diversity in rhizosphere enhance plant growth. *Biological Fertile Soils* 47:907–16.
103. *Tombesi, Sergio; Frioni, Tommaso; Poni, Stefano; Palliotti, Alberto (2018). "Effect of water stress "memory" on plant behavior during subsequent drought stress". Environmental and Experimental Botany.* ***150****: 106–114.* doi:10.1016/j.envexpbot.2018.03.009. ISSN 0098-8472.
104. Ulfat Samreen, Muhammad Ibrar, Lalbadshah, Shahida Naveed, Imran and Inayat Khatak. 2016. Ethnobotanical study of subtropical hills of Darazinda, Takht-e-Suleman range F.R D.I. Khan, Pakistan. Pure and Applied Biology. Vol. 5, Issue 1, pp149-164. http://dx.doi.org/10.19045/bspab.2016.50020
105. Van Hoorn, J. W., K. Nader, A. Hamdy, and M. Mastrorilli. 2001. Effect of salinity on yield and nitrogen uptake of four grain legumes

and on biological nitrogen contribution from the soil. *Agriculture Water Management* 51:87–98.

106. Imran, Amanullah, Abdel Rehman Al Tawaha et al. 2021. Use of Biochar in Agriculture: An Inspiring Way in Existing Scenario. Organic Farming for Sustainable Development. CRC Press, Taylor and Frances, USA.
107. Imran, Amanullah, Abdel Rehman Al Tawaha, et al. 2021. Organic Amendments: Right Way to Soil Sustainability. Organic Farming for Sustainable Development. CRC Press, Taylor and Frances, USA.
108. Imran, Amanullah and Abdel Rahman Al-Tawaha. 2021. Carbon Sources Application Increase Wheat Yield And Soil Fertility. *Communication in soil science and plant analysis DOI: 10.1080/00103624.2020.1865397.*
109. *Wang, W.; Vinocur, B.; Altman, A. (2007). "Plant responses to drought, salinity and extreme temperatures towards genetic engineering for stress tolerance". Planta.* ***218*** *(1): 1–14.* doi:10.1007/s00425-003-1105-5. PMID 14513379.
110. Wang, G., and Z. Xu. 2013. The effects of Nano-black carbon on germination and growth of wheat in different saline-alkali soil. *Asian Agricultural Research* 5:116–9.
111. Wang. H., and J. Jin. 2007. Effects of zinc deficiency and drought on plant growth and metabolism of reactive oxygen species in maize (*Zea mays* L.). *Agricultural Sciences in China*6:988–95.
112. Wolfe, A. 2007. “Patterns of biodiversity.” Ohio State University, 10.1016/j.agee.2006.12.013.
113. Xu, D., J. Huang, S. Guo, X. Yang, Y. Bao, H. Tang, and H. Zhang. 2008. Over-expression of a TFIII-type zinc finger protein gene ZFP252 enhances drought and salt tolerance in rice (*Oryza sativa*L.). *Federation of European Biochemical Societies Letters* 582:1037–43.
114. Yeo, A. R., M. E. Yeo, S. A. Flowers, and T. J. Flowers. 1990. Screening of rice (*Oryza sativa* L.) genotypes for physiological characters contributing to salinity resistance, and their relationship to overall performance. *Theoretical Applied Genetics* 79:377–84.
115. Younis, U., M. Athar, S. A. Malik, M. H. RazaShah, S. Mahmood.2015. Nano-black carbon impact on physiological and biochemical attributes of spinach *Spinacia oleracea* (L.) in nickel contaminated soil. *Global Journal Environment Science Management* 1:245–54.

116. Zakaria, M., S. Daniel, V. Murphy, and L. K. Abbott. 2012. Nano-black carbons influence seed germination and early growth of seedlings. *Plant Soil* 353:273–87.
117. *Zargar, Sajad Majeed; Nagar, Preeti; Deshmukh, Rupesh; Nazir, Muslima; Wani, Aijaz Ahmad; Masoodi, Khalid Zaffar; Agrawal, Ganesh Kumar; Rakwal, Randeep (2017). "Aquaporins as potential drought tolerance inducing proteins: Towards instigating stress tolerance". Journal of Proteomics.* ***169****: 233–238.*
118. doi:10.1016/j.jprot.2017.04.010. ISSN 1874-3919.
119. Zhan, H. X., Z. J. Chang, A. L. Wei, X. J. Zhang, and X. Li. 2011. Impact of drought to wheat physiological index. *Journal of Shanxi Agriculture Sciences* 39:1049–51.
120. A.R. M. Tawaha, Z. Farrokhi, N. Yoga, P. Roshan, Imran, Amanullah, et al.2021. Weed Management in Organic Cropping Systems. Organic Farming for Sustainable Development. CRC Press, Taylor and Frances, USA
121. Imran, Amanullah and Abdel Rehman Al-Tawaha.2020. The Productivity of subsequent wheat enhanced with residual carbon sources and phosphorus under improved irrigation system. *Communication in soil science and plant analysis*, DOI:10.1080/00103624.2020.1763387
122. Imran, Amanullah, Asad Ali khan, Tariq Mahmood, Abdel Rahman Al Tawaha and Samia Khanum[4]. 2020. Adequate Fertilization, Application Method and Sowing Techniques Improve Maize yield and related traits. *Communication in soil science and plant analysis*.
123. Mohammad Ilyas, Gohar Ayub, Imran, Azmat Ali Awan and Manzoor Ahmad.2020. Effect of Calcium and Boron on Production and Quality of Autumn Potato under Chilling Temperature. *Communication in soil science and plant analysis*. DOI: 10.1080/00103624.2020.1854286
124. Abdel Rahman M. Al-Tawaha, Nusrat Jahan…Imran, et al 2020. Growth, Yield and Biochemical Responses in Barley to DAP And Chitosan Application Under Water Stress. *Journal of Ecological Engineering*. 21(6). 86-93. Doi. 10.12911/22998993/123251.
125. Imran, Amanullah, Muhammad Arif, Zahir Shah and Abdul Bari. 2020. Integration of Peach (*Prunus persica* L.) residues, beneficial microbes and phosphorus enhance phenology, growth and yield of soybean. ISSN 1068-3674, *Russian Agricultural Sciences*, 2020, Vol. 46, No. 3, pp. 223–230. © Allerton Press, Inc., 2020.
126. Imran, Amanullah, Muhammad Arif, Zahir Shah and Abdul Bari. 2020. Soil application of *Trichoderma* and Peach (*Prunus persica* L.) residues possesses biocontrol potential for weeds and enhances growth

and profitability of soybean (Glycine max). *Sarhad Journal of Agriculture* 36 (1) 10-20. DOI | http://dx.doi.org/10.17582/journal.sja/2020/36.1.10.20

127. Imran, Amanullah and Abdel Rahman M. Al Tawaha. 2021a. Management of Nano-black carbon, Phosphorous and Bio Fertilizer improve Soil organic carbon and ensilage biomass of soybean and maize. *Communication in Soil Science and Plant Analysis.* DOI.10.1080/00103624.2021.1966439
128. Imran, Amanullah and Abdel Rehman Altawaha. 2021b. Carbon Assimilation and Dry Matter Partitioning in Soybean Ameliorates with the Integration Of Nano-Black Carbon, Along With Beneficial Microbes and Phosphorus Fertilization. *Journal of Plant Nutrition*
129. Imran. 2021. The bioavailability of Phosphorus in composite vs. Hybrid maize differ with phosphorus and boron fertilization. *Phosphorus, Sulfur, and Silicon and the Related Elements. https://doi.org/10.1080/10426507.2021.1920588.*
130. Imran, Amanullah, Asad Ali khan, Tariq Mahmood, Abdel Rahman Al Tawaha and Samia Khanum[4]. 2021c. Adequate Fertilization, Application Method and Sowing Techniques Improve Maize yield and related traits. *Communication in soil science and plant analysis.* DOI:10.1080/00103624.2021.1925688
131. Imran and Amanullah· 2021. Phosphorus and Boron Application Optimizing Biofortification Of phosphorus and Productivity of French Bean (*Phaseolus Vulgaris* L.). *Communication In Soil Science and Plant Analysis.*
132. Imran, Amanullah and Abdel Rahman Al-Tawaha. 2021d. Carbon Sources Application Increase Wheat Yield And Soil Fertility. *Communication in soil science and plant analysis DOI: 10.1080/00103624.2020.1865397.*
133. Imran, Amanullah and Abdel Rehman Al-Tawaha.2020b. The Productivity of subsequent wheat enhanced with residual carbon sources and phosphorus under improved irrigation system. *Communication in soil science and plant analysis*, DOI:10.1080/00103624.2020.1763387
134. Imran, Amanullah, Muhammad Arif, Zahir Shah and Abdul Bari. 2020a. Soil application of *Trichoderma* and Peach (*Prunus persica* L.) residues possesses biocontrol potential for weeds and enhances growth and profitability of soybean (Glycine max). *Sarhad Journal of Agriculture* 36 (1) 10-20. DOI | http://dx.doi.org/10.17582/journal.sja/2020/36.1.10.20

135. Mohammad Ilyas, Gohar Ayub, Imran*, Azmat Ali Awan and Manzoor Ahmad.2020. Calcium and Boron effect on Production and Quality of Autumn Potato under Chilling Temperature. *Communication in soil science and plant analysis*.
DOI: 10.1080/00103624.2020.1854286

CHAPTER NINE

STRATEGIES FOR IMPROVING CROP YIELD UNDER SOIL, WATER AND ENVIRONMENT CRISES

IMRAN[1,2,3]

[1]DEPARTMENT OF AGRONOMY, THE UNIVERSITY OF AGRICULTURE, PESHAWAR

[2]AGRICULTURE OFFICER (EXTENSION WING), OFFICE OF THE DISTRICT DIRECTOR AGRICULTURE EXTENSION, DISTRICT KURRAM, MERGED DISTRICTS KHYBER PAKHTUNKHWA, PAKISTAN

[3]RHIZOSPHERE LABORATORY, UNIVERSITY OF CUKUROVA, FACULTY OF AGRICULTURE, DEPARTMENT OF SOIL SCIENCE AND PLANT NUTRITION, ADANA, TURKEY,

Abstract

To increase crop productivity various strategies should be applied. Different techniques like crop rotation and proper cropping pattern, suitable crop verities and cultivars, organic farming, brown and green manuring should be applied sole or as integrated approach, to get maximum benefits of soil resources on sustainable bases. Crop rotation can increase diversity of crop species, ensure effective nutrient uptake and Preserve soil fertility. Conservation tillage or reduce tillage can decrease soil pollution because more tillage cause soil erosion which increase water pollution and decrease soil productivity and crop yield. Combination with chemical with organic selecting suitable time and source can enhanced productivity of soil. Green manuring is also a good method for maintaining soil fertility by the incorporation of crops in the soil at full vegetative growth stage. Organic farming enhancing soil productivity on sustainable bases incorporation of different organic manures like FYM, composting etc. Application of Beneficial microbes decrease dependence on chemical fertilizers, and hence decrease soil and air pollution and also reduced the emission of GHGs, thereby help to mitigate and reduced the risk of climate change.

Keywords: Climate change, cropping pattern, organic amendments, green house gases, soil fertility, nutrients eplenishment

1. Introduction

Integrated crop management is method of agriculture farming that fulfill the farming need and requirements of a profitable business with sustainable and safety to the environment. This system includes practices of managing waste, increasing efficacy of energy and decrease soil and air pollution. It is the combination of both, technology of the modern period along with good foaming principles, for ensuring crop productivity, their sustainability and safety to the environment. ICM include crop rotation and proper cropping pattern, suitable crop verities and cultivars, enhancing soil organic matter, reducing soil and air pollution, land scape maintenance, decreasing dependency on chemical inputs (fertilizers and pesticides), integrated nutrients management, tillage system and advances farming techniques like organic farming, green and brown manuring, enhancing soil biodiversity and using of climate smart agriculture practices.

1.1. Cropping pattern and rotation

For improving soil structure and health crop rotation is good agronomic practices for example deep rooted crops should be followed by shallow rooted crops and exhaustive crops should be followed by restorative crops, non-legumes crops should be followed by legumes crops. Tosti et al. (2014) stated that intercropping decrease nitrogen losses ac compared with monocropping. Pure vetch was reported highest N source potential as green manure. Crop rotation have beneficial effects on crop and soil status (1) increase diversity of crop species to prevent disease and pest carry over from crop to crop; (2) Ensure effective nutrient uptake by scheduling crops with different nitrogen demands in the correct sequences; (3) Preserve soil fertility, structure and minimize erosion by ensuring adequate crop cover, good rooting depth and reduction of compaction. Including pulses in the maize-wheat rotation improved soil organic carbon (SOC) and plant available macronutrients being higher in maize-wheat-mungbean (*Vigna radiata* L.) rotation. Inclusion of mungbean to maize-wheat rotation enhanced the nitrogen (33.9%), phosphorus (46.4%), potassium (36.3%), and sulphur (55.5%) uptake in maize crop; likewise, alternate-year chickpea inclusion increased the uptake of these nutrients by 18.2, 19.1, 21.7, 32.1%, respectively (Venkatesh et al., 2017). Co-cropping of restorative (legume) and exhaustive (cereal) crops paved its way again in modern agriculture as an environment friendly alternative to fertilizers (Akhtar et al., 2010). In context of both the cost and environmental impact of chemical fertilizers, excessive reliance on the chemical fertilizers is not viable strategy in long run because of the cost, both in domestic resources and foreign exchange, involved in setting up of fertilizer plants and sustaining the production. In this context, organic fertilizers and bio-fertilizers would be the viable option for farmers to increase productivity per unit area (Mahdi et al., 2010)

1.2. Tillage practices

Reduce tillage is tool for reducing soil and air pollution. Conservation tillage or reduce tillage can decrease soil pollution because more tillage cause soil erosion which increase water pollution and decrease soil productivity and crop yield. More tillage also harmful for the soil beneficial microorganism inside the soil. Conservation tillage improving soil health and crop productivity by preventing soil erosion as compared with conventional tillage system. Conservation tillage system can be defined as tillage or planting combination which left about 30% of plant residue on soil surface (Soil Science Society of America, 2008). Conservation system of tillage have many advantages in system of agriculture farming like conservation of soil, economic advantages, decrease crop failure and crop establishment time, conserve energy use, decrease soil erosion and air pollution and also reduces the rate of organic matter decomposition as result improve soil quality (Carter, 1994; Fawcett, 2008; Dumanski and Peiretti 2013; Derpsch et al. 2014). Zero tillage system not only decrease soil erosion but also improve soil fertility (Sullivan, 2004). Conservation tillage increase soil moisture retention and reduces runoff of water and retain organic matter on soil surface for long time. Zero tillage also help to increase soil aeration due loosen soil materials because it not compacts by the heavy instruments. In contrast, more and deep tillage system considerably decrease soil worm population.

1.3. Integrated nutrients management: climate change mitigation

Chemical fertilizers excessive use particularly N and P are among the major agriculture pollution factors (Roba, 2018). Application of improve agriculture practice can enhance organic carbon of soil, productivity of soil and water holding capacity (Manna et al., 2005). Application of INM practices decrease dependence on synthetic fertilizers FAO, 2018a). INM practice is more useful to bring sustainability and ecofriendly agriculture management (Panhwar et al., 2019). Combination with chemical with organic selecting suitable time and source can enhanced productivity of soil (Choudhary et al., 2018) also can improving growth of plants, portioning of dry matter and photosynthesis (Bharali et al., 2017) by providing a healthy and productive soil to the farmers, without affecting ecosystem (Tian et al., 2015; Venkatesh et al., 2017). INM permits workable farms to prevent the excessive use of chemical fertilizers. Application of organic and inorganic fertilizers can enhanced accumulation of carbon and soil quality(Bharali et al., 2017) Benbi et al., 2012. Moreover, INP techniques can increase NPK uptake of plant as compared with single chemical fertilizer application (Dilshad et al., 2010). Combine application of both inorganic and organic can enhanced crop yield and soil carbon. Application of FYM and NPK fertilizers will be good choice for enhancing the yield of crops (Panhwar et al., 2019) by enhancing soil biodiversity (Kumar et al., 2017).

1.4. Sowing methods

Proper sowing of crops enhancing soil health and maintaining soil fertility by maintaining suitable plant to plant and row to row distance, very dense plant population leads to all nutrients reduction in soil for supporting good crop growth. Intensive farming practices for example over grazing and intensive farming without proper management increase soil erosion and soil fertility. Egg plants intercropped with S. nigrum enhanced the growth of crop and decrease uptake of Cd (Tang et al., (2017). Intercropping of maize/fababean had high uptake of nitrogen as compared with wheat/maize with faba bean (Li et al., 2011). Crops of different maturity dates grown together increase nitrogen use efficacy and decrease accumulation of soil mineral nitrogen.

1.5. Green manuring

Green manuring is also a good method for maintaining soil fertility by the incorporation of crops in the soil at full vegetative growth stage, green manuring crops is also called fertility building crops, these crops are grown for the maintaining and enhancing fertility of the soil (Rayns and Rosenfeld, 2018). Green manuring crops mostly grown for supply nitrogen to the soil, in the system of organic farming it is consider is a main source of nitrogen. Green manuring of forage and mustard crops improving soil status by increasing soil respiration rate and decrease heavy metals availability in the soil (Foucault et al., 2013). Green manuring supplying crops nutrients from organic sources, but it has got some restrictions as it needs a period of about 45-60 days for decomposition with proper temperature and optimum moisture conditions after incorporation. The viable option left behind that is brown manuring as a tool for integrated nutrient management (Maitra and Zaman, 2017).

1.6. Brown manuring

Like green manuring, brown manuring is also a good techniques for improving plant and soil nutrients status by growing green manuring crops like *sesbina, dhaincha* and *sunhemp* etc., as intercrops or mixed crops and then destring them by the postemergence herbicide for manuring purposes. Due to the application of herbicide the color of these crops become brown due the loss of chlorophyll, that I why this process is called brown manuring (Tanwar et al., 2010). Application of brown manuring have many benefits for the us by improving soil organic matter, enhancing soil nitrogen and other plant nutrients in the soil and decreasing soil bulk density and soil erosion (Iliger et al., 2017). Application of brown manuring and decrease the use of chemical fertilizer by 25% (Iliger et al., 2017). Raped decomposition of *Sesbania* leaves fell on soil surface increase nitrogen supply, organic matter in the soil and also decrease weed density and their prolific growth as result decrease crops weed competition (Mondal et al., 2012).

Brown manuring is new version of no tillage green farming can b used for maize, sugarcane and rice etc. by using herbicide.

1.7. *Drought resistant varieties of crops*

Selecting suitable crops verities is essential for good crop productivity under changing climate. Those area which are prone to drought plant of drought resistant crops and their suitable verities are very helpful to get economic yield and to reduce climate change vulnerability. For example, crops like wheat require considerable lees amount of water than dry season rice. Growing of drought resistant and tolerant crop verities are used as effective tool as method for climate change adaptation in Burkina, Nigeria and Ghana (Ngigi, 2009).

1.8. *Crop diversification*

Diversification of crop is very important to increase crop productivity and farmers income. On priority base diversification of crop is range first in both irrigated and non-irrigated areas. For example, in wastarun sudan Statas like Kordofan and, food crops have substituted cash crop, and introduction of high resilient verities of crop was made towards high value crops is feasible in (DFID, 2004). Diversion of crops can decrease and spread the crops failure risk. Diversification of crop decrease the negative effects and risks of variability in rainfall (Orindi and Eriksen 2005; Adger *et al.*, 2003).

1.9. *Change in cropping pattern and calendar of planting*

Change in climate negatively affects production of crops due to change in rainfall pattern as result pattern and calendar of cropping and farm operation is also change like crop sowing time, maturity, harvesting, and irrigation schedule. Agriculture system diversity is a real operative source of reducing climate change risks and growing of farmer income. Diversification of crops is equally important either on highly potential areas or marginal area. Development of crop verities which have low requirement of crops is another good opportunity, to grow the crops and fodder for human and animals, respectively. For example, spineless cactus is grown as ruminants' fodder in Tunisia.

1.10. *Organic farming: crop productivity and soil health*

Application organic fertilizer is useful for enhancing fertility of soil, soil organic carbon and water holding capacity and porosity (Han et al., 2016). Soil having enough organic matter can reduced the risk of climate change (Paris Agreement, 2015). Organic crop management can increase soil carbon than conventional farming (Gattinger et al., 2013; Tuomisto et al., 2012). Due to the exponential growing of population our agriculture soil is forcing to a tremendous pressure to produce food and feed for the human and their animals, respectively (Tsiafouli et al., 2016). Therefore, to enhance soil productivity on sustainable bases incorporation of different organic manures like FYM, composting, green and brown manuring in soils are urgently (Fan et al. 2011). Maintaining carbon-rich soils on sustainable base can play a vital role in addressing the surety of food security, food systems adaptation and people to change in climate (Paris Agreement, 2015). Soil microbial activity and their population can be improving with application of Organic manures which subsequently increase soil productivity (Francioli et al., 2016). in semiarid climate condition low soil organic carbon is the major problem o low crop and soil productivity. Low organic matter in these soils cause depletion of essential plant nutrients. Different sources of soil organic matter such aa animal and plant residues) in various forms (ash, biochar, dry and fresh etc can enhanced fertility and productivity of soil, income of the growers. Biochar application in combination with organic or inorganic fertilizer has been testified to have a outstanding effect on crop yield (Adekiya et al. 2019). Organic carbon of the soil is important pointer of a healthy soil due to its important functions that affect soil fertility, productivity, and the environment (Merante et al., 2017; Fang et al., 2018).

1.11. Biofertilizers

Application of biofertilizers is new techniques for higher production on sustainable base (Naher et al., 2016). Application of Beneficial microbes decrease dependence on chemical fertilizers, and hence decrease soil and air pollution and also reduced the emission of GHGs, thereby help to mitigate and reduced the risk of climate change. Application of different strain of microorganism like bacteria and Fungai can increase availability of certain nutrient in soil which subsequent increase nutrient concentration in the soil like phosphorus solubilizing bacteria, zinc and potassium solubilizing bacteria. Additionally, application of biofertilizers are very ecofriendly, it increases soil microorganism population and soil biodiversity as well (Kannaiyan, 2002). Micorgansim which are free living in the soil and fix nitrogen for the plant can account of 25% to 40% throughout their life cycle (Panhwar et al., 2012. P availability and crop yield can be enhanced with application of Phosphate solubilizing bacteria (Wani et al., 2007; Henri et al. 2008)

Conclusions

Integrated crop management is method of agriculture farming that fulfill the farming need and requirements of a profitable business with sustainable and safety to the environment through different techniques like crop rotation and proper cropping pattern, suitable crop verities and cultivars, enhancing soil organic matter, reducing soil and air pollution, land scape maintenance, Crop rotation can increase diversity of crop species, ensure effective nutrient uptake and Preserve soil fertility. Conservation tillage or reduce tillage can decrease soil pollution because more tillage cause soil erosion which increase water pollution and decrease soil productivity and crop yield. Combination with chemical with organic selecting suitable time and source can enhanced productivity of soil. Green manuring is also a good method for maintaining soil fertility by the incorporation of crops in the soil at full vegetative growth stage.

References

1. Adekiya A, Agbede T, Aboyeji C, Dunsin O, Simeon V. 2019. Effects of biochar and poultry manure on soil characteristics and the yield of radish. Scientia Horticulturae 243:457-463.
2. Bharali A, Baruah KK, Bhattacharyya P, Gorh D. 2017. Integrated nutrient management in wheat grown in a northeast India soil: impacts on soil organic carbon fractions in relation to grain yield. Soil and Tillage Research 168, 81-91. doi:
3. Choudhary, M., Panday, S.C., Meena, V.S., Singh, S., Yadav, R.P., Mahanta, D., Mondal, T., Mishra, P.K., Bisht, J.K., Pattanayak, A., 2018. Long-term effects of organic manure and inorganic fertilization on sustainability and chemical soil quality indicators of soybean-wheat cropping system in the Indian mid-himalayas. Agric. Ecosyst. Environ. 257, 38–46.
4. Dilshad, M., Lone, M., Jilani, G., Malik, M.A., Yousaf, M., Khalid, R., et al., 2010. Integrated plant nutrient management (IPNM) on maize under rainfed condition. Pak J Nutr 9 (9), 896_901.
5. Fan M, Shen J, Yuan L, Jiang R, Chen X, Davies WJ, Zhang F. 2011. Improving crop productivity and resource use efficiency to ensure food security and environmental quality in China. Journal of Experimental Botany 63:13-24.
6. Fang, J., Zhan, L., Ok, Y.S., Gao, B., 2018. Minireview of potential applications of hydrochar derived from hydrothermal carbonization of biomass. J. Ind. Eng. Chem. 57, 15–21.

7. FAO (2018a). Soil Pollution: a hidden reality. Rome, FAO. 142 pp. (ISBN 978-92-5-130505-8).
8. Francioli, D., Schulz, E., Lentendu, G., Wubet, T., Buscot, F., Reitz, T., 2016. Mineral vs. Organic amendments: microbial community structure, activity and abundance of agriculturally relevant microbes are driven by long-term fertilization strategies. Frontiers in Microbiology 7, 1–16. http://dx.doi.org/10.3389/fmicb.2016.01446.
9. Gattinger, A., Muller, A., Haeni, M., Skinner, C., Fliessbach, A., Buchmann, N., Mäder, P., Stolze, M., Smith, P., Scialabba, N.E.-H., Niggli, U., 2012. Enhanced top soil carbon stocks under organic farming. Proc. Natl. Acad. Sci. U. S. A. 109, 18226–18231.
10. Han, S.H., Young, J., Hwang, J., Kima, S.B. and Parka, B. (2016) the Effects of Organic Manure and Chemical Fertilizer on the Growth and Nutrient Concentrations of Yellow Poplar (Liriodendron tulipifera Lin.) in a Nursery System. Forest Science and Technology, 12, 137-143. https://doi.org/10.1080/21580103.2015.1135827
11. Henri, F., Laurette, N.N., Annette, D., John, Q., Wolfgang, M., Franc¸ois-Xavier, E., et al., 2008. Solubilization of inorganic phosphates and plant growth promotion by strains of Pseudomonas fluorescens isolated from acidic sols of Cameroon. Afr. J. Microbiol. Res. 2, 171_178.
12. Kannaiyan, S., 2002. Biotechnology of biofertilizers: Alpha Science Int'l Ltd. UK, pp. 1_375.
13. Merante, P., Dibari, C., Ferrise, R., Sanchez, B., Iglesias, A., Lesschen, J.P., Kuikman, P., Yeluripati, J., Smith, P., Bindi, M., 2017. Adopting soil organic carbon management practices in soils of varying quality: implications and perspectives in Europe. Soil Tillage Res. 165, 95–106.
14. Naher, U.A., Panhwar, Q.A., Radziah, O., Razi, I.M., Zulkarami, B., 2016. Biofertilizer as a supplement
15. Panhwar, Q.A., A. Ali, U.A. Naher, and M.Y. Memon. 2019. Fertilizer Management Strategies for Enhancing Nutrient Use Efficiency and Sustainable Wheat Production. In Organic Farming. Elsevier. 17-39.
16. Panhwar, Q.A., Othman, R., Rahman, Z.A., Meon, S., Ismail, M.R., 2012. Isolation and characterization of phosphate-solubilizing bacteria from aerobic rice. Afr J Biotechnol 11 (11), 2711_2719.
17. Roba, T.B., 2018. Review on: the effect of mixing organic and inorganic fertilizer on productivity and soil fertility. Open Access Library J. 5, e4618. https://doi.org/10.4236/oalib.1104618.
18. Venkatesh, M.S., K.K. Hazra, P.K. Ghosh, and B.L. Khuswah et al., 2017. Long–term effect of crop rotation and nutrient management on

soil–plant nutrient cycling and nutrient budgeting in Indo–Gangetic plains of India. Arch. Agron. Soil. Sci. https://doi.org/10.1080/03650340.2017.1320392.

19. Wani, P.A., Khan, M.S., Zaidi, A., 2007. Chromium reduction, plant growth promoting potentials and metal solubilization by Bacillus sp. isolated from alluvial soil. Curr. Microbiol. 54, 237_243.

CHAPTER TEN

CARBON FARMING: A SOLUTION TO SOIL AND ENVIRONMENT CRISES

IMRAN[1,2,3]

[1]DEPARTMENT OF AGRONOMY, THE UNIVERSITY OF AGRICULTURE, PESHAWAR

[2]AGRICULTURE OFFICER (EXTENSION WING), OFFICE OF THE DISTRICT DIRECTOR AGRICULTURE EXTENSION, DISTRICT KURRAM, MERGED DISTRICTS KHYBER PAKHTUNKHWA, PAKISTAN

[3]RHIZOSPHERE LABORATORY, UNIVERSITY OF CUKUROVA, FACULTY OF AGRICULTURE, DEPARTMENT OF SOIL SCIENCE AND PLANT NUTRITION, ADANA, TURKEY

Abstract

Organic matter incorporation is an ancient age agriculture practices to enhance soil aeration, nutrients supply and water infiltration rate along with water holding capacity. In modern day agriculture practices, the organic matter amendments is proven beyond doubts in respect of their diverse role for soil health. This study revealed that recycle of farm leftovers and its incorporation in soil can replenish soil nutrients, soil health, microbial activity and soil productivity. Previous study results revealed that phenological and Morphological traits of both cropping system (maize-wheat and soybean-wheat) responded positively and ameliorated with peach biochar organic amendments. Pre and post-harvest soil analysis of the maize and soybean field showed that soil health and nutrients status was significantly boosted with organic matter addition.

Keywords: carbon sequestration, soil health, climate change mitigation, microbial action, maximum plant biomass production

Introduction

Organic matter incorporation is an ancient age agriculture practices to enhance soil aeration, nutrients supply, soil porosity, soil health, soil productivity and water infiltration rate along with water holding capacity (Imran et al., 2015). It has been reported by Imran et al. (2016) and Imran and Asad. (2017), that initially organic amendments was used as supplement of chemical fertilizer to provide nutrients in abundance for crop growth and maintenance. Many researchers examined that soil reclamation and soil productivity could only be achieved with soil organic matter incorporation (Imran and Asad. 2015; Naveed et al., 2016; Imran et al., 2015; Khan et al., 2015; Baqa et al., 2015 and Iqbal et al., 2015). Various organic sources (manure and composted manure) brings a lot of beneficial microorganisms that can act as PGPR. All manures are not the same manure which gives the quick response. These are low in ammonia but have good nitrate concentration. In addition they are high is soluble phosphate and the complete complement of needed nutrition including its excellent ability to supply Calcium. Imran et al. (2020) reported that the use of a thermophilic turned leaf compost when amended with 10 to 20% castings gives season long and early growth stimulus. It also saves on application since it is useful for early and later stage nutrition without unnecessary multiple lay by applications. Plant residues are environmentally friendly materials has many favorable physicochemical characteristics, making it suitable for mixture in substrates including high porosity, good aeration, drainage and water holding capacity(Edwards and Burrows, 1988).

Vermicomposting

Vermicomposting, in contrast to conventional compost, is the product of an accelerated biooxydation of organic matter by the use of high densities of earthworm populations without passing a thermophilic stage Iqbal et al. (2016); Samreen et al. (2016); Anwar et al. (2016); Khan et al. (2016); and Imran et al. (2016). Different earthworm species are able to consume a wide range of organic residues such as sewage sludge, animal wastes (Edwards et al., 1985);, crop residues; (Orozco et al., 1996) and industrial wastes. The earthworm-processed organic wastes are finely divided peat-like materials with high porosity, aeration, drainage, and water-holding capacity sludge Compared to conventional compost which passes a thermophilic stage, vermicompost usually has a much finer structure and larger surface area providing strong absorbability and retention of nutrients. Based on all these characteristics, earthworm-processed organic waste would have a great commercial potential in the horticultural industry as container media for growing bedding and vegetable plants. There is strong scientific evidence that vermicompost can significantly influence the growth and productivity of plants(Edwards, 1998). All these characteristics of soil known as soil health.

Soil Reclamation by organic amendments

For degraded soil reclamation organic matter is mostly incorporated acts mainly as a source of nutrients, and improves some physical properties. On strongly degraded soils, mainly by acidity (pH in KCl even of 3,2, in water probably about 4) the effect of single dose of farmyard manure last even 3 years or more and consists mainly on softening of negative effects of acidity (probably complexing of exchangeable Al). It also depends on crop. It has been reported that, during 3 years the crop uptake (apparent) from farmyard manure was about of 25-40% of N, 20% P, 37-55%K and 10% of Mg introduced with farmyard manure Iqbal et al. (2016); Samreen et al. (2016); Anwar et al. (2016); Khan et al. (2016); and Imran et al. (2016). Organic matter application has a central role in nutrients supply, maintaining soil functions and preventing soil degradation (Zerihun and Deressa, 2017).

Additions of plant organic sources significantly increase the reactivity of the indigenous phosphate rock because of the different organic acids produced by the compost microflora (Nadeem et al., 2016). It increase soil characteristics like, mobilization of nutrients by mineralization, sorption capacity, short term immobilization into soil organisms (Prihastuti et al., 2015). Organic matter is mostly used in various form like fresh application of the farm residues incorporated in the soil, composted form of the farm residues application and sometime partially burn by a process of pyrolysis and applying known as biochar application (Khaim et al., 2013; Ahmad et al., 2014). Organic sources combined application with chemical fertilizers could be a promising technique to increase the production of soybean and maize. Intensive use of organic matter with chemical fertilizer for crop production increase grain yield, chemical composition of plant and soil with the advantage of maximizing economic return (Yagoub et al., 2012; Khaim et al., 2013).

Organic Manures application

The addition of organic manure improves soil fertility and increase crops yield (Maltas *et al*., 2013). When organic fertilizer is continually applied to the soil it decrease soil bulk density and increase porosity and buffering capacity of the soil (Edmeades, 2003). Soil amendments have both agronomically and economically important and help in availability of nutrients and enhance soil fertility (Opala *et al*., 2010). Addition of organic manure to the soil provide the essential macro and micro nutrients and improves soil health and thereby increase the crop growth and yield (Garcia-Franco, 2015). This integration can improve the yield and also sustain soil productivity (Chukwu *et al*., 2012). Organic fertilizer help in the enhancing soil characteristics by obtaining high yield as compared to inorganic fertilizer (Cezar, 2004).

The organic amendments improve soil physical, chemical and biological properties (Chukwu *et al*., 2012). Farmyard manure reduces soil erosion, nutrients leaching, and losses through evaporation and had residual effects for next crops. Farmyard manure remain for longer time in the soil, and provide the nutrients over longer period upon mineralization (FAO, 2004). However, the inappropriate application of manure causes crop and soil problem including weakening of soil structure, water and environmental pollution (Nyamangara *et al*., 2009). Different sources of organic manure farmyard manure, green manure, and poultry manure provide organic material to the soil (Muhammad *et al.,* 2011). Poultry manure add essential nutrient to the soil which improve soil fertility and thus improve maize productivity (Farhad *et al.*, 2011). Crop yield and yield component of maize was significantly increased by the application of PM (Javeed *et al.*, 2014).

Soil Health

To ensure soil quality, improved nutrients supply and potential crop production along with farmer's income there is a need of revising the current agricultural management practices with a view of recycling plant biomass and organic matter application (Arif et al., 2017). Recycling of nutrient from organic sources into the soil is a sustainable approach for improving soil chemical, physical and biological properties (Diep et al., 2016). Reduction in soil quality and poor nutrients use efficiencies is the major fact consistently declining agriculture productivity. Several management practices are adopted for a variety of soil types, cropping patterns and farmers resources, to reduce the magnitude of soil fertility degradation. Regardless of soil types and crop seasons, organic amendments are generally more effective to increase soil C and organic matter status as compared to mineral fertilizers (Fageria et al., 2013). To improve soil quality and fertility status sustainable fertilizer management practices have been developed by integrating them with a large number of organic nutrient sources (Yagoub et al., 2012). Application of organic amendments has been advocated to sustain soil organic matter, improve soil biological functioning, aeration and moisture retention, reduce compaction, pollutant attenuation and enhance soil nutrient supply (Prihastuti et al., 2015). These numerous benefits associated with soil application of organic amendments depend strongly on the quantity and nature of organic wastes (Khaim et al., 2013). Organic amendments for such purpose are available in huge quantities and originate from diverse resources including crop residues, peach residues, fruit plants wastes, green manures, industrial wastes, animal wastes and household waste (Arif et al., 2017).

Nutrients Replenish with organic matter

Low P availability to plants is a global problem limiting crop production, especially in arid and semi-arid regions of the world (Haque and Khan, 2012). As a result, a number of agronomic practices have been proposed to enhance P availability and use by crops under diverse climatic conditions (Munda et al., 2018). However, no studies have focussed on the use of peach wastes and beneficial microbes to increase P utilization from organic and inorganic P fertilizers (Ahmad et al., 2014).

Phosphorus is an important element for seeds development, roots elongation, roots nodulation in legumes, N_2 fixation and its deficiency limiting crop production and proper seed development (Munda et al., 2018). Phosphorus is needed by most of bacteria for conversion of atmospheric N (N_2) into an ammonium (NH_4). *Rhizobium* bacteria convert atmospheric N to useable form. Nitrogenase enzyme catalyzed N_2 into two molecules of ammonia (NH_3). These enzymes are synthesized by *Rhizobium*. It is well known that symbiotic N_2 fixation is a high P demanding process, hence nodule formation and N_2 fixation are generally limited by low P availability which adversely affects nodule number and mass, as well as nitrogenase activity (Begum et al., 2015). Symbiotic N_2 fixation has high ATP requirements for nodule development, function and need additional P for signal transduction and membrane biosynthesis (Fageria et al., 2013; Sharif et al., 2013; Prihastuti et al., 2015; Diep et al., 2016). Phosphorus is partially mobile in soil and interaction of maize and soybean with phosphorus along with beneficial microbes increase crop productivity and profitability (Ali et al., 2018). Combine use of phosphorus fertilizers with organic matter and beneficial microbes provide frequent supply of nutrients to plants and improve plant production and biomass with advantage of reasonable nutrients use efficiency.

Nutrients availability and Beneficial Microbes

To ensure phosphorus and other nutrients availability to plant and minimize adsorption on soil particles, beneficial microbe's application is more productive to work as a natural chelate (Wang et al., 2015; Zhongjun et al., 2017; Munda et al., 2018). Beneficial microbes (fungi, bacteria, protozoa, algae and viruses) are containing living cells commonly used in agriculture known as bio fertilizers. Some of the bio fertilizer products, used for nutrients availibilty in agronomic crops are Mycorrhizae, Rhizobia, Azospirillum, Pseudomonas, Bacillus, *Trichoderma*, Streptomyces species etc (Zarei et al., 2012; Bargaz et al., 2018). Bio fertilizer is a wide term which includes a diverse category of bio inoculants such as nitrogen fixers, phosphate solubilizers, phosphate mobilizers and plant growth promoting rhizobacteria (Suppadit et al., 2012; Henri et al., 2014). The mineral mobilizer, arbuscular mycorrhizal fungus has been found to be quite effective for growth improvement and saving in phosphorus nutrition in agronomic crops such as maize, rice, wheat and soybean under field conditions (Argaw, 2012; Haque and Khan, 2012; Keswani et al., 2014). The phosphate solubilizers (PSB, *Trichoderma*) is generally solubilize phosphate by the release of low molecular weight organic acids (formic acids, fumeric acids, acetic acids, propionic acids and succinin acids), which through their hydroxyl and carboxyl groups chelate the cations bound to phosphate, thereby converting it into soluble forms (Xiang et al., 2012). Further vesicular arbuscular mycorrhizal (VAM) fungi grow extensively in soil to form a well-developed hyphal network that absorbs inorganic phosphate several centimeters away from the root surface. Phosphorus is translocated rapidly to the roots overcoming the slow diffusion that occurs in the soil solution. It has been reported that soybean and wheat grain yield increases with combines application of phosphate solubilizers (Mahanta et al., 2014; Munda et al., 2018).

Integrated Approach and soil sustainability

Combined use of organic and mineral fertilizers along with beneficial microbes' inoculation increases each other's efficiency. Interaction of organic, chemical and biological sources not only sustain crop productivity and soil health by effective management but they also optimize chemical fertilizers utilization for different crops in different cropping systems (Prabha et al., 2013; Zhongjun et al., 2017). Integrated P management is essential for the improvement of crop growth, seed yield, nodulations in lugumes and human health (Arif et al., 2017; Baghdadi et al., 2018; Sara et al., 2018). Application of organic sources increases soil organic matter, P availability and soybean productivity (Bibi et al., 2010; Frank ct al., 2017). Use of proper organic and inorganic fertilizer combination, is a key factor in crop production for sustainable agriculture (Hidayatullah et al., 2013; Iqbal et al., 2015; Oloo et al., 2016). Arif et al. (2017) concluded that biochar application enhanced crop productivity, soil properties and phosphorus use efficiency (PUE) when applied with chemical fertilizer along with other organic sources.

In many parts of the world have been plagued by saturated soils from a very early time. Imran et al (2020) reported that we continued to get record rainfall from April, when most farmers would think about planting, all the way through today. This has caused not only late-planted soybeans, but soybeans that in a lot of cases were planted into less than ideal soil conditions and are experiencing continued saturated soil conditions. This has caused soybeans to take a long time to emerge and they seem to have struggled to grow ever since. Recent warm weather has helped pushed beans along a little, but with saturated soils we're still seeing soybeans with slow growth.

The most common symptoms observed have been slow emergence, reduced early vigor, yellowing of lower leaves, necrosis or death, poor root systems and lack of nodule formation. The main cause for these symptoms can be attributed to saturated soils, which has caused a lack of oxygen and poor root development. We then saw problems like nutrient deficiencies and lack of nodulation. Determining how to overcome these problems may require practices and products that we don't typically use.

Nitrogen: When we see yellow and stunted crops our minds automatically think nitrogen (N), and rightfully so. Applying nitrogen to soybeans is still a somewhat controversial topic. We know that soybeans will begin fixing their own nitrogen and rarely recommend adding N to this crop. There is concern that giving the soybean N early on will make it lazy and not able to produce the nitrogen it needs later in the season. However, I think we are seeing a lack of nitrogen availability in small soybeans that are not yet fixing N and seeing issues with larger soybeans being able to fix their own nitrogen. I think the addition of a small amount of N may give the soybean just the boost it needs. Be cautious on applying too much, though.

Key nutrients: A few other key nutrients of interest are sulfur (S) and molybdenum (Mo). More soybeans are receiving planned applications of sulfur than ever before. With less sulfur available through acid rain we are seeing a larger need for sulfur applications. Sulfur helps promote nodulation in soybeans, is required for chlorophyll production and aids in seed production. Sulfur is mobile in the soil, so it may have been leached out with all the rain. Molybdenum is vital for fixing nitrogen by rhizobia bacteria. It can also move through the soil profile and may have leached out. The addition of S and Mo are likely to help soybeans that are struggling to produce nodules and begin fixing N.

Foliar feeding: Another idea would be an application of foliar fertilizers in the POST herbicide trip. Before making a fertilizer application it's important to identify which nutrients are lacking in the plant. A tissue sample, along with a soil sample, will help determine the best route to take. I believe that we may not be lacking nutrients in the soil, but because of poor root development we are seeing nutrient deficiencies in the plant. Giving the soybeans the needed nutrients through either a foliar fertilizer with herbicides or a topdress application could be extremely beneficial this year. Make sure the product you use has a high concentration of the needed nutrients because those with only small concentrations won't last long if conditions don't improve.

Aeration: One highly unpopular recommendation would be to cultivate if you are able. This would help to dry and put oxygen back in the soil. However, I realize that most growers probably don't own a rotary hoe or row crop cultivator in an era of reduced and no-till, with herbicides to control weeds.

Disease risk: Disease management will play an important role in protecting the soybean crop we have. If we continue to receive wet weather, we will likely see foliar diseases like white mold and frogeye leaf spot develop. Scout fields and be prepared to make foliar fungicide applications at R3 to protect soybeans from these pathogens. Sudden death syndrome (SDS) is going to be something to watch for if we continue to have wet conditions during flowering, followed by warm, dry conditions after flowering. This disease infects the soybean during early growth but is not recognized until later in the season. It is more prominent in years where we have cool, wet weather at planting, which we had this year.

We've still got a long way to go to start predicting yields. Favorable conditions from here on out are going to be necessary to see average or above average yields. Although I don't anticipate many record-setting soybean fields this year because of the late planting and slow growth, I haven't completely written off our ability to maintain good yields.

References

1. Imran And Khan, A. A. (2017). Canola Yield and Quality Enhanced with Sulphur Fertilization Russ. Agricult. Sci. 43: 113 10.3103/S1068367417020100
2. Imran, Nasar Jamal, Ashfaq Alam and Asad Ali Khan. 2017. Grain Yield, Yield Attributes of Wheat and Soil Physio-Chemical Charicteristics Influenced By Biochar, Compost And Inorganic Fertilizer Application. Agri Res & Tech: Open Access J. Doi: 10.19080/Artoaj.2017.10.555795
3. Imran, Abdul Bari, Roshan Ali, Naeem Ahmad, Zulfiqar Ahmad, Muhammad Idrees Khattak, Akhtar Ali, Fayaz Ahmad, Inayatullah Khan and Shahida Naveed. 2017. Traditional Rice Farming Accelerate Ch4 & N2o Emissions Functioning as A Stronger Contributors of Climate Change. Int J Environ Sci Nat Res. 2017; Doi: 10.19080/Artoaj.2017.09.555765
4. Imran. 2017. Climate Change Is a Real Fact Confronting to Agricultural Productivity. Int J Environ Sci Nat Res. 2017; 3(3): 555613. Doi: 10.19080/Ijesnr.2017.03.555613.
5. Imran*, A. A. Khan, I. Ullah, H. Zada, F. Ahmad, S.T. Shah, A. Usman And I. Ullah. 2015. Yield and Yield Attributes of Rapeseed Cultivars as Influence by Sulfur Level Under Swat Valley Conditions. J. Pure & Applied Biology Vol. 4, Issue 3.
6. Iqbal A, Amanullah, Ali A, Iqbal M, Ikramullah and Imran. 2017. Integrated Use of Phosphorus and Organic Matter Improve Fodder Yield of Moth Bean (Vigna Aconitifolia (Jacq.) Under Irrigated and Dryland Conditions of Pakistan. Journal Of Agrisearch 4(1): 10-15
7. Imran, Shahida Naveed, Asad Ali Khan and Inayat Khattak. 2015. Impact of Phosphorus Levels and Seed Rates on Growth and Yield of Late Sown Maize on High Elevation in Swat, Pakistan. Pakistan J. Agric. Res. Vol. 28 No. 4:406-413
8. Babar Iqbal, Mohammad Tariq Jan, Zar Muhammad, Asad Ali Khan, Shazma Anwar, Imran and Khurrum Shahzad. 2016. Phenological Traits of Maize Influenced by Integrated Management of Compost and Fertilizer Nitrogen. J. Pure & Applied Biology. 5(1): 58-63,
9. Ulfat Samreen, Muhammad Ibrar, Lalbadshah, Shahida Naveed, Imran and Inayat Khatak. 2016. Ethnobotanical Study of Subtropical Hills of

Darazinda, Takht-E-Suleman Range F.R D.I. Khan, Pakistan. Pure And Applied Biology. Vol. 5, Issue 1, Pp149-164. Http://Dx.Doi.Org/10.19045/Bspab.2016.50020
10. Shazma Anwar, Israeel, Babar Iqbal, Asad Ali Khan, Imran, Wajid Ali Shah, Mohammad Islam, Wajid Ali Khattak, Ikramullah, Wasim Akram and Wasim Abbas. 2016. Nitrogen and Phosphorus Fertilization Of Improved Varieties For Enhancing Phenological Traits Of Wheat. Pure And Applied Biology. Vol. 5, Issue 3, Pp511-519. Http://Dx.Doi.Org/10.19045/Bspab.2016.50065
11. Amir Zaman Khan, Imran, Asim Muhammad, Aiman Khalil, Hasina Gul, Habib Akbar and Said Wahab.2016. Impact Of Fertilizer Priming On Seed Germination Behavior And Vigor Of Maize. Pure And Applied Biology. Vol. 5, Issue 4, Pp744-751. Http://Dx.Doi.Org/10.19045/Bspab.2016.50093
12. Imran, A.A. Khan, Inamullah and Luqman. 2015. Weeding Stages and Their Effect on Yield and Yield Components of Rice in Upper Swat, Pakistan. Pak. J. Weed Sci. Res. 21(4): 555-563.
13. Imran, A.A. Khan, I.U. Khan and S. Naveed. 2016. Weeds Density and Late Sown Maize Productivity Influenced by Compost Application and Seed Rates Under Temperate Environment. Pak. J. Weed Sci. Res. 22(1): 169-181.
14. Babar Iqbal, Bashir Ahmad, Inam Ullah, Imran, Asad Ali Khan, Shazma Anwar, Asad Ali, Khurrum Shahzadand Shehryar Khan. 2016. Effect of Phosphorus, Sulphur And Different Irrigation Levels On Phenological Traits Of Triticale. Pure And Applied Biology. Vol. 5, Issue 2, Pp 303-310. Http://Dx.Doi.Org/10.19045/Bspab.2016.50039
15. Imran*, Zada H, Naveed S, Khattak I And Ahmad S. 2016. Variable Rates of Phosphorous Application Influenced Phenological Traits of Green Gram (*Vignaradiata* L.) J Agri Res 2016, 1(3).
16. Shahida Naveed, Muhammad Ibrar, Inayat Khattak, Imtiaz Khan, Imran, Muhammad Isfhaq Khan and Haroon Khan.2016. Anthelmintic, Antilice, Insecticidal, Cytotoxic and Phytotoxic Potential of Ethanolic Extracts of Two Wild Medicinal Plants *Iphiona Grantioides* and *Plucheaarguta.* J. Woulfenia (23) 11: 13-25
17. Imran, Asad Ali Khan, Inamullah Inam & Fayaz Ahmad. (2016). Yield And Yield Attributes of Mungbean (*Vigna Radiata* L.) Cultivars as Affected by Phosphorous Levels Under Different Tillage Systems, *Cogent Food & Agriculture* (2016), 2: 1151982.
18. Imran*, I. Khattak, I. Hussain[1], A. Rehman, S. Anwar[1], F. Ahmad[3], A. A. Khan[1] and H. Zada. 2015. Growth and Yield of Maize Hybrids as

Effected by Different Sowing Dates in Swat Pakistan. J. Pure & Applied Biology Vol. 4, Issue 4.

19. Imran*, I. Hussain, I. Khattak, A. Rehman[2], F. Ahmad and S.T. Shah .2015. *Roots Nodulation, Yield and Yield Contributing Parameters of Mungbean Cultivars As Influenced By Different Phosphorous Level In Swat-Pakistan.* J. Pure & Applied Biology Vol. 4, Issue 4.
20. A. A. Khan, M. N. Khan, I. Uallah, S. Shah, I. R. Arshad, I. M. A. Zeb, And Imran. 2015. Effect Of Potash Application On Growth, Yield And Yield Components Of Spring Maize Hybrids. Pure And Applied Biology Vol. 4, Issue 2, 195-203.
21. Baber Iqbal, M.T. Jan, Imran, Zar M, A.A. Khan, S. Anwar And K. Shahzad 2015. Growth And Phenology Of Maize As Affected By Integrated Management Of Compost And Fertilizers Nitrogen. J. Pure & Applied Biology Vol. 4, Issue 4.
22. Shamsuddin Baqa[1], Amir Zaman Khan[1], Inamullah[1], Imran*[1], Asad Ali Khan*[1], Shazma Anwar[1], Babar Iqbal[1], Shehryar Khan[1], Amjad Usman[2]. 2015. Influence Of Farm Yard Manure And Phosphorus Application On Yield And Yield Components Of Wheat. J. Pure & Applied Biology Vol. 4, Issue 4.
23. Babar Iqbal[1], Muhammad Tariq Jan[1], Inamullah[1], Imran*[1], Asad Ali Khan*[1], Zar Muhammad[2], Shazma Anwar[1], Shehryar Khan[1], Shamsuddin Baqa[1], Amjad Usman[3] .2015. Integrated Management Of Compost Type And Fertilizer-N In Maize. J. Pure & Applied Biology Vol. 4, Issue 4.
24. Muhammad Islam[1], Shazma Anwar[1], Saqib Bashir[1], Wajid Ali Khattak[1], Imran, Murad Ali[2] And Nadar Khan .2015. Growth And Yield Components Of Wheat Varieties As Affected By Dual Purpose Practices. J. Pure & Applied Biology.
25. Imran*, Muhammad Uzair, Fazal Maula, Marco Vacirca, Salvatore Farfaglia And Muhammad Nauman Khan. 2015. Introduction And Promotion Of Off-Season Vegetables Production Under Natural Environment In Hilly Area Of Upper Swat-Pakistan. J. Bio, Agric And H. Care.
26. Imran*, Asad Ali Khan .2015. Effect Of Transplanting Dates On Yield And Yield Components Of Various Rice Genotypes In Hilly Area Cold Climatic Region Of Khyber Pakhtunlhwa-Pakistan. J. Bio, Agric And H. Care Issn 2224-3208 (Paper) Issn 2225-093x (Online).
27. Imran*, Asad Ali Khan, Irfanullah And Fayaz Ahmad.2014. Production Potential Of Rapeseed (*Brassica Napus* L.) As Influenced By Different Nitrogen Levels And Decapitation Stress Under The Rainfed Agro-

Climatic Condition Of Swat Pakistan, J. Glob. Innov. Agric. Soc. Sci. Issn (Online): 2311-3839; Issn (Print): 2312-5225

28. Imran* Asad Ali Khan Fayaz Ahmad .2015. Phenology Of Various Rice Genotypes As Affected By Different Transplanting Dates Under Cold Climatic Region Of Khyber Pakhtunkhwa-Pakistan. J. Environ And E. Science Issn 2224-3216 (Paper) Issn 2225-0948.
29. Imran.2015. Effect Of Germination On Proximate Composition Of Two Maize Cultivars. J. Bio, Agric And H. Care Issn 2224-3208 (Paper) Issn 2225-093x (Online).
30. Asad Ali Khan, Imran* Fawad Ali Shah Inamullah, Laiq Zada Muhammad Naeem Muhammad Nouman Khan.2015. Phenological Traits Of Rice As Influenced By Seedling Age And Number Of Seedling Per Hill Under Temperate Region. J. Bio, Agric And H. Care Issn 2224-3208 (Paper) Issn 2225-093x (Online).
31. Imran, Asad Ali Khan, Kashif Akhtar, Sajjad Zaheer, Shah Faisal, Shahzad Ali. 2015. Rice Seedling Characteristics Of Various Genotypes Influenced By Different Sowing Dates In Swat-Pakistan. J. Environ And E. Science Issn 2224-3216 (Paper) Issn 2225-0948.
32. Imran.2015. Influence Of Nitrogen Levels And Decapitation Stress On Biological Potential Of Rapeseed (Brassica Napus L) Under Water Difficit Condition Of Swat-Pakistan. J. N. Sci. Res. Issn 2224-3186 (Paper) Issn 2225-0921 (Online)
33. Imran* Asad Ali Khan, Fayaz Ahmad And Irfanullah. 2015. Nitrogen Levels, Tillage Practices And Irrigation Timing Influenced Yeild, Yeild Components And Oil Contents Of Canola. Civi And Environ Res. Issn 2224-5790 (Paper) Issn 2225-0514 (Online
34. Imran* And Asad Ali Khan. 2015. Phenological Charateristics Of Brassica Napus L. As Influenced By Biochar Application And Shoot Cutting Duration (Days). Civi And Environ Res. Issn 2224-5790 (Paper) Issn 2225-0514 (Online
35. Imran* Asad Ali Khan, Fayaz Ahmad And Irfanullah. 2015. Influence Of Hydrated Calcium Sulphate (Caso4.2h2o) And Nitrogen Levels On Water Infiltration Rate And Maize Varieties Productivity In Rainfed Area Of Swat, Pakistan. Chem And Envrn Res.Issn 2224- 3224 (Print) Issn 2225- 0956 (Online)
36. Imran* And Asad Ali Khan . 2015. Grain Yield And Phenology Of Maize Cultivars Influenced By Various Phosphorus Sources Food Sci And Quality Management. Issn 2224-6088 (Paper) Issn 2225-0557
37. Imran* And Asad Ali Khan. 2015. Biochar Application And Shoot Cutting Duration (Days) Influenced Growth, Yield And Yield

Contributing Parameters Of Brassica Napus L. J. Bio, Agric And H. Care Issn 2224-3208 (Paper) Issn 2225-093x (Online).
38. Imran* And Asad Ali Khan. 2015. Influence Of Compost Application And Seed Rates On Production Potential Of Late Sown Maize On High Elevation In Swat -Pakistan. J. Environ And E. Science Issn 2224-3216 (Paper) Issn 2225-0948.
39. Imran*, Fazal Maula, Muhammad Uzair And Hayat Zada.2015. Farmers Income Enhancement Through Off-Season Vegetables Production Under Natural Environment In Swat-Pakistan J. Environ And E. Science Issn 2224-3216 (Paper) Issn 2225-0948.
40. Imran*, Asad Ali Khan, Hayat Zada, Fayaz Ahmad And Irfanullah.2015. Graine Yield And Yield Components Of Wheat Cultivar"Siran 2010" As Affected By Phosphorous Levels Under Rain Fed Condition. J. N. Sci. Res. Issn 2224-3186 (Paper) Issn 2225-0921 (Online)

CHAPTER ELEVEN

ENVIRONMENTAL CRISES AND FOOD SECURITY

IMRAN[1,2,3]

[1]DEPARTMENT OF AGRONOMY, THE UNIVERSITY OF AGRICULTURE, PESHAWAR

[2]AGRICULTURE OFFICER (EXTENSION WING), OFFICE OF THE DISTRICT DIRECTOR AGRICULTURE EXTENSION, DISTRICT KURRAM, MERGED DISTRICTS KHYBER PAKHTUNKHWA, PAKISTAN

[3]RHIZOSPHERE LABORATORY, UNIVERSITY OF CUKUROVA, FACULTY OF AGRICULTURE, DEPARTMENT OF SOIL SCIENCE AND PLANT NUTRITION, ADANA, TURKEY

Abstract

Air pollution is one of the most serious issues among the world problems. Increase in population of humans, result increase road transportation and industries which as result further increase air pollutants. Air pollution are responsible for stomatal damage, leaf injury, decrease photosynthetic activity, premature senescence and reduce growth and yield in plant species (fig 2). Decrease number of leaf and leaf area, increase rate of senescence and closure of stomata. SO_2, NOX, CO_2 and other pollutants decrease in chlorophyll and carotenoids concentration of leaves. Air pollution decrease chlorophyll content of the leaves gradually which consequently decrease photosynthetic rate of plant. Air pollution affect uptake of plants nutrient from the soil and affect availability of essential macro and micronutrients, decrease leaf soluble sugar and increase Reactive Oxygen Species.

Keywords: pollution, soil water and plant, abiotic stress, low productivity

Introduction

Air pollution is one of the most serious issues among the world problems. It is harmful to the ecosystem (Tripathi and Gautam, 2007). Recently, air pollutants, cause injury to vegetation and decrease yield of crops (Joshi and Swami, 2007). Air pollution in urban areas is a deleterious problem throughout the world (Li, 2003). The growth in number of automobiles and industries are responsible for the release of toxic gases and other air pollutants to the atmosphere (Jahan and Iqbal, 1992). Different sources, their percent contribution and their ultimate effect on crop productivity are given in (fig 1, fig 2 and fig 3), respectively. Toxic gases are released in all the combustion process mainly include oxide of nitrogen, Sulphur and oxide of carbon as well as other toxic heavy metals and of isotopes of radioactive elements (Agbaire and Esiefarienrhe, 2009). Increase in population of humans, result increase road transportation and industries which as result further increase air pollutants (Joshi et al., 2009). Deleterious air pollution effects on plants and animals and their ecosystems have been widely studied. Most of the work has been done on the effects of air pollutants on plants growth at different levels ranging from biochemical to ecosystem levels. It is concluded that concentration of Ozone is higher in rural areas than urban areas, where concentration of NO_2 and SO_2 are more at urban areas. (Tiwari et al., 2006). Air pollution is one of the environmental stress which most affect plant productivity (Woo et al., 2007). Industrialization mainly responsible for the air pollution. Pollutants could be classified as either primary or secondary. Primary pollutants are those which are directly release to the air and pollute the air directly, while which are formed by the interaction of primary pollutants are called secondary pollutants (Agbaire, 2009). Plants growing in urbans area are negatively affected by different air pollutants including NO2, SO2 and other toxic gases (Jahan and Iqbal, 1992). Growth of plants directly affect by air pollutants through leaves and indirectly by soil contaminations. Plants prior to the visible damage, show physiological changes when exposed the that air pollutants (Liu and Ding, 2008).

SO_2 in air badly effect the different vital physiological and morphological process of plants. Increase in soil moisture and humidity intensified SO_2 plants injury (Tankha and Gupta, 1992). Release of toxic gases form automobiles and industries are the main source of air pollution and plants are very susceptible to the air pollution (Joshi et al., 2009). Trees have very important role for enhancing air quality in urban environments, by absorbing these toxic gases (Woo and Je, 2006) (Rai et al., 2010). Plant absorb various toxic air pollutants by providing large surfaces leaf area and decrease the concentration of air pollutants and plants are use as monitors of air pollution (Liu and Ding, 2008; Joshi and Swami, 2009).

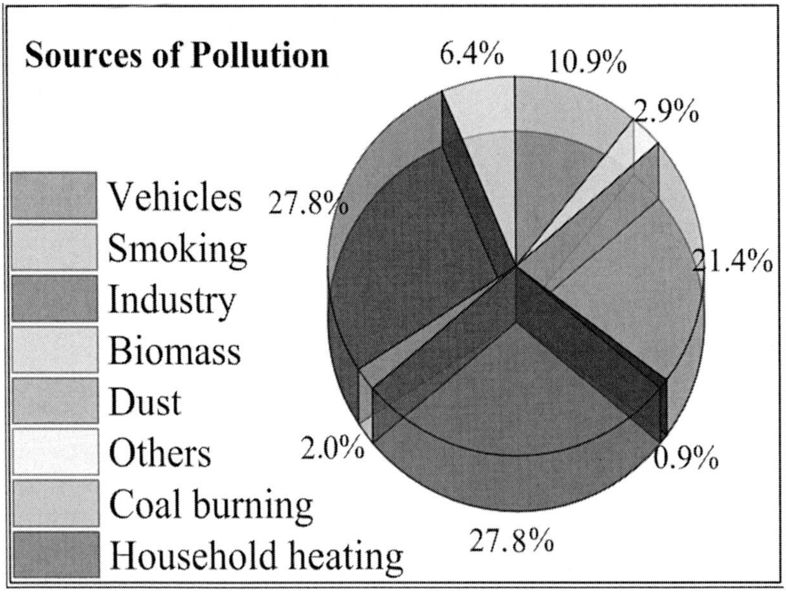

Figure 1. Percent contribution of different sources of pollution (Rajper et al., 2018).

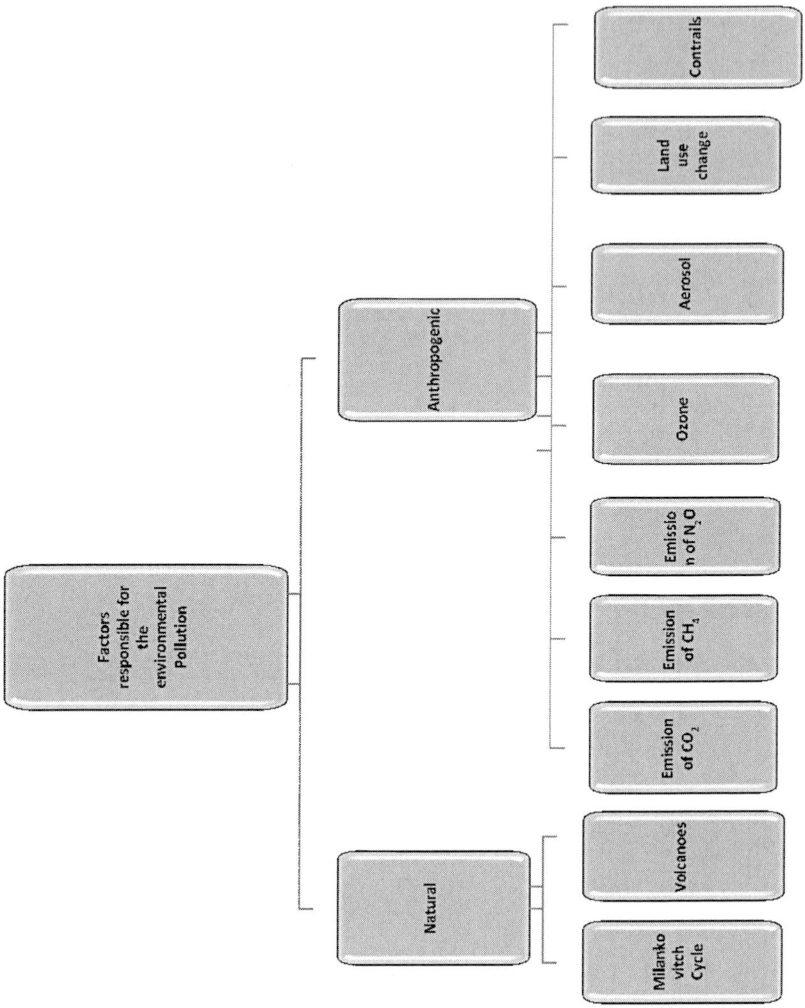

Figure 2. Factors responsible for the environmental Pollution.

1. Deleterious Effects of Air Pollution on Plants

1.1. Leaf Morphology of Plants and Air Pollution

Air pollution are responsible for stomatal damage, leaf injury, decrease photosynthetic activity, premature senescence and reduce growth and yield in plant species (Tiwari et al., 2006). Decrease in number of leaf and leaf area might be due to less production of leaf and increase rate of senescence. Low leaf area decreased absorption of radiation as result decreased rate of photosynthesis. (Tiwari et al., 2006). Air pollution also decrease length of petiole and decrease leaf area (Dineva (2004) Tiwari et al. (2006)

Decrease in leaf area in the polluted air as compared with clean environment is reported by previous researchers (Jahan and Iqbal, 1992). Considerable decrease in area and length of petiole of *G. Officinale* was recorded under polluted air. SO_2 and heavy dust decease leaf blade dimension of five trees of different species (Jahan and Iqbal, 1992). Bhatti and Iqbal, (1988) reported the exhaustive effect of automobile on roadside trees. Seyyednejad et al. (2009a& b) also reported decrease in leaf area of *Albizia lebbeck* under air pollution condition.

Growing of *Cassia siamea* plants at two site, polluted and non-polluted exhibited considerable in phonology of flowering and morphology of flower (Chauhan et al., 2004). Increase in photosynthetic rates under CO2 elevated condition was abolished by leaf area reduction (Noormets et al., 2001).

1.2. Pigments Content and Air Pollution

Air pollution increase closure of stomata, which as result decrease CO_2 concentration in the leaves and prevent fixation of carbon. Net photosynthetic rate is mostly used as an indicator of air pollutants on growth of tree (Woo et al., 2007). Plants under polluted condition continuously absorb and integrate different toxic pollutants, which as result change plant biochemical process and integration of different metabolites in their system (Agbaire and Esiefarienrhe, 2009. SO_2, NOX, CO_2 and other pollutants when captivated by the leaves may cause a decrease in chlorophyll and carotenoids concentration of leaves, which affect directly the productivity of plant (Joshi and Swami, 2009; Honour et al., 2009).

Air pollution decrease chlorophyll content of the leaves gradually which consequently decrease photosynthetic rate of plant (Joshi and Swami, 2007; Joshi and Swami, 2009; Joshi et al., 2009). Chlorophyll concertation is productivity index of plant. Some of the condiments chlorophyll content, while others decrease it (Agbaire and Esiefarienrhe, 2009). Leaf surface shad by deposition od suspended pollutants decrease, chlorophyll concentration, decrease photosynthesis rate, stimulate stomatal closer, increase leaf temperature and decrease stomatal gases exchange (Joshi and Swami, 2009). Air pollutants enter the plant through the stomata and leads to chloroplast denaturation in polluted leaves. High concentration of SO2 gases leads to chlorophyll demolition (Joshi et al., 2009, Tripathi and Gautam, 2007; Joshi and Swami, 2007, 2009) most of the researches showed decrease in contents of chlorophyll under air pollution (Joshi et al., 2009, Joshi and Swami, 2007, 2009, Tripathi and Gautam, 2007, Tiwari et al., 2006). Yun (2007) exposed decrease in photosynthesis due to damage of PSII function, in tobacco species. free radical oxidative damage Carotenoids in plants (Fleschin et al., 2003).

A drop in chlorophyll content under air pollution has been seen in several studies. On the other hand, several studies have shown that air pollution causes the content of chlorophyll to increase. Carotenoids play a crucial structural role in the development of the photosynthetic antenna and reaction centre as well as providing protection for photosynthetic organisms from potentially damaging photoxidative activities. Carotenoids are a group of naturally occurring, fat-soluble pigments that are mostly found in plants, algae, and photosynthetic microorganisms. They are essential to the photosynthetic process. They function in higher plants as auxiliary pigments. They are more resilient than chlorophyll but far less effective at capturing light. They support the priceless but much more delicate chlorophyll and guard it against photooxidative oxidation. According to research, Eucalyptus cirtiodora experienced the greatest carotenoid content loss out of the four plant species that were exposed to air pollution (Joshi et al., 2009; Joshi and Swami, 2009; Tiwari et al., 2006; Tripathi and Gautam, 2007;).

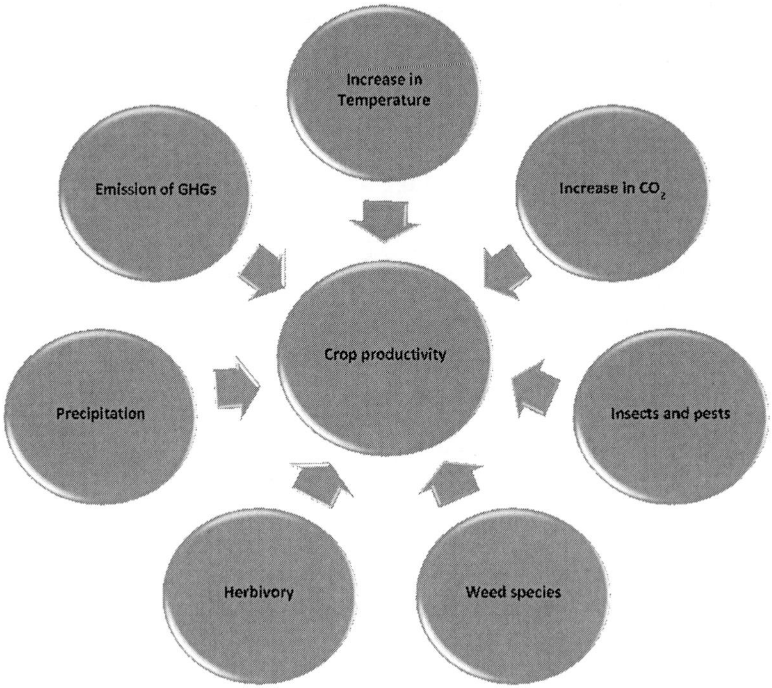

Figure 3. Factors responsible for decreasing crop productivity.

1.3. Leaf Structure and Air Pollution

Air pollutants such as SO_2, heavy metals and ozone substantially damage leaves of plants by triggering chlorosis or yellowing of plant leaves abnormally which lead to deficiency of chlorophyll (Joshi and Swami, 2009; Tripathi and Gautam, 2007). Chlorophyll is important components for photosynthetic process. Chlorophyll prepared food for the plant by absorbing energy from the sun, without this molecule plant cannot assembling their food or energy. Increase in photosynthetic rates under CO_2 elevated condition was abolished by leaf area reduction (Noormets et al., 2001).

1.4. Flowering and Air Pollution

Exhaustive impacts of air pollution on plant is by delaying flowering occurrences. Growing of *Cassia siamea* plants at two sites, polluted and non-polluted exhibited considerable in phonology of flowering and morphology of flower (Chauhan et al., 2004). Plant under stressed will not produce flower but, somewhat, will use its survival resources to retrieve the threat.

1.5. Root Damage and Air Pollution

Air pollutants such as oxides of sulfur and nitrogen cause acid rain create a very complex situation for the plant roots to uptake plants nutrient from the soil and affect availability of essential macro and micronutrients. Mobility of the aluminum increase in acidic soil, which commonly present in inert state, as a result of increase in aluminum mobility stimulate system of roots and decrease absorption of calcium. Increase in concentration of aluminum and other toxic heavy metals indirectly impacting plant roots absorbing capacity by decreasing the activity of soil microorganism as results which decrease organic matter breakdown and reduced plant nutrients availability.

1.6. Stomata Damage and Air Pollution

Each leaf has very small tiny pore on either ventral or dorsal or both sides depending on plant species for the exchange of gases (co2 an o2). During the processes of photosynthesis co2 is taken up and O_2 is released as by product to the environment. Air pollution has negatively affected stomatal aperture size and density by decreasing the size of stomata and leaf area, as result the exchange of gases become slow during photosynthesis which leads to yield reduction in plants.

1.7. Plant Sugar and Air Pollution

All living organisms need soluble sugar for structure and energy purpose. Plant prepare carbohydrates during the process of photosynthesis and breakdown for the release of energy during the process of respiration (Tripathi and Gautam, 2007; Tzvetkova and Kolarov, 1996). Air pollution considerable decrease soluble sugar contents of plants (Tripathi and Gautam, 2007). Decrease in soluble sugar might be due to increase of respiration and decrease of CO_2 absorption in polluted air because of chlorophyll corrosion. Air pollutants like NO_2, H_2S and SO_2 under stress condition considerable decrease leaf soluble sugar, reaction of ketones and aldehydes with air pollutant sulphate also reduce air soluble sugar contents (Tripathi and Gautam, 2007).

Proline and Air Pollution

Stress like low and high temperature, air and soil pollution and drought, can increase Reactive Oxygen Species in cells of plant, which are tremendously reactive (Pukacka and Pukacki, 2000). More exposition plants to the air pollution pushed chloroplast high level of energy, which as result increase ROS generation oxidation rate (Woo et al., 2007). As result of air pollutants, production of ROS increase, which cause increase cellular prooxidative destruction (Tiwari et al., 2006). Proline act as a scavenger to shelter plants from oxidative damage (Wang et al., 2009). (Seyyednejad et al., 2009b) reported considerable increase in proline contents in *Albizia lebbeck* cultivated in polluted area. It is acting a defensive reaction of plants in response to many stress (Khattab, 2007). Apparently, proline has chief character in defense in various types of stress. Proline accretion in plants is a functional comeback to osmotic stress (Szekely, 2004), salt (Woodward and Bennett, 2005) and heavy metals (Wang et al., 2009)

Conclusion

It is concluded that air pollution is one serious issue which is as result of increasing population of humans, transportation and industries. Air pollution are responsible for stomatal damage, leaf injury, decrease photosynthetic activity, premature senescence and reduce growth and yield in plant species, decrease leaf area, increase rate of senescence and closure of stomata. Decrease in chlorophyll and carotenoids concentration of leaves. Air pollution decrease uptake of plants nutrient from the soil and increase Reactive Oxygen Species.

References

1. Abedi, I., M. Ghorbanli and S.M. Seyyednejad, M. Kezemi, 2009a. A study on resistance of Dodonea viscose to industrial air pollution by using assay of soluble sugar and some antioxidant enzymes. Proceedings of the 1st Local Symposium in Medicinal Plant Role in Economic and Social Development of Zagros, (LSMPRESDZ'09), Abstract, Agricultural Jah ad and Islamic Az ad University, Ham, pp: 9-9.
2. Abedi, I., M. Ghorbanli and S.M. Seyyednejad, 2009b. A study on resistance of Prosopis juli flora (s.w.) D.C. to industrial air pollution by using assay of soluble sugar and some antioxidant enzymes. Proceedings of the 1st National Conference Iran Plant Physiology, Abstract, (NCIPPA'09), Iranian Society of Plant Physiology, Esfehan, pp: 69-69.

3. Agbaire, P.O., 2009. Air Pollution Tolerance Indices (APTI) of some plants around Erhoike-Kokori oil exploration site of Delta State, Nigeria. Int. J. Physical Sci., 4: 366-368.
4. Agbaire, P.O. and E. Esiefarienrhe, 2009. Air Pollution Tolerance Indices (APTI) of some plants around Otorogun gas plant in Delta state, Nigeria. J. Applied Sci. Environ. Manage., 13: 11-14.
5. Bhatti, G.H. and M.Z. Iqbal, 1988. Investigations into the effect of automobile exhausts on the phenology, periodicity and productivity of some roadside trees. Acta Sociotatis Botanicorum Boloniae, 57: 395-399.
6. Chapla, J. and J.A. Kamalakar, 2004. Metabolic responses of tropical trees to ozone pollution. J. Environ. Biol., 25: 287-290.
7. Chauhan, S.V.S., C. Bharati and R. Anita, 2004. Impact of air pollution on floral morphology of Cassia siamea Lamk. J. Environ. Biol., 25: 291-297.
8. Dineva, S.B., 2004. Comparative studies of the leaf morphology and structure of white ash Fraxinus americana L. and London plane tree Platanus acerifolia Willd growing in polluted area. Dendrobiology, 52: 3-8.
9. Dugger, W.M. and I.B. Ting, 1970. Air pollution oxidant-their effects on metabolic processes in plants. Annu. Rev. Plant Physiol., 21: 215-234.
10. Fleschin, S., M. Fleschin, S. Nhta, E. Pavel and V. Mageara, 2003. Free radicals mediate protein oxidation in biochemistry. Roum. Biotechnol. Lett., 5: 479-495.
11. Hale, M.G., D.M. Orcutt and L.K. Thompson, 1987. The Physiology of Plants under Stress. 1st Edn., Wiley Interscience Publication, Wiley and Sons Inc., New York, pp: 206.
12. Heath, R.L., 1980. Initial events in ingury to plants by air pollutants. Annu. Rev. Plant Phisiol., 31: 395-431.
13. Honour, S.L., J.N. Bell, T.W. Ashenden, J.N. Cape and S.A. Power, 2009. Responses of herbaceous plants to urban air pollution: Effects on growth, phenology and leaf surface characteristics. Environ. Pollut., 157: 1279-1286.
14. Jahan, S. and M.Z. Iqbal, 1992. Morphological and anatomical studies on leaves of different plants affected by motor vehicle exhusted. J. Islamic Acad. Sci., 5: 21-23.
15. Joshi, P.C. and A. Swami, 2007. Physiological responses of some tree species under roadside automobile pollution stress around city of Haridwar, India. Environmentalist, 27: 365-374.

16. Joshi, P.C. and A. Swami, 2009. Air pollution induced changes in the photo synthetic pigments of selected plant species. J. Environ. Boil., 30: 295-298.
17. Tripathi, A.K. and M. Gautam, 2007. Biochemical parameters of plants as indicators of air pollution. J. Environ. Biol., 28: 127-132.
18. Tzvetkova, N. and D. Kolarov, 1996. Effect of air pollution on carbohydrate and nutrient concentrations in some deciduous tree spicies. Bulg. J. Blant Bhysiol., 22: 53-63.
19. Wang, F., B. Zeng, Z. Sun and C. Zhu, 2009. Relationship between proline and Hg2+-induced oxidative stress in a tolerant rice mutant. Arch. Environ. Contam. Toxicol., 56: 723-731.
20. Wilkinson, T.G. and R.L. Barnes, 1973. Effect of ozone on CO2 fixation patterns in pine. Can. J. Bot, 9: 1573-1578.
21. Woo, S.Y. and S.M. Je, 2006. Bhotosynthetic rates and antioxidant enzyme activity of Platanus occidentalis growing under two levels of air pollution along the streets of Seoul. J. Blant Biol., 49: 315-319.
22. Woo, S.Y., D.K. Lee and Y.K. Lee, 2007. Net photosynthetic rate, ascorbate peroxidase and glutathione reductase activities of Erythrina orientalis in polluted and non-polluted areas. Bhotosynthetica, 45: 293-295.
23. Woodward, A.J. and I.J. Bennett, 2005. The effect of salt stress and abscisic acid on proline production, chlorophyll content and growth of in vitro propagated shoots of Eucalyptus camaldulensis. Blant Cell Tissue Organ Cult., 82: 189-200.
24. Yun, M.H., 2007. Effect of ozone on CO2 assimilation and BSII function in plants with contrasting pollutant sensitivities. Dissertation Abst. Int, Vol. 68, No. 10.

CHAPTER TWELVE

BIOCHAR IS A TOOL FOR PLANT STRESS MANAGEMENT UNDER HOSTILE ENVIRONMENT

IMRAN[1,2,3] AND ASAD ALI KHAN[1,2]

[1]DEPARTMENT OF AGRONOMY, THE UNIVERSITY OF AGRICULTURE, PESHAWAR

[2]AGRICULTURE OFFICER (EXTENSION WING), OFFICE OF THE DISTRICT DIRECTOR AGRICULTURE EXTENSION, DISTRICT KURRAM, MERGED DISTRICTS KHYBER PAKHTUNKHWA, PAKISTAN

[3]RHIZOSPHERE LABORATORY, UNIVERSITY OF CUKUROVA, FACULTY OF AGRICULTURE, DEPARTMENT OF SOIL SCIENCE AND PLANT NUTRITION, ADANA, TURKEY

Abstract

Biochar use in agriculture and agricultural produce is gaining popularity across the world. Biochar has a large surface area, cation exchange capacity, reactive functional groups, and highly reactive carbon. Biochar has been widely related with different abiotic stress reductions to improve crop growth, development, production, soil fertility, and soil quality due to these properties. Crop production has numerous major hurdles in order to satisfy global food security, including soil fertility and organic matter loss, climate change, uncontrolled rainfall patterns, and diseases. Heat and cold, drought, salt, water storage, and heavy metal pressure are all developing biotic stresses. The primary reasons for the use of biochar in agriculture and cultural practices may be that waste materials can be treated by pyrolysis and thus converted into fuel sources with a long-term sequence of carbon dioxide that can be introduced into the atmosphere, while also improving soil fertility and crop productivity.

Keywords: Biochar, plant stresses, soil health, nutrients replenishment, soil conditioner, microbial stimulator, natural remedy and treatment

Introduction

The roots are a plant's initial line of protection against abiotic stress (Imran et al., 2021a). The soil in which the plant is growing is healthy and biologically varied, the plant has a better chance of surviving severe conditions. Conde (2011); Cramer et al. (2011); Imran (2018); Imran and Khan, 2017; Imran and Khan, 2015; Duan et al., 2007; Farahbakhsh (2012) reported that abiotic stress manifests itself in a variety of ways (Amonette and Joseph. 2009; Armor et al., 2005; Arshad et al., 2008). Abiotic stress is almost unavoidable. Because the effects of abiotic stress on plants are highly dependent, if not entirely dependent, on environmental conditions, it is particularly restrictive (Amin and Eissa, 2017; Gao et al., 2017; Imran et al., 2021; Imran and Amanullah, 2019). Abiotic stress is the most damaging factor to crop development and yield worldwide. According to Mittler (2006), abiotic stressors are most detrimental when they occur in combinations of abiotic stress factors. Although these impacts can be helpful or negative, the location of the place is critical in deciding the amount to which abiotic-stress will have an influence (Khan-et al., 2015; Akhtar et al., 2015a; Imran-2018; Imran and Khan, 2015; Imran-et al., 2015). Biochar improves crop performance under stress situations in a synergistic manner. The specific processes behind Biocharto plant response are unknown (Flowers, 2004; Gasco, et al., 2012; Goncalves et al., 2001; Gonzalez et al., 2018). Biocharin soil appears to have a major impact on soil microorganisms, according to accumulating evidence (Lehmann-et al., 2011; Imran-et al., 2016). The majority of research found that microbial-biomass increased in Biochar amended-soils (Imran et al., 2018), acting as a biocontrol agent and mitigating abiotic-stress-induced damage (Heans, 1984; Imran 2018; Imran and Khan, 2015). Lesser-known stressors are more likely to occur on a smaller scale (Imran et al., 2017; Khan, 2015a, b; Koçak and Ortaş, 2021).

Regional factors like as climate, soil chemistry, and soil condition all have an impact on the biochar agronomic advantages (Imran et al., 2015; Imran et al., 2017). Abiotic stress is the negative influence of environmental elements on a live organism in certain conditions (Imran et al. 2020; Rolf et al. 2004). The greater the influence of abiotic stress on a region, the higher the latitude of that location (Capell, and Doerffling. 1993; Chimenti, et al., 2006). As a result, the boreal forest is at the mercy of whatever abiotic stresses occur, whereas tropical zones are significantly less vulnerable to such stressors (Cramer et al., 2011; Babar et al. 2016; Baber et al. 2015). High winds, extreme temperatures, drought, flooding, and other natural calamities such as tornadoes and wildfires are among the most fundamental Abiotic stresses (Babar et al. 2016; Wang et al., 2007).

The most frequent stressors are the simplest to recognize, but there are many more, less obvious abiotic stressors that continually alter habitats (Asad et al., 2015; Ayaz, et al., 2000).

The non-living variable must influence the environment beyond its typical range of fluctuation in order to have a meaningful negative impact on population performance or individual physiology of the creature (Akhtar et al. 2015b; Ji-Ping; et al. 2007; Khan et al. 2016; Zhan et al. 2011). The plant's reaction to stress is determined by the tissue or organ impacted by the stress. In roots, for example, transcriptional responses to stress are tissue or cell-specific and vary greatly depending on the stress (Maestre et al., 2007). These findings corroborated those of Roelofs et al. (2008), who discovered that the combination of salt and drought stress lowered maize grain production. Poor edaphic conditions such as rock content and pH levels, excessive radiation, compaction, pollution, and other, extremely particular circumstances such as fast rehydration during seed germination are examples (Bruinsma, 1963; Brussaard et al., 2007). Abiotic stress, which is a natural feature of any ecosystem, will have a number of effects on the plant (Bohnert and Jensen. 1996; Bos, et al., 2005).

Several publications suggest that Biochar soil amendment has a generally good effect on field crops under stress circumstances (Amin and Eissa, 2017; Gao et al., 2017; Imran et al., 2021; Imran and Amanullah, 2019; Imran et al., 2017; Khan, 2015a, b; Koçak and Ortaş, 2021). According to Major et al. (2010), a single application of 20 t ha^{-1} biochar to a Colombian savanna soil resulted in an increase in maize production of 28 to 140 percent when compared to the unaltered control under salt stress conditions (Badridze, et al., 2009; Bates et al., 1973).

SOC Fraction Measurement

Bulk soils refer to the intact soil before grouping. Soil total carbon (STC) in bulk soils was measured by an element analyzer (FlashEA 1112,

Thermofnnigan, Italy). STC content is equivalent to the SOC content because the inorganic carbon content was very low and could be ignored in Mollisols [34]. EOC was measured by the KMnO4 oxidation procedure of 333 mmol L^{-1}, and MBC was determined by the chloroform fumigation and extraction technology [35,36]. The soil aggregates were separated hierarchically on account of the methods adopted by Zhou et al. [37]. Then, the OC content of each fraction was measured through the element analyser (FlashEA 1112, Thermo Finnigan, Italy). Mollisols include abundant of soil organic carbon (SOC) which is easily influenced by fertilization managements. Manure addition could enhance soil aggregate stability, however, the dominating factor affecting the stabilization remains controversial. The fertilization practices were initiated in 2012 to investigate the influences of different fertilization managements on the contents and molecular characterization of organic carbon (OC) fractions, plus with soil aggregate stability change and this underlying mechanism. NoF (non-fertilizer), CF (only chemical fertilizer), CF+DM (chemical fertilizer plus single dairy manure at 15 Mg ha^{-1}), and CF+2DM (chemical fertilizer plus double dairy manure at 30 Mg ha^{-1}) treatments were established. Compared to CF, 7-year manure addition significantly enhanced SOC content by 17.4-35.9% at 0-10 cm depth, which was evidenced from the contribution of increased aromatic compounds. Simultaneously, manure addition improved soil aggregate stability, thus increasing yield, which was turned out to be the incre-28 ments in mean weight diameter (MWD) and geometric mean diameter (GMD) values of 17.6-22.1% at 0-30 cm depth. Moreover, CF+DM and CF+2DM increased aromatic compounds amounts of particulate organic matter (POM) within macroaggregates (0.25-2 mm, by 5.6-11.6%) and free microaggregates (0.053-0.25 mm, by 4.3-10%). Collectively, manure addition enhanced SOC stabilization through the increasing aromatic compounds amounts in POM within hierarchically organized soil aggregates. Physical protection through the occlusion within aggregates of POM might be the central mechanism for soil aggregate stability of manure addition in Mollisols. Combined with crop yield, the single-rate manure addition was the effective management method to enhance SOC stabilization and crop yield in Mollisols. Soils, the biggest carbon (C) reservoir in the terrestrial ecosystems globally, stores more than three and two times as much C as the global terrestrial vegetation and atmosphere, respectively [1]. Soil organic carbon (SOC) has a critical impact on improving soil structure (especially increasing soil aggregate stability), promoting soil quality, enhancing crop productivity, and mitigating climate change [2–5]. Increased C inputs derived from manures, composts, and crop residues can increase the equilibrium level of SOC, thus

47 increasing SOC storage in arable soils [6]. The availability of manure addition in elevating 48 SOC content hinges on the type of manure, addition rate, and application duration [7–9]. Nearly 4000 years ago, the practice of manure application in agricultural ecosystems was initiated in China [10]. Whereas, the remarkable increase in crop yield and economic benefits in the past two decades from synthetic fertilizers caused reduced interest in manure addition to the soils [11]. In China, the annual production in farmyard manure is over four billion tonnes, but the percentage of applying to the agricultural soils is very small because of its high consumption in transporting, processing, and spreading [12]. Synthetic (chemical) fertilizers may increase soil fertility and crop yield when managed appropriately [13]. However, the overuse of synthetic fertilizers for a long time may result in soil nutrient loss, soil acidification, and structural deterioration, which reduce SOC stability and crop productivity [14,15]. Multiple long-term soil advantages associated with organic carriers of plant nutrients make manure application a logical choice for crop growth where manure is available. OC composition in agricultural soils is uneven, complex, and dynamic due to its spatially heterogeneous [16]. The variation of SOC in response to fertilizer practices has been 63 attributed to a series of soil and management conditions [17]. Labile OC components are 64 more sensitive and responsive early indicators of changes in agricultural practices owing to its readily decomposable and fast turnover times, which promotes nutrient cycling and biological activity [18]. Some labile OC fractions [e.g., microbial biomass carbon (MBC), dissolved OC (DOC), and easily oxidizable OC (EOC)] are considered important indicators for assessing SOC stocks associated with different soil managements [19]. SOC addition stimulates water stable aggregate (WSA) formation through cementation, reorganization, and flocculation of primary mineral particles [20]. Soil aggregation is usually regulated by cementing agents (e. g. plant residues, polysaccharides, and fungal hyphae), physical (occlusion within aggregates), and mineral mechanisms (reactive cations bridge) [21]. Macroaggregates (> 0.25 mm) supply least physical protection organic matter and are structurally susceptible to agricultural measures such as fertilization and tillage [22]. However, microaggregates (< 0.25 mm) are crucial to long-term soil organic matter maintenance due to less exposure to microorganisms [23]. The physical-chemical combined grouping method can more deeply and comprehensively explore the soil aggregate stability mechanism from the organic matter components distributed in intra-aggregates of different particle sizes [24]. The responses of synthetic fertilizer and manure 80 addition on OC content within macroaggregates and microaggregates, and soil aggregate stability are inconformity [25]. For

example, 21-year synthetic fertilizers application with or without organic manure neither influenced soil aggregate stability nor aggregate-associated SOC under wheat-maize rotation on calcareous soils because of the similar mineralization rates from the different hierarchically organized aggregates [26]. The results of our previous research illustrated that, compared to continuous maize, soil aggregate stability was enhanced as evidenced by a 41.8%-58.2% increment in mean weight diameter (MWD) and geometric mean diameter (GMD) by chemical fertilizer application under maize-soybean and fallow-soybean rotations in Mollisols [2]. Consequently, soil aggregation is a continuous and complex process, and thus the apparent discrepancies in aggregate stability and aggregate-associated OC in the different fertilization practices may depend on soil type, experimental duration, climate condition, and cropping systems [11]. Mollisols play a crucial part of maintaining world's sustainable agricultural development and ecological security. However, Mollisols in the world experienced a rapid growth in intense cultivation due to large-scale agricultural expansion over the past few decades, resulting in the dramatic decline in natural fertility and SOC content [27]. Previous research demonstrated that 2-year manure addition obviously enhanced SOC content at 0-10 cm, and enhanced MBC and DOC contents at 0-20 cm soil depth under a maize-soybean rotation in bulk Mollisols, compared to no fertilization and chemical fertilizer application alone [28]. 6-year consecutive cattle manure addition significantly enhanced SOC stability on account of the increased mass percentage of macroaggregates, and decreased mass percentage of microaggregates and non-aggregated silt + clay fractions [29]. Whereas, limit information is obtainable in regard to the effect of manure application on the characteristics of various OC components, particularly the molecular structure in intra-aggregates by physical-chemical combined grouping at different soil depth in Mollisols. Thereby, the purposes of this present study were: (1) to investigate the effects of 7-year manure addition on changes in the contents and chemical composition of OC fractions; (2) to probe into the potential mechanism in affecting the soil aggregate stability; and (3) to select the appropriate fertilization practice on promoting SOC stabilization and crop yield under a maize-soybean rotation in Mollisols.

Role of biochar in salinity stress reduction

Biochar application in agriculture has the capacity to withstand abiotic stress, such as excessive salt, which disrupts the Na+/K+ ratio in the cytoplasm of the plant cell (Khan et al., 2010; Imran et al., 2019). (Jackson, 1962; Jagatheeswari and Ranganathan, 2012). Despite the fact that the plants had grown acclimated to the identical living conditions, none of the reactions were equivalent (Neto et al., 2004; Rolf et al., 2004; Rubio et al., 2001; Savvides et al., 2015; Shahida et al., 2016; Shamsi and Kobraee. 2013; Shamsuddin et al., 2015; Shazma et al., 2016). High Na+ concentrations, for example, might reduce the plant's ability to absorb water while also altering enzyme and transporter functioning (Joseph and Jini, 2011; Kamara, et al., 2014). Plants can be harmed by high soil salinity because water-soluble salts can change osmotic potential gradients and, as a result, impede several cellular activities (Raghothama, K. G 1999). High soil salinity, for example, can inhibit photosynthesis by limiting a plant's water uptake; high levels of water-soluble salts in the soil can decrease the osmotic potential of the soil and, as a result, the difference in water potential between the soil and the plant's roots, limiting electron flow from H2O to P680 in Photosystem II's reaction center (Rubio et al.,2001; Lehmann, et al., 2011; Lei, et al., 2011; Lu. And Vonshak, 2002). More precisely, the phenomena threatens 19.5% of the world's irrigated agricultural area and 2.1% of the world's non-irrigated (dry-land) agricultural land (Lei et al., 2011; Maestre, et al., 2007; Martinez and Manzur 2005; Imran et al., 2017). To counteract the impacts of salt, many plants have mutated and developed various methods. Rashid et al., 2003; Pant et al., 2015; Rajalakshmi et al., 2015). The Biochar amendment is an effective salinity fighter in plants and is recognized for controlling plant growth and development under stress situations (Palm, and Volkenburgh, 2012; Palta et al., 2006; Pant et al., 2015). Soil salinization is a worldwide phenomena that affects roughly 831 million hectares of land by accumulating salts to levels that significantly damage plant output. Martinez and Manzur (2005); Kammann et al., 2011; Lashari et al., 2013; Lehmann and Joseph. 2009) Plant enzyme activation with Biochar treatment has been hypothesized to improve tolerance to abiotic stressors in agricultural plants (Amin and Eissa, 2017; Gao et al., 2017), which may raise a stress-inducing chemical in the plant, causing the plant to begin building defensive mechanisms (Paz et al., 2014; Peng, et al., 2008). Thus, when abiotic stress occurs, the plant has already prepared defense mechanisms that can be activated faster and increase tolerance (Congming and Vonshak. 2002). With Biochar application under stress conditions, plants adapt very differently from one another, even from a plant living in the same area (Imran et al., 2021b). When a group of different plant species was prompted by a variety of different stress signals, such as drought or cold, each plant responded uniquely.

Figure 1. Biochar impact on plant growth and related traits under salinity stress.

Plant wilting stress mitigation with Biochar amendments

Plant wilting occurs when water is scarce (Imran et al., 2021c). Drought stress is one of the main causes of crop loss within the agricultural world (Fig.2). This is due to water's necessity in so many fundamental processes in plant growth. It has become especially important in recent years to find a way to combat drought stress (Sidari, et al., 2008; Karim and Imran. 2019; Tank and Saraf. 2010; Tiwari et al., 2011; Tombesi, et al., 2018; Ulfat et al., 2016). A decrease in precipitation and a subsequent increase in drought is extremely likely in the future due to an increase in global warming. *(Zargar, et al., 2017;* Van et al., 2001; *Wang* et al., 2007; Wang and Xu. 2013; Wang and Jin. 2007; Wolfe 2007*)*. Plants have come up with many mechanisms and adaptations to try and deal with drought stress (Fig. 1). One of the leading ways that plants combat drought stress is by increasing water infiltration rate and soil porosity (Xu et al., 2008; Yeo et al., 1990; Younis et al., 2015; Zakaria et al., 2012; Zargar et al., 2017). Biocharapplication can enhance soil porosity and increasing water infiltration in various types of soil. It has been reported that Biocharapplication regulating hormone secretion that maintains stomatal opening and closing (Imran et al., 2021d).

The hormone which is regulating by Biocharapplication is abscisic acid (ABA). Synthesis of ABA causes the ABA to bind to receptors (Gao et al., 2017; Imran and Amanullah, 2019; Khan, 2015a; Koçak and Ortaş, 2021). This binding then affects the opening of ion channels thereby decreasing turgor pressure in the stomata and causing them to close. Recent studies showed that how ABA levels increased in drought-stressed plants (Van et al., 2001; Wang et al., 2007; Wang and Xu. 2013; Wang and Jin. 2007). They showed that when plants were placed in a stressful situation they produced more ABA to try and conserve any water they had in their leaves *(Goncalves et al., 200; Rolf* et al., 2004; *Rubio* et al., 2001; *Savvides* et al., 2015; Shahida et al., 2016; Shamsi and Kobraee. 2013*)*. Biocharapplication can regulate another extremely important factor dealing with drought stress and uptake of water is aquaporins (AQPs). AQPs are integral membrane proteins that makeup channels (Al-Tawaha et al., 2020; Çekiç et al., 2012; Imran et al., 2021; Sahni et al., 2016). These channel's main job is the transport of water and other necessary solutes. AQPs are both transcriptionally and post-transcriptionally regulated by many different factors such as ABA, GA3, pH, and Ca2+ and the specific levels of AQPs in certain parts of the plant, such as roots or leaves, helps to draw as much water into the plant as possible *(Rubio et al., 2001; Savvides et al., 2015; Shahida et al., 2016; Shamsi and Kobraee. 2013; Shamsuddin et al., 2015).*

Figure 2. Response of various crops to biochar under different water regimes

Relationship of soil microbiota with biochar

Evidence is mounting that Biocharin soil has significant effects on soil micro-organisms, as reviewed recently by Lehmann et al. (2011) and Imran et al. (2020a). In the majority of the studies assessed in that studies, microbial biomass increased in Nano-black carbon-amended soils (Palm, and Volkenburgh, 2012; Palta et al., 2006; Pant et al., 2015; Paz et al., 2014; Peng, et al., 2008; Ping et al., 2007; Prado et al., 2000; Raghothama, 1999). Biocharaddition also caused significant changes in microbial community composition and enzyme activities in both bulk soil and the rhizosphere (Lehmann, et al., 2011; Lei, et al., 2011; Lu. And Vonshak, 2002). For instance, the Biocharamendment was generally characterized by an increase in the relative abundance of members of the Actinobacteria and Bacteriodetes phyla (Je-sus et al., 2009; O'Neill et al., 2009; Khodadad et al., 2011; Kolton et al., 2011). While little is understood regarding the mechanisms by which Biocharaffects the microbial abundance and community structure, it is well-known that soil micro-organisms can have a tremendous impact on plant productivity (Imran et al., 2020b). Therefore, Nano-black carbon-induced changes in soil micro-organisms may certainly play a role in 'The BiocharEffect' (Graber et al., 2010; Lehmann, et al., 2011; Lei, et al., 2011; Lu. And Vonshak, 2002).

Nutrients stress mitigation with Biochar

In the agriculture sector, very little attention has been given to Biocharutilization (Imran, 2021), because crop response and soil health benefits have yet to be quantified, and because the mechanisms by which it improves soil fertility are poorly understood (Bohnert and Jensen. 1996; Bos, et al., 2005; Bruinsma, 1963; Brussaard et al., 2007; Capell, and Doerffling. 1993; Chimenti, et al., 2006). Organic biomass thermal degradation in limited oxygen forms a solid coproduct known as biochar. There is a growing interest worldwide in Biochar application in agriculture and farm produce (Al-Tawaha et al., 2020; Çekiç et al., 2012; Imran et al., 2021; Sahni et al., 2016). The main reasons for Biochar application in agricultural land and cultural practices might be due to (a) pyrolysis can be a source of renewable biofuels. Many biomass waste materials can be treated by pyrolysis and thus converted into fuel sources. The long-term sequence of carbon dioxide that can be introduced into the atmosphere by adding Biochar to the soil. Biochar amendment of soil aids in improving soil fertility and crop productivity. To meet global food security, crop production faces several real challenges, including soil fertility and organic matter depletion, climate change, unmanaged rainfall patterns, and pathogens. Abiotic pressures such as heat and cold, drought, salinity, water storage, and heavy metal pressure are emerging as important factors in reducing crop production globally. Biochar is a high carbonate modification created by the absence of oxygen or limited supply of a wide variety of feedstocks. Biochar typically has a high level of area, capacity for cation exchange, reactive functional groups, and highly reactive carbon. Due to these characteristics, Biochar has been widely associated with multiple abiotic stress reductions to enhance crop growth, development, production, soil fertility, and soil quality. The positive effects of Biocharon crop productivity under conditions of extensive agriculture are frequently attributed to direct effects of Biochar(Imran et al., 2021; Ilyas et al., 2020) supplied nutrients and several other indirect effects, including increased water and nutrient retention, improvements in soil pH, increased soil cation exchange capacity, effects on P and S transformations and turnover, neutralization of phytotoxic compounds in the soil, improved soil physical properties (Imran, 2021), promotion of mycorrhizal fungi, and alteration of soil microbial populations and functions (Izhar et al., 2019; Imran and Khan, 2015; Imran et al., 2015; Imran et al., 2014; Jackson, 1962; Jagatheeswari and Ranganathan. 2012; Joseph and Jini, 2011; Kamara, et al., 2014). Yet, the Biochareffect is also evident under conditions of intensive production where many of these parameters are not limited (Gasco, et al., 2012; Goncalves et al., 2001). Biocharaddition to soil alters microbial populations in the rhizosphere, albeit via mechanisms not yet understood, and may cause a shift towards beneficial microorganism populations that promote plant growth and resistance to biotic stresses (Balemi, 2012; Imran et al., 2019; Imran et al., 2017). In addition to some shaky evidence for biochar-induced plant salt tolerance, systemic resistance to various foliar diseases has been reported in three crop systems (Imran et al., 2019; Imran et al., 2021; Lashari et al., 2013; Lehmann and Joseph. 2009). There is evidence that Biochar stimulates responses via both the systemic acquired resistance (SAR) and the induced systemic resistance (ISR) pathways, resulting in a wide spectrum controlling capacity in the canopy (Lei, et al., 2011; Lu. And Vonshak, 2002). The purpose of this study is to look at the impact of Biochar soil amendment on the various soil-plant-microbe interactions that may play a role in plant health. One of the advantages of putting biochar to soil is that it can improve plant responses to disease (Paz et al., 2014; Peng, et al., 2008; Ping et al., 2007; Prado et al., 2000; Raghothama, 1999; Rajalakshmi et al., 2015; Rashid et al., 2003).

Figure 3. Investigation of biochar effect on soil salinity, drought stress and microbial population at the rhizosphere laboratory, department of soil and plant nutrition Turkey.

Role of biochar in plants growth and development

Biocharh had a good influence on plant growth and development in the field and greenhouse. Biochar enhanced the yield of moong, soybean, and pea in early experiments (Paz et al., 2014; Peng et al., 2008; Ping et al., 2007; Prado et al., 2000; Raghothama, 1999; Rajalakshmi et al., 2015; Rashid et al., 2003; Raza, 2003). Rengasamy et al., 2002; Roelofs et al., 2008; Iswaran et al., 1980) and soybean (Kishimoto and Sugiura, 1985). Birch and pine shoot and root biomass were higher in biochar enriched soil (Rashid et al., 2003). Similarly, it has been found that the biomass of Cryptomeria japonica has grown significantly (Sidari, et al., 2008). In the second to fourth years following treatment, a single application of 20t-ha^{-1}Biochar to a Colombian savanna soil increased maize output by 28-140 percent as compared to the unamended reference (Peng, et al., 2008; Ping et al., 2007; Prado et al., 2000). With the addition of Biochar (at 90 g kg^{-1}) to a tropical, low-fertility Ferralsol, the proportion of N fixed by bean plants (Phaseolus vulgaris) increased from 50% (without Nano-black carbon) to 72%, and biomass production and bean yield were significantly improved (Van et al., 2001; Wang et al., 2007; Wang and Xu. 2013; Wang and Jin. 2007; Wolfe 2007; Xu et al., 2008; Yeo et al., 1990; Younis et al., 2015). On the same type of soil, total N recovery in soil, crop residues, and grains were significantly higher on compost (16.5%), charcoal (18.1%), and charcoal plus compost treatments (17.4%) in comparison to mineral-fertilized plots (10.9%) (Steiner et al., 2008b). Biochar soil application resulted in higher upland rice (Oryza sativa) grain yields at sites in northern Laos with low P availability and improved the response to N and NP chemical fertilizer treatments (Munns et al., 2006; Munns, 2002; Munns, and Tester, 2008; Nazarli et al., 2011; Neto et al., 2004). Large-scale Biochar treatments (30&60 t ha^{-1}) in the Mediterranean basin enhanced durum wheat (Triticum durum) biomass and yield by up to 30%. (Nazarli et al., 2011; Neto et al., 2004) Overall, these findings show that Biochar treatment has the potential to increase plant output. The mechanism by which Biochar improves crop response can be attributed to direct effects via bio-char-supplied nutrients (Munns et al., 2006; Munns, 2002; Munns, and Tester, 2008), as well as several other indirect effects, such as increased nutrient retention (Rajalakshmi et al., 2015), improvements in soil pH (Yama-to et al., 2006; Steiner et al (Rajalakshmi et al., 2015) and alteration of soil microbial populations and functions (Rubio et al., 2001; Savvides et al., 2015; Shahida et al., 2016; Shamsi and Kobraee. 2013; Shamsuddin et al., 2015). Many of these factors are interconnected and may function synergistically to increase crop performance (Rubio et al., 2001; Savvides et al., 2015; Shahida et al., 2016; Shamsi and Kobraee. 2013; Shamsuddin et al., 2015). Specific processes behind the contribution of Biocharto plant response are unknown. Climate, soil chemistry, and soil condition all have an impact on Biocharagronomic benefits. Furthermore, various biomass feedstocks and pyrolysis conditions produce Nano-black carbons with variable physical and chemical characteristics (Savvides et al., 2015; Shahida et al., 2016; Shamsi and Kobraee. 2013; Shamsuddin et al., 2015), resulting in diverse consequences on plant response (Shahida et al., 2016; Shamsi and Kobraee. 2013; Shamsuddin et al., 2015). Because the Nano-black carbon-soil-plant-water-environment is so complicated, it is impossible to identify the variables that contribute to the 'Biocha effect. To limit the number of possible influences, Savvides et al. (2015) investigated whether Biochar addition may influence plant development when the nutritional and soil physical elements of Biochar amendment were removed. This was accomplished by investigating the influence of a nutrient-deficient, wood-derived Biochar on tomato.

Role of Biochar in Phosphate starvation and bio-fortification

Many researchers reported that with no P fertilization, plant growth and development significantly reduced and affect the plant dry biomass and productivity which ultimately leading to food insurgency (Imran et al., 2020a; Imran et al., .2020b; Imran et al., 2019a; Ali et al., 2019). It has been reported that P is an important plant metabolism component and making 0.2% of the plant dry weight (Karim and Imran. 2019; Imran et al., 2018;

Imran 2018a). Therefore P fertilization is needed as it is an important major plant nutrient after N. In limited P application or no P fertilization, plants cannot grow as it is a key component of nucleic acids, phospholipids, and ATP, and without these components, the plant cannot survive (Imran 2018b; Imran. 2018c; Imran 2018d). Phosphorus is involved in controlling and regulating several enzymatic functions and reactions in the metabolic pathway and thus regulating plant growth and development and ultimately plant productivity and yield (Imran and Amanullah. 2018; Imran. 2018e; Imran and Khan 2017). Other researchers reported that P is very rarely available in soil or only available in the soil out of the rhizosphere which cannot support the plant for high growth rate and development (Imran et al., 2020c; Imran et al., 2017a). Phosphorus application is necessary for agriculture systems to ensure crop productivity and food security (Imran et al., 2017b; Imran 2017; Imran et al., 2015a). Limited P supply in bulk soil affect P uptake by plants as it is only moving by the process of diffusion and do not uptake or very low amount is uptake by the crop plants which is an alarming threat for low productivity (Imran, et al., 2015b; Imran et al., 2019b). Phosphorus (P) is an essential macronutrient required for plant growth and development, but most of the world's soil is limited in this important plant nutrient (Palta et al., 2006; Pant et al., 2015; Paz et al., 2014; Peng, et al., 2008; Ping et al., 2007; Prado et al., 2000). Plants can utilize P mainly in the form of soluble inorganic phosphate (Pi) but are subjected to abiotic stress of P-limitation when there are not sufficient soluble PO_4 available in the soil. Phosphorus forms insoluble complexes with Ca and Mg in alkaline soils and Al and Fe in acidic soils that makes it unavailable for plant roots (Imran, 2018; Imran, 2015; Iqbal et al., 2017; Raza 2012; Rengasamy, 2002; Roelofs et al., 2008; Rolf et al., 2004; Rubio et al., 2001; Savvides et al., 2015). When there is limited bioavailable P in the soil, plants show extensive abiotic stress phenotype such as short primary roots and more lateral roots and root hairs to make more surface available for Pi absorption, exudation of organic acids, and phosphatase to release Pi from complex P containing molecules and make it available for growing plants organs (Imran, 2018; Imran, 2015; Iqbal et al., 2017; Raza 2012; Rengasamy, 2002; *Roelofs* et al., 2008; *Rolf* et al., 2004; *Rubio* et al., 2001; *Savvides* et al., 2015). Organic biomass thermal degradation in limited oxygen forms a solid coproduct known as biochar. There is a growing interest worldwide in Biochar application in agriculture and farm produce. The main reasons for Biochar application in agricultural land and cultural practices might be due to (a) pyrolysis can be a source of renewable biofuels (Balemi, 2012; Imran et al., 2019; Imran et al., 2017). (ii) Many biomass waste materials can be treated by pyrolysis and thus converted into fuel sources. The long-term sequence of carbon dioxide that can be introduced into the atmosphere by adding Biochar to the soil (Amin and Eissa, 2017; Imran et al., 2021; Imran et al., 2017; Khan, 2015b). Biochar amendment of soil aids in improving soil fertility and crop productivity. To meet global food security, crop production faces several real challenges, including soil fertility and organic matter depletion, climate change, unmanaged rainfall patterns, and pathogens (Amin and Eissa, 2017; Gao et al., 2017; Imran et al., 2021; Imran and Amanullah, 2019; Imran et al., 2017; Khan, 2015a, b). Abiotic pressures such as heat and cold, drought, salinity, water storage, and heavy metal pressure are emerging as important factors in reducing crop production globally. Biochar is a high carbonate modification created by the absence of oxygen or limited supply of a wide variety of feedstocks. Biochar typically has a high level of area, capacity for cation exchange, reactive functional groups, and highly reactive carbon (Amin and Eissa, 2017; Gao et al., 2017; Imran et al., 2021; Imran and Amanullah, 2019; Imran et al., 2017; Khan, 2015a, b; Koçak and Ortaş, 2021). Due to these characteristics, Biochar has been widely associated with multiple abiotic stress reductions to enhance crop growth, development, production, soil fertility, and soil quality (Amin and Eissa, 2017; Gao et al., 2017; Imran et al., 2021; Imran and Amanullah, 2019; Imran et al., 2017; Khan, 2015a, b; Koçak and Ortaş, 2021). It has been shown that Biochar application is a master regulator of P-starvation response in plants *(Gonzalez et al., 2018)*. Biochar amended soil has been shown to regulate extensive secondary metabolites during phosphorus limitation stress *(Bos, et al., 2005; Bruinsma, 1963; Brussaard et al., 2007; Capell, and Doerffling. 1993).*

Figure 4. Investigation of root response to biochar and its effect on root length, root diameter, root colonization, root weight and microbial population

Figure 5. Biochar chamber (pyrolyser) and various type of biochar prepared on various temperature (Cukurova University Turkey).

Exogenous application of mycorrhiza and beneficial microbes

Commercial use of beneficial bacteria and fungi is gaining ground as part of integrated pest management strategies. Belchim's Contans contains Coniothyrium minitans, which helps control soilborne pathogens such as Sclerotinia sclerotiorum and S. minor (Hasanuzzaman, 2019). According to specialist Fargro technical and energy director Paul Sopp: "Growers are experimenting with biopesticides such as Serenade (Bacillus subtilis QST713) because this widens their options for pest and disease control (Fig 5). This is especially the case with diseases such as downy mildew on brassicas and onions (Hussain et al., 2013). Several formulations of beneficial microbes antagonistic to pathogens are available including Agrobacterium radiobacter active against A. tumefaciens (crown gall, NoGall), Pseudomonas fluorescens active against Sclerotinia, Rhizoctonia and Pythium (Biomonas), Streptomyces griseoviridis active against Fusarium, Phytophthora, Pythium, and Rhizoctonia (Mycostop), Trichoderma atroviride active against Sclerotium cepivorum (Plantmate) and T. harzianum active against Pythium, Rhizoctonia and Fusarium (Rootshield, PlantShield).

Biofertilizers and Pulses

Mungbean (green gram) inoculation with VAM gave the highest green forage, dry matter, and crude protein. It has been reported that the highest number of plants m^{-2}, plant height (cm), number of leaves $plant^{-1}$, pods length (cm), numbers of seeds pod^{-1}, number of pods plants-^{1}, thousand seed weight, seed yield, and harvest index was noted with the inoculation of biofertilizer (PSB) (Imran et al., 2016). Alam et al. (2007) studied beneficial microbe inoculation with three different cultivars of mung bean and reported that growth and yield components were highest in cv. NM-92 with inoculation of Trichoderma. Recently the author (Imran 2020a) conducted two years (2016 and 2017) consecutive experiments on soybean, maize, and subsequent wheat to evaluate soybean, maize, and subsequent wheat response to biofertilizers (PSB, Trichoderma) and their interactions with organic sources and phosphorous levels. The author not yet published the work and will be submitted to the higher education commission of Pakistan. He reported that soybean phenological and growth characterizes were holistically improved with the inoculation of Trichoderma. Dry matter portioning was much higher with PSB inoculation. He concluded that Trichoderma is favored by the presence of high levels of plant roots, which they colonize readily. Some strains are highly rhizosphere competent, i.e., able to colonize and grow on roots as they develop. The most strongly rhizosphere competent strains can be added to soil or seeds by any method (Ansari and Sukhraj 2010). Once they come into contact with roots, they colonize the root surface or cortex, depending on the strain. Thus, if added as a seed treatment, the best strains will colonize root surfaces even when roots a meter or more below the soil surface and they can persist at useful numbers up to 18 months after application. However, most strains lack this ability. By fixing atmospheric nitrogen and solubilizing phosphates, pulses contribute to reducing the need for synthetic fertilizers and, in doing so, greatly contribute to reducing the risk of soil and water pollution, supporting soil biodiversity, and combating and building resilience to climate change (Aziz et al. 2016). Biological nitrogen fixation is particularly important for global agricultural productivity and might be considered one of the most important biological processes on the planet. It provides *circa* 100 million metric tonnes of N which leads to an annual saving of around USD10 billion in N fertilizer (Arancon et al. 2005). Lentils alone could fix nitrogen in the range of 35-100 kg ha^{-1}. Furthermore, the reduced need for (or use of) synthetic fertilizers indirectly reduces the number of greenhouse gases released into the atmosphere. Pulses also promote soil carbon sequestration and, ultimately, reduce soil erosion when included in intercropping farming systems and/or used as cover crops (Ancha et al. 1997). Furthermore, due to their high nutritional value, pulses are also valuable allies in fighting hunger worldwide.

Phosphate Solubilizing Bacteria and Crop Production

Phosphate rock minerals are often too insoluble to provide sufficient P for crop uptake. The use of PSMs can increase crop yields up to 70 percent (Carsky et al. 2001). Combined inoculation of arbuscular mycorrhiza and PSB give better uptake of both native P from the soil and P coming from the phosphatic rock (Carvalho et al. 2011). Higher crop yields result from solubilization of fixed soil P and applied phosphates by PSB (Cassman et al. 1981). Microorganisms with phosphate solubilizing potential increase the availability of soluble phosphate and enhance plant growth by improving biological nitrogen fixation (Chaturvedi 2006). *Pseudomonas* spp. enhanced the number of nodules, dry weight of nodules, yield components, grain yield, nutrient availability, and uptake in soybean crop (Chauhan et al. 1992; Chela et al. 1993). Phosphate solubilizing bacteria enhanced the seedling length of *Cicer arietinum* while co-inoculation of PSM and PGPR reduced P application by 50 % without affecting corn yield. Inoculation with PSB increased sugarcane yield by 12.6 percent (Chen et al. 2006). Sole application of bacteria increased the biological yield, while the application of the same bacteria along with mycorrhizae achieved the maximum grain weight (Chen et al. 2008). Single and dual inoculation along with P fertilizer was 30-40 % better than P fertilizer alone for improving grain yield of wheat, and dual inoculation without P fertilizer improved grain yield up to 20 % against sole P fertilization. Mycorrhiza along with *Pseudomonas putida* increased leaf chlorophyll content in barley. Rhizospheric microorganisms can interact positively in promoting plant growth, as well as N and P uptake. Seed yield of green gram was enhanced by 24 % following triple inoculation of *Bradyrhizobium* + *Glomus fasciculatum* + *Bacillus subtilis* (Chiezy and Odunze 2012). Growth and phosphorus content in two alpines Carex species increased by inoculation with *Pseudomonas Fortini.* Integration of half dose of NP fertilizer with biofertilizer gives crop yield as with full rate of fertilizer, and through reduced use of fertilizers the production cost is minimized and the net return maximized (Crew 1993). Mostly on farm crops using compost may have some limitations. For example, compost could be effectively used during the establishment of the nursery in the field to increase the stock and maintain the health of the plant (Ahmed et al., 2008). Crop nutrient management through an integrated approach is an important aspect provided mainly by bio, organic and chemical fertilizers. However, it is widely accepted that the application of a balanced fertilizer with effective use of organic sources and beneficial microbes is key to achieving higher crop production and net return (Imran et al. 2016). Agriculture scientists exploring an alternative source to increase farm products with low input cost and maintain soil health without affecting soil and plant environment. Plants obtain nutrients from two natural sources, organic matter, and minerals or chemical fertilization. Farmers only rely on chemical fertilizer which not only deteriorates the quality of soil but also increases the cost of production. Soil amendments and bio fertilization is an integrated approach to reduce dependency on chemical fertilizers with the advantage of the low cost of production (Imran 2017). Crop response to biological fertilizers depends largely on crop species/strains of microbe and application method along with climatic conditions. It has been reported that using beneficial microbes reduces the requirement of nitrogen 50-70 percent and increases yield up to 20%. Bacterial inoculation not only provides nitrogen, phosphorus, and growth hormones but also makes the plant healthy and less susceptible to pathogens. Using beneficial microbes along with the optimal dose of fertilizers to save about half the recommended dose of chemical fertilizers (Imran et al. 2016; Imran and Asad 2015).

Conclusion

Biochar is an activated carbon soil conditioner that can alleviate the negative impacts of salinity, drought, heat and cold, water-logging, and heavy metal stresses. Drought and salt stress negatively affect soil fertility and plant growth. The application of Biochar ameliorates the negative effects of drought and salt stress on plants. The Biochar application increased the plant growth, biomass, and yield under either drought or salt stress and also increased photosynthesis, nutrient uptake, and modified gas exchange characteristics in drought and salt-stressed plants. Under drought stress, Biochar increased the water holding capacity of the soil and improved the physical and biological properties of soils. Under salt stress, Biochar decreased Na^+ uptake, while increased K^+ uptake by plants. Nano-black carbon-mediated increase in salt tolerance of plants is primarily associated with improvement in soil properties, thus increasing plant water status, reduction of Na^+ uptake, increasing uptake of minerals, and regulation of stomatal conductance and phytohormones. This chapter highlights both the potential of Biochar in alleviating abiotic stresses in plants and the prospect of the role of Biocharunder abiotic stresses in plants.

References

1. Imran, Zada H, Naveed S, Khattak I And Ahmad S. 2016. Variable Rates of Phosphorous Application Influenced Phenological Traits of Green Gram (*Vignaradiata* L.) J Agri Res 2016, 1(3).
2. Imran, A.A. Khan, I.U. Khan and S. Naveed. 2016. Weeds density and late sown maize productivity influenced by compost application and seed rates under temperate environment. Pak. J. Weed Sci. Res. 22(1): 169-181.
3. A. Khan, M. N. Khan, I. uallah, S. Shah, I. R. Arshad, I. M. A. Zeb, and Imran. 2015. Effect of potash application on growth, yield and yield components of spring maize hybrids. Pure and Applied Biology Vol. 4, Issue 2, 195-203.
4. Akhtar, S. S., M. N. Andersen, and F. Liu. 2015a. Biocharmitigates salinity stress in potato. *Journal of Agronomy and Crop Sciences* 201:321–400.
5. Akhtar, S. S., M. N. Andersen, and F. Liu. 2015b. Residual effects of Biocharon improving growth, physiology and yield of wheat under salt stress. *Agricultural Water Management* 158:61–8.
6. Imran, A.A. Khan, Inamullah and Luqman. 2015. Weeding stages and their effect on yield and yield components of rice in upper Swat, Pakistan. Pak. J. Weed Sci. Res. 21(4): 555-563.
7. Imran, Abdul Bari, Roshan Ali, Naeem Ahmad, Zulfiqar Ahmad, Muhammad Idrees Khattak, Akhtar Ali, Fayaz Ahmad, Inayatullah Khan and Shahida Naveed. 2017. Traditional Rice Farming Accelerate CH4 & N2O Emissions Functioning As A Stronger Contributors Of Climate Change. Int J Environ Sci Nat Res. 2017; DOI: 10.19080/ARTOAJ.2017.09.555765
8. Imran, Amanullah, A. Bari and R. Ali. 2018. Peach sources, phosphorous and beneficial microbes enhance productivity of soybean. Soy. Res. 16(2): 39-48.
9. Akhtar, S. S., M. N. Andersen, M. Naveed, Z. A. Zahir, and F. Liu. 2015. Interactive effect of Biocharand plant growth-promoting bacterial endophytes on ameliorating salinity stress in maize. *Functional Plant Biology* 42:770–81.
10. Amir Zaman Khan, Imran, Asim Muhammad, Aiman Khalil, Hasina Gul, Habib Akbar and Said Wahab.2016. Impact of fertilizer priming on seed germination behavior and vigor of maize. Pure and Applied Biology. Vol. 5, Issue 4, pp744-751. http://dx.doi.org/10.19045/bspab.2016.50093

11. Amonette, J. E., and S. Joseph. 2009. Characteristics of Nano-black carbon: Microchemical properties. In *Biocharfor environmental management: Science and technology*, eds. J. Lehmann and S. Joseph, 33–52. Earthscan, London: Sterling.
12. Armor, N. B., K. B. Hamed, A. Debez, and C. Grignonand Crabbedly. 2005. Physiological and antioxidant responses of perennial halophyte *Crithmum maritimum* to salinity. *Plant Science*4:889–99.
13. Arshad, M., B. Shaharoona, and T. Mahmood. 2008. Inoculation with *Pseudomonas* spp. containing ACC-deaminase partially eliminates the effects of drought stress on growth, yield, and ripening of pea (*Pisum sativum* L.). *Pedosphere* 18:611–20.
14. Asad Ali Khan, Imran* Fawad Ali Shah Inamullah, Laiq Zada Muhammad Naeem Muhammad Nouman Khan.2015. Phenological Traits of Rice as Influenced by Seedling Age and Number of Seedling per Hill under Temperate Region. J. Bio, Agric and H. care ISSN 2224-3208 (Paper) ISSN 2225-093X (Online).
15. Ayaz, F. A., A. Kadioglu, and T. R. Turgut. 2000. Water stress effects on the content of low molecular weight carbohydrates and phenolic acids in *Cienanthe setosa*. *Candian Journal of Plant Sciences* 80:373–8.
16. Babar Iqbal, Bashir Ahmad, Inam Ullah, Imran, Asad Ali Khan, Shazma Anwar, Asad Ali, Khurrum Shahzadand Shehryar Khan. 2016. Effect of phosphorus, sulphur and different irrigation levels on phenological traits of Triticale. Pure and Applied Biology. Vol. 5, Issue 2, pp 303-310. http://dx.doi.org/10.19045/bspab.2016.50039
17. Babar Iqbal, Mohammad Tariq Jan, Zar Muhammad, Asad Ali Khan, Shazma Anwar, Imran and Khurrum Shahzad. 2016. Phenological traits of Maize influenced by integrated management of compost and fertilizer Nitrogen. J. Pure & Applied biology. 5(1): 58-63.
18. Babar Iqbal[1], Muhammad Tariq Jan[1], Inamullah[1], Imran*[1], Asad Ali Khan*[1], Zar Muhammad[2], Shazma Anwar[1], Shehryar Khan[1], Shamsuddin Baqa[1], Amjad Usman[3] .2015. Integrated Management Of Compost Type And Fertilizer-N In Maize. J. Pure & Applied biology Vol. 4, Issue 4.
19. Baber Iqbal, M.T. Jan, Imran, Zar M, A.A. Khan, S. Anwar and K. Shahzad 2015. Growth and phenology of maize as affected by integrated management of compost and fertilizers Nitrogen. J. Pure & Applied biology Vol. 4, Issue 4.
20. Badridze, G., A. Weidner, F. Asch, and A. Borner. 2009. Variation in salt tolerance within Georgian wheat germplasm collection. *Genetic Resources and Crop Evolution* 56:1125–30.

21. Bates, L. S., R. P. Waldren, and I. D. Teare. 1973. Rapid determination of free proline for water stress studies. *Plant and Soil* 39:205–7.
22. Bohnert, H. J., and R. G. Jensen. 1996. Strategies for engineering water stress tolerance in plants. *Trends in Biotechnology* 14:89–97.
23. Imran, Shahida Naveed, Asad Ali Khan And Inayat Khattak. 2015. Impact Of Phosphorus Levels And Seed Rates On Growth And Yield Of Late Sown Maize On High Elevation In Swat, Pakistan. Pakistan J. Agric. Res. Vol. 28 No. 4:406-413
24. Imran. 2018. Ecological Environmental Variability Influence Growth and Yield Potential of Rice under Northern Climatic Scenario. Russian Agricultural Sciences, 2018, Vol. 44, No. 1, pp. 18–24.
25. Bos, C., B. Juillet, H. Fouillet, L. Turlan, S. Dare, C. Luengo, R. Ntoun da, R. Benamouzig, N. Gausseres, V. Tome, and C. Gaudichon. 2005. Postprandial metabolic utilization of wheat protein in humans. *The American Journal of Clinical Nutrition* 81:87–94.
26. Bruinsma, J. 1963. The quantitative analysis of chlorophyll a and b in plant extract. *Photochemistry and Photobiology* 2:241–9.
27. *Brussaard, Lijbert; de Ruiter, Peter C.; Brown, George G. (2007). "Soil biodiversity for agricultural sustainability". Agriculture, Ecosystems and Environment.* ***121*** *(3): 233–244.* doi:10.1016/j.agee.2006.12.013.
28. Capell, B., and K. Doerffling. 1993. Genotype specific difference in chilling tolerance of maize in relation to chilling induce changes in water status and abscisic acid accumulation. *Physiologia Plantarum* 88:638–46.
29. Chimenti, C. A., M. Marcantonio, and J. A. Hall. 2006. Divergent selection for osmotic adjustment results in improved drought tolerance in maize (*Zea mays* L.) in both early growth and flowering phases. *Field Crops Research* 95:305–15.
30. *Conde, Artur (2011).* "Membrane Transport, Sensing and Signaling in Plant Adaptation to Environmental Stress" (PDF). *Plant & Cell Physiology.* ***52*** *(9): 1583–1602.* doi:10.1093/pcp/pcr107. PMID 21828102 – *via Google Scholar.*
31. *Cramer, Grant R; Urano, Kaoru; Delrot, Serge; Pezzotti, Mario; Shinozaki, Kazuo (2011-11-17).* "Effects of abiotic stress on plants: a systems biology perspective". *BMC Plant Biology.* ***11****:163.* doi:10.1186/1471-2229-11-163. ISSN 1471-2229. PMC 3252258. PMID 22094046.
32. Duan, D. Y., W. Q. Li, X. J. Liu, H. P. Ouyang, and P. An. 2007. Seed germination and seedling growth of *Suaeda salsa* under salt stress. *Annales Botanici Fennici* 44:161–9.

33. Dubois, M., K. Gilles, J. K. Hammiltron, P. A. Robers, and F. Smith. 1951. A colorimetric method for the determination of sugars. *Nature* 168:167–8.
34. Farahbakhsh, H. 2012. Germination and seedling growth in un-primed and primed seeds of fennel as affected by reduced water potential induced by NaCl. *International Research Journal of Applied and Basic Sciences* 3:737–44.
35. Imran 2018. Phosphorous fertilization influenced weeds attributes and phenological characteristics Of mungbean cultivars (vigna radiata l.) Russian Agricultural Sciences, 2018, Vol. 44, No. 3, pp. 229–234.
36. Imran and Khan, A. A. (2017). Canola Yield And Quality Enhanced With Sulphur Fertilization Russ. Agricult. Sci. 43: 113 10.3103/S1068367417020100
37. Flowers, T. J. 2004.Improving crop salt tolerance. *Journal of Experimental Botany* 55:307–19.
38. Gasco, G., Paz-Ferreiro, and A. Mendez. 2012. Thermal analysis of soil amended with sewage sludge and Biocharfrom sewage sludge pyrolysis. *Journal of Thermal Analysis and Calorimetry*108:769–75.
39. Giannopolitis, C. N., and S. K. Ries. 1977. Superoxide dismutase occurance in higher plants. *Plant Physiology* 59:304–14.
40. *Goncalves-Alvim, Silmary J.; Fernandez, G. Wilson (2001). "Biodiversity of galling insects: historical, community and habitat effects in four neotropical savannas". Biodiversity and Conservation. **10**: 79–98.* doi:10.1023/a:1016602213305
41. Imran, A. A. Khan, I. Ullah, H. zada, F. Ahmad, S.T. Shah, A. Usman and I. Ullah. 2015. Yield and yield attributes of rapeseed cultivars as influence by sulfur level under swat valley conditions. J. Pure & Applied biology Vol. 4, Issue 3.
42. *Gonzalez-Villagra, Jorge; Rodrigues-Salvador, Acácio; Nunes-Nesi, Adriano; Cohen, Jerry D.; Reyes-Díaz, Marjorie M. (2018). "Age-related mechanism and its relationship with secondary metabolism and abscisic acid in Aristotelia chilensis plants subjected to drought stress". Plant Physiology and Biochemistry. **124**: 136–145.*
43. doi:10.1016/j.plaphy.2018.01.010. ISSN 0981-9428. PMID 29360623.
44. Imran, I. Hussain, I. khattak, A. Rehman[2], F. Ahmad and S.T. shah .2015. *Roots nodulation, yield and yield contributing parameters of mungbean cultivars as influenced by different phosphorous level in swat-Pakistan.* J. Pure & Applied biology Vol. 4, Issue 4.
45. Heans, D. L. 1984. Determination of total organic carbon in soils by an improved chromic acid digestion and spectrophotometric procedure. *Communications in Soil Science and Plant Analysis* 15:1191–213.

46. Imran, I. Khattak, I. Hussain, A. Rehman, S. Anwar[1], F. Ahmad, A. A. Khan[1] and H. Zada. 2015. Growth and Yield of Maize Hybrids as effected by different Sowing Dates in Swat Pakistan. J. Pure & Applied biology Vol. 4, Issue 4.
47. Imran, Amanullah, A.A Khan, Abdul bari….. 2019. Production statistics and modern technology of maize cultivation in Khyber Pakhtunkhwa Pakistan. Plant sci. Arc.2 (2) 1-12.
48. Imran, Amanullah. 2018. Review. Global impact of climate change on water, soil resources and threat towards food security: evidence from Pakistan. Adv Plants Agric Res. 2018;8(5):350–355.
49. DOI: 10.15406/apar.2018.08.00349
50. Imran, Asad Ali Khan, Inamullah Inam & Fayaz Ahmad. (2016). Yield and yield attributes of Mungbean (*Vigna radiata* L.) cultivars as affected by phosphorous levels under different tillage systems, *Cogent Food & Agriculture* (2016), 2: 1151982.
51. Imran. 2018. Organic Matter Amendments Improve Soil Health, Productivity and Profitability of Maize and Soybean. Ann Rev Resear. 1(3): 555564.
52. Iqbal A, Amanullah, Ali A, Iqbal M, Ikramullah and Imran. 2017. Integrated use of phosphorus and organic matter improve fodder yield of moth bean (Vigna aconitifolia (Jacq.) under irrigated and dryland conditions of Pakistan. Journal of AgriSearch 4(1): 10-15
53. ISTA. 2005. *International rules for seed testing*. Zurich, Switzerland: International Seed Testing Association.
54. Izhar Ali, Asad Ali Khan, Imran, inamullah, Aman Khan …. 2019. Humic acids and Nitrogen levels optimizing productivity of green gram (vigna radiate L.). Russian Agriculture Sciences, 45(1), 43-47.
55. Jackson, M. 1962. *Soil chemical analysis*, 496. London: Constable and Company Limited.
56. Jagatheeswari, D., and P. Ranganathan. 2012. Influence of Mercuric Chloride on seeds germination, seedling growth and biochemical analysis of green gram (*Vigna radiata* (L.) Wilczek. Var. Vamban-3). *International Journal of Pharmaceutical and Biological Archives* 3:291–5.
57. Joseph B., and D. Jini. 2011. Development of salt stress-tolerant plants by gene manipulation of antioxidant enzymes. *Asian Journal of Agricultural Research* 5:17–27.
58. Kamara, A., A. Kamara, M. M. Mansaray, and P. A. Sawyerr. 2014. Effects of Biocharderived from maize stover and rice straw on the germination of their seeds. *American Journal of Agriculture and Forestry* 2:246–9.

59. Kammann, C. I., S. Linsel, J. W. Gobling, and H. Koyro. 2011. Influence of Biocharondrought tolerance of *Chenopodium quinoa* willd and on soil-plant relation. *Plant and Soil* 345:195–210.
60. Khan, M. A., E. Islam, M. U. Shirazi, S. Mumtaz, S. M. Mujtaba, M. A. Khan, A. Shereen, M. Y. Ashrafand, and G. M. Kaleri. 2010. Physiological responses of various wheat genotypes to salinity. *Pakistan Journal of Botany* 42:3497–505.
61. Lashari M.S., Y. Liu, W. Li, J. Pan, G. Pan, J. Zheng, G. Zheng, X. Zh ang, and X. Yu. 2013. Effect of amendment of Biocharmanure compost in conjunction with pyroligneous solution on soil quality and wheat yield of a salt stressed cropeland from central China Great Plain. *Food Crop Research* 144:113–8.
62. Lehmann J., and S. Joseph. 2009. Biocharfor environmental management: An introduction. In *Biocharfor environmental management: Science and technology*, eds. J. Lehmann and S. Joseph, 1–12. London: Earthscan.
63. Lehmann, J., M. C. Rillig, J. Thies, C. A. Masiello, W. C. Hockaday, and D. Crowley. 2011. Biochareffects on soil biota. *Soil Biology and Biochemistry* 43:1812–36.
64. *Lei, Gang; Shen, Ming; Li, Zhi-Gang; Zhang, Bo; Duan, Kai-Xuan; Wang, Ning; Cao, Yang-Rong; Zhang, Wan-Ke; Ma, Biao (2011). "EIN2 regulates salt stress response and interacts with a MA3 domain-containing protein ECIP1 in Arabidopsis". Plant, Cell & Environment.* ***34*** *(10):1678–1692.*
doi:10.1111/j.13653040.2011.02363.x. ISSN 1365-3040.
65. Lu. Congming, A. Vonshak. (2002). Effects of salinity stress on photosystem II function in cyanobacterial Spirulina platensis cells. Physiol. Plant 114 405-413.
66. *Maestre, Fernando T.; Cortina, Jordi; Bautista, Susana (2007). "Mechanisms underlying the interaction between Pinus halepensis and the native late-successional shrub Pistacia lentiscus in a semi-arid plantation". Ecography.* ***27*** *(6): 776–786.*
doi:10.1111/j.0906-7590.2004.03990.x.
67. Martinez-Beltran J, Manzur CL. (2005). Overview of salinity problems in the world and FAO strategies to address the problem. Proceedings of the international salinity forum, Riverside, California, April 2005, 311–313.
68. Mazhar, R., N. Ilyas, M. Saeed, F. Bibi, and N. Batool. 2016. Biocontrol and salinity tolerance potential of *Azospirillum lipoferum* and its inoculation effect in wheat crop. *International Journal Of Agriculture and Biology* 3:494–500.

69. Ministry of Food, Agriculture and Livestock, 2007. *Agricultural statistics of Pakistan*. Islamabad: Ministry of Food, Agriculture and Livestock, Government of Pakistan.
70. Miri, Y., and S. A. Mirjalili. 2013. Effect of salinity stress on seed germination and some physiological traits in primary stages of growth in purple coneflower (*Echinacea purpurea*). *International Journal of Agronomy and Plant Production* 4:142–6.
71. *Mittler, Ron (2006). "Abiotic stress, the field environment and stress combination". Trends in Plant Science.* ***11*** *(1): 15–19.* doi:10.1016/j.tplants.2005.11.002. PMID 16359910.
72. Muhammad Islam[1], Shazma Anwar[1], Saqib Bashir[1], Wajid Ali Khattak[1], Imran, Murad Ali[2] and Nadar Khan .2015. Growth and Yield Components of Wheat Varieties As Affected By Dual Purpose Practices. J. Pure & Applied biology.
73. Munns, R. R., A. James, and A. Lauchli. 2006. Approaches to increasing the salt tolerance of wheat and other cereals. *Journal of Experimental Botany* 57:1025–43.
74. Munns, R. 2002. Comparative physiology of salt and water stress. *Plant Cell Environment*25:239–50.
75. Munns, R., and M. Tester. 2008. Mechanisms of salinity tolerance. *Annual Review of Plant Biology* 59:651–81.
76. Nazarli, H., F. Faraji, and M. R. Zardashti. 2011. Effect of drought stress and polymer on osmotic adjusment and photosynthetic pigments of sunflower. *Carcetriari Agronomica in Maldova* 1:45.
77. *Neto, Azevedo; De, André Dias; Prisco, José Tarquinio; Enéas-Filho, Joaquim; Lacerda, Claudivan Feitosa de; Silva, José Vieira; Costa, Paulo Henrique Alves da; Gomes-Filho, Enéas (2004-04-01). "Effects of salt stress on plant growth, stomatal response and solute accumulation of different maize genotypes". Brazilian Journal of Plant Physiology.* ***16****(1): 31 38.* doi:10.1590/S1677-04202004000100005. ISSN 1677-0420.
78. Olsen, S. R., and L. E. Sommers. 1982. Phosphorous. In *Methods ofsoil analysis: Chemical and microbiological properties*, ed. D. Bunton, 403–27. 2nd ed. Madison, WI: Soil Science Society of America.
79. *Palm, Brady; Van Volkenburgh (2012). "Serpentine tolerance in Mimuslus guttatus does not rely on exclusion of magnesium". Functional Plant Biology.*
80. Palta, Jiwan P. and Farag, Karim. September 2006. "Methohasds for enhancing plant health, protecting plants from biotic and abiotic stress related injuries and enhancing the recovery of plants injured as a result of such stresses." United States Patent 7101828.

81. *Pant, Bikram Datt; Burgos, Asdrubal; Pant, Pooja; Cuadros-Inostroza, Alvaro; Willmitzer, Lothar; Scheible, Wolf-Rüdiger (2015).* "The transcription factor PHR1 regulates lipid remodeling and triacylglycerol accumulation in Arabidopsis thaliana during phosphorus starvation". *Journal of Experimental Botany. **66** (7): 1907–1918.* doi:10.1093/jxb/eru535. ISSN 0022-0957. PMC 4378627. PMID 25680792.
82. *Pant, Bikram-Datt; Pant, Pooja; Erban, Alexander; Huhman, David; Kopka, Joachim; Scheible, Wolf-Rüdiger (2015). "Identification of primary and secondary metabolites with phosphorus status-dependent abundance in Arabidopsis, and of the transcription factor PHR1 as a major regulator of metabolic changes during phosphorus limitation". Plant, Cell & Environment. **38** (1): 172–187.* doi:10.1111/pce.12378. ISSN 1365-3040. PMID 24894834.
83. Paz-Ferreiro, J., S. Fu, A. Mendez, and G. Gasco. 2014. Interactive effects of Biocharand the earhworm *Pontoscolex corethrurus* on plant productivity and soil enzymes activities, *Journal of Soils Sediments*, 14:483–94.
84. Peng, Y., Z. Gao, Y. Gao, G. Liu, L. Shengand D. Wang. 2008. Ecophysiological characteristics of alfalfa seedlings in response to various mixed salt-alkaline stresses. *Journal of Integrative Plant Biology* 50:29–39.
85. *Ping Ji, Geo K. Ron M. (2007). "Understanding Abiotic Stress Tolerance Mechanisms: Recent Studies on Stress Response in Rice". Journal of Integrative Plant Biology. **49** (6): 742–750.* doi:10.1111/j.1744-7909.2007.00495.x.
86. Prado, F. E., C. Boero, M. Gallarodo, and A. J. Gonzalez. 2000. Effect of NaCl on germination, growth and soluble sugar content in *Chenopodium quinoa* wil seeds. *Botanical Bulletin - Academia Sinica Taipei* 14:27–34.
87. *Raghothama, K. G. (1999). "Phosphate Acquisition". Annual Review of Plant Physiology and Plant Molecular Biology. **50** (1): 665–693.* doi:10.1146/annurev.arplant.50.1.665. PMID 15012223.
88. Rajalakshmi, A. G., S. K. Kumar, C. D. Bharathi, R. Karthika, K. D. Visalakshi, R. Meera, and S. Mohanapriya. 2015. Effect of Biocharin seed germination: *In-vitro* study. *International Journal of Biosciences and Nanosciences* 2:132–6.
89. Rashid, A., Q. Saleem, A. Nazir, and H. S. Kazım. 2003. Yield potential and stability of nine wheat varieties under water stress conditions. *International Journal of Agriculture and Biology* 5:7–9.

90. Raza, A. 2012. Grain and field annual 2012 gain report. Global Agricultural Information Network. *USDA Foreign Agrilcultural services* 1–3.
91. Rengasamy, P. 2002. Transient salinity and subsoil constraints to dryland farming in Australian sodic soils: An overview. *Australian Journal of Experimental Agriculture* 42:351–61.
92. *Roelofs, D, Jiwan P and kwangy I.. (2008). "Functional ecological genomics to demonstrate general and specific responses to abiotic stress". Functional Ecology. **22**: 8–18.* doi:10.1111/j.1365-2435.2007.01312.x.
93. *Rolf D, Vinebrooke K., Mittler J. (2004). "Impacts of multiple stressors on biodiversity and ecosystem functioning: the role of species co-tolerance". OIKOS. **104** (3): 451–457.* doi:10.1111/j.0030-1299.2004.13255.x.
94. *Rubio, Vicente; Linhares, Francisco; Solano, Roberto; Martín, Ana C.; Iglesias, Joaquín; Leyva, Antonio; Paz-Ares, Javier (2001).* "A conserved MYB transcription factor involved in phosphate starvation signaling both in vascular plants and in unicellular algae". *Genes & Development. **15** (16): 2122–2133.* doi:10.1101/gad.204401. ISSN 0890-9369. PMC 312755. PMID 11511543.
95. *Savvides, Andreas (2015).* "Chemical Priming of Plants Against Multiple Abiotic Stresses: Mission Possible?". *Trends in Plant Science. **21** (4): 329 340.* doi:10.1016/j.tplants.2015.11.003. hdl:10754/596020.
96. Scholander, P. F., H. T. Hammel, E. D. Bradstreet, and E. A. Hemminogsen. 1965. Sap pressure in vascular plants: Negative hydrostatic pressure can be measured in plants. *Plant Sciences*148:339–46.
97. Shahida Naveed, Muhammad Ibrar, Inayat Khattak, Imtiaz Khan, Imran, Muhammad Isfhaq Khan And Haroon Khan.2016. Anthelmintic, Antilice, Insecticidal, Cytotoxic And Phytotoxic Potential Of Ethanolic Extracts Of Two Wild Medicinal Plants *Iphiona Grantioides* And *Plucheaarguta.* J. Woulfenia (23) 11: 13-25
98. Shamsi, K., and S. Kobraee. 2013. Biochemical and physiological responses of three wheat cultivars (*Triticum aestivum* L.) to salinity stress. *Annals of Biological Research* 4:180–5.
99. Shamsuddin Baqa[1], Amir Zaman Khan[1], Inamullah[1], Imran*[1], Asad Ali Khan*[1], Shazma Anwar[1], Babar Iqbal[1], Shehryar Khan[1], Amjad Usman[2]. 2015. Influence Of Farm Yard Manure And Phosphorus

Application On Yield And Yield Components Of Wheat. J. Pure & Applied biology Vol. 4, Issue 4.

100. Shazma Anwar, Israeel, Babar Iqbal, Asad Ali Khan, Imran, Wajid Ali Shah, Mohammad Islam, Wajid Ali Khattak, Ikramullah, Wasim Akram and Wasim Abbas. 2016. Nitrogen and phosphorus fertilization of improved varieties for enhancing phenological traits of wheat. Pure and Applied Biology. Vol. 5, Issue 3, pp511-519. http://dx.doi.org/10.19045/bspab.2016.50065
101. Sidari, M., C. Mallamaci, and A. Muscolo. 2008. Drought, salinity and heat differently affect seed germination of Pinus pinea. *Journal of Forest Research* 13:326–330.
102. Syed Arshad Karim and Imran. 2019. Reduction in substrate moisture content reduce final yield of wheat. Biomed J. Sci & Tec Res. 1-5.
103. Tank, N, and M. Saraf. 2010. Salinity-resistant plant growth promoting rhizobacteria ameliorates sodium chloride stress on tomato plants. *Journal of Plant Interaction* 5:51–8.
104. Tiwari S., P. Singh, R. Tiwari, K. K. Meena, M. Yandigeri, D. P. Singh, and D. K. Arora. 2011. Salt-tolerant rhizobacteria-mediated induced tolerance in wheat (*Triticum aestivum*) and chemical diversity in rhizosphere enhance plant growth. *Biological Fertile Soils* 47:907–16.
105. *Tombesi, Sergio; Frioni, Tommaso; Poni, Stefano; Palliotti, Alberto (2018). "Effect of water stress "memory" on plant behavior during subsequent drought stress". Environmental and Experimental Botany. **150**: 106–114.* doi:10.1016/j.envexpbot.2018.03.009. ISSN 0098-8472.
106. Ulfat Samreen, Muhammad Ibrar, Lalbadshah, Shahida Naveed, Imran and Inayat Khatak. 2016. Ethnobotanical study of subtropical hills of Darazinda, Takht-e-Suleman range F.R D.I. Khan, Pakistan. Pure and Applied Biology. Vol. 5, Issue 1, pp149-164. http://dx.doi.org/10.19045/bspab.2016.50020
107. Van Hoorn, J. W., K. Nader, A. Hamdy, and M. Mastrorilli. 2001. Effect of salinity on yield and nitrogen uptake of four grain legumes and on biological nitrogen contribution from the soil. *Agriculture Water Management* 51:87–98.
108. Imran, Amanullah, Abdel Rehman Al Tawaha et al. 2021. Use of Biochar in Agriculture: An Inspiring Way in Existing Scenario. Organic Farming for Sustainable Development. CRC Press, Taylor and Frances, USA.

109. Imran, Amanullah, Abdel Rehman Al Tawaha, et al. 2021. Organic Amendments: Right Way to Soil Sustainability. Organic Farming for Sustainable Development. CRC Press, Taylor and Frances, USA.
110. Imran, Amanullah and Abdel Rahman Al-Tawaha. 2021. Carbon Sources Application Increase Wheat Yield And Soil Fertility. *Communication in soil science and plant analysis DOI: 10.1080/00103624.2020.1865397.*
111. *Wang, W.; Vinocur, B.; Altman, A. (2007). "Plant responses to drought, salinity and extreme temperatures towards genetic engineering for stress tolerance". Planta.* ***218*** *(1): 1–14.* doi:10.1007/s00425-003-1105-5. PMID 14513379.
112. Wang, G., and Z. Xu. 2013. The effects of Biocharon germination and growth of wheat in different saline-alkali soil. *Asian Agricultural Research* 5:116–9.
113. Wang. H., and J. Jin. 2007. Effects of zinc deficiency and drought on plant growth and metabolism of reactive oxygen species in maize (*Zea mays* L.). *Agricultural Sciences in China*6:988–95.
114. Wolfe, A. 2007. "Patterns of biodiversity." Ohio State University, 10.1016/j.agee.2006.12.013.
115. Xu, D., J. Huang, S. Guo, X. Yang, Y. Bao, H. Tang, and H. Zhang. 2008. Over-expression of a TFIII-type zinc finger protein gene ZFP252 enhances drought and salt tolerance in rice (*Oryza sativa*L.). *Federation of European Biochemical Societies Letters* 582:1037–43.
116. Yeo, A. R., M. E. Yeo, S. A. Flowers, and T. J. Flowers. 1990. Screening of rice (*Oryza sativa* L.) genotypes for physiological characters contributing to salinity resistance, and their relationship to overall performance. *Theoretical Applied Genetics* 79:377–84.
117. Younis, U., M. Athar, S. A. Malik, M. H. RazaShah, S. Mahmood. 2015. Biocharimpact on physiological and biochemical attributes of spinach *Spinacia oleracea* (L.) in nickel contaminated soil. *Global Journal Environment Science Management* 1:245–54.
118. Zakaria, M., S. Daniel, V. Murphy, and L. K. Abbott. 2012. Nano-black carbons influence seed germination and early growth of seedlings. *Plant Soil* 353:273–87.
119. *Zargar, Sajad Majeed; Nagar, Preeti; Deshmukh, Rupesh; Nazir, Muslima; Wani, Aijaz Ahmad; Masoodi, Khalid Zaffar; Agrawal, Ganesh Kumar; Rakwal, Randeep (2017). "Aquaporins as potential drought tolerance inducing proteins: Towards instigating stress tolerance". Journal of Proteomics.* ***169****: 233–238.* doi:10.1016/j.jprot.2017.04.010. ISSN 1874-3919.

120. Zhan, H. X., Z. J. Chang, A. L. Wei, X. J. Zhang, and X. Li. 2011. Impact of drought to wheat physiological index. *Journal of Shanxi Agriculture Sciences* 39:1049–51.
121. A.R. M. Tawaha, Z. Farrokhi, N. Yoga, P. Roshan, Imran, Amanullah, et al.2021. Weed Management in Organic Cropping Systems. Organic Farming for Sustainable Development. CRC Press, Taylor and Frances, USA
122. Imran, Amanullah and Abdel Rehman Al-Tawaha.2020. The Productivity of subsequent wheat enhanced with residual carbon sources and phosphorus under improved irrigation system. *Communication in soil science and plant analysis*, DOI:10.1080/00103624.2020.1763387
123. Imran, Amanullah, Asad Ali khan, Tariq Mahmood, Abdel Rahman Al Tawaha and Samia Khanum[4]. 2020. Adequate Fertilization, Application Method and Sowing Techniques Improve Maize yield and related traits. *Communication in soil science and plant analysis*.
124. Mohammad Ilyas, Gohar Ayub, Imran, Azmat Ali Awan and Manzoor Ahmad.2020. Effect of Calcium and Boron on Production and Quality of Autumn Potato under Chilling Temperature. *Communication in soil science and plant analysis*. DOI: 10.1080/00103624.2020.1854286
125. Abdel Rahman M. Al-Tawaha, Nusrat Jahan…Imran, et al 2020. Growth, Yield And Biochemical Responses In Barley To DAP And Chitosan Application Under Water Stress. *Journal of Ecological Engineering*. 21(6). 86-93. Doi. 10.12911/22998993/123251.
126. Imran, Amanullah, Muhammad Arif, Zahir Shah and Abdul Bari. 2020. Integration of Peach (*Prunus persica* L.) residues, beneficial microbes and phosphorus enhance phenology, growth and yield of soybean. ISSN 1068-3674, *Russian Agricultural Sciences*, 2020, Vol. 46, No. 3, pp. 223–230. © Allerton Press, Inc., 2020.
127. Imran, Amanullah, Muhammad Arif, Zahir Shah and Abdul Bari. 2020. Soil application of *Trichoderma* and Peach (*Prunus persica* L.) residues possesses biocontrol potential for weeds and enhances growth and profitability of soybean (Glycine max). *Sarhad Journal of Agriculture* 36 (1) 10-20. DOI | http://dx.doi.org/10.17582/journal.sja/2020/36.1.10.20
128. Imran, Amanullah and Abdel Rahman M. Al Tawaha. 2021a. Management of Nano-black carbon, Phosphorous and Bio Fertilizer improve Soil organic carbon and ensilage biomass of soybean and maize. *Communication in Soil Science and Plant Analysis*. DOI.10.1080/00103624.2021.1966439

129. Imran, Amanullah and Abdel Rehman Altawaha. 2021b. Carbon Assimilation and Dry Matter Partitioning in Soybean Ameliorates with the Integration Of Nano-Black Carbon, Along With Beneficial Microbes and Phosphorus Fertilization. *Journal of Plant Nutrition*
130. Imran. 2021. The bioavailability of Phosphorus in composite vs. Hybrid maize differ with phosphorus and boron fertilization. *Phosphorus, Sulfur, and Silicon and the Related Elements. https://doi.org/10.1080/10426507.2021.1920588.*
131. Imran, Amanullah, Asad Ali khan, Tariq Mahmood, Abdel Rahman Al Tawaha and Samia Khanum[4]. 2021c. Adequate Fertilization, Application Method and Sowing Techniques Improve Maize yield and related traits. *Communication in soil science and plant analysis.* DOI:10.1080/00103624.2021.1925688
132. Imran and Amanullah· 2021. Phosphorus and Boron Application Optimizing Biofortification Of phosphorus and Productivity of French Bean (*Phaseolus Vulgaris* L.). *Communication In Soil Science and Plant Analysis.*
133. Imran, Amanullah and Abdel Rahman Al-Tawaha. 2021d. Carbon Sources Application Increase Wheat Yield And Soil Fertility. *Communication in soil science and plant analysis DOI: 10.1080/00103624.2020.1865397.*
134. Imran, Amanullah and Abdel Rehman Al-Tawaha.2020b. The Productivity of subsequent wheat enhanced with residual carbon sources and phosphorus under improved irrigation system. *Communication in soil science and plant analysis*, DOI:10.1080/00103624.2020.1763387
135. Imran, Amanullah, Muhammad Arif, Zahir Shah and Abdul Bari. 2020a. Soil application of *Trichoderma* and Peach (*Prunus persica* L.) residues possesses biocontrol potential for weeds and enhances growth and profitability of soybean (Glycine max). *Sarhad Journal of Agriculture* 36 (1) 10-20. DOI | http://dx.doi.org/10.17582/journal.sja/2020/36.1.10.20
136. Mohammad Ilyas, Gohar Ayub, Imran*, Azmat Ali Awan and Manzoor Ahmad.2020. Calcium and Boron effect on Production and Quality of Autumn Potato under Chilling Temperature. *Communication in soil science and plant analysis.* DOI: 10.1080/00103624.2020.1854286
137. Al-Tawaha, A. R. M., Jahan, N., Odat, N., Al-Ramamneh, E. A.-D., Al-Tawaha, A. R., Rauf, A., Wedyan, M., Shariati10, M. A., Qaisi11, A. M., and Imran, K. T. (2020). Growth, Yield and Biochemical

Responses in Barley to DAP and Chitosan Application Under Water Stress. *Journal of Ecological Engineering* **21**, 86-93.

138. Amin, A. E.-E. A. Z., and Eissa, M. A. (2017). Biochar effects on nitrogen and phosphorus use efficiencies of zucchini plants grown in a calcareous sandy soil. *Journal of soil science and plant nutrition* **17**, 912-921.
139. Balemi, T. (2012). Effect of integrated use of cattle manure and inorganic fertilizers on tuber yield of potato in Ethiopia. *Journal of soil science and plant nutrition* **12**, 253-261.
140. Çekiç, F. Ö., Ünyayar, S., and Ortaş, İ. (2012). Effects of arbuscular mycorrhizal inoculation on biochemical parameters in Capsicum annuum grown under long term salt stress. *Turkish Journal of Botany* **36**, 63-72.
141. Gao, L., Wang, R., Shen, G., Zhang, J., Meng, G., and Zhang, J. (2017). Effects of biochar on nutrients and the microbial community structure of tobacco-planting soils. *Journal of soil science and plant nutrition* **17**, 884-896.
142. Imran, A., Khalid, Shah, Arif, M., Fahad, S., and Al-Tawaha, A. R. M. (2021). The Application of Biochar for the Mitigation of Abiotic Stress-Induced Damage. *In* "Sustainable Soil and Land Management and Climate Change", pp. 45-56. CRC Press.
143. Imran, A., Khalid, Shah, Khan, H. A., Arif, M., Altawaha, A. R., Adnan, M., Fahad, S., and Parmar, B. (2019). Organic matter management in cereals based system: Symbiosis for improving crop productivity and soil health. *In* "Sustainable Agriculture Reviews 29", pp. 67-92. Springer.
144. Imran, I., and Amanullah, A. (2019). Integration of Peach (Prunus persica L) Biochar, Compost and Peach Residues along with Beneficial Microbes and Phosphorus Improve Agronomic Efficiency, Phosphorus use Efficiency, Partial Factor Productivity and Soil P in Soybean vs Maize Crops. *Preprints*.
145. Imran, N. J., Ashfaq, A., and Asad, A. (2017). Grain yield, yield attributes of wheat and soil physiochemical characteristics influenced by biochar, compost and inorganic fertilizer application. *Agric Res Tech Open Access J. https://doi. org/10.19080/ARTOAJ* **555795**.
146. Khan, A. (2015a). Biochar application and shoot cutting duration (days) influenced growth, yield and yield contributing parameters of Brassica napus L. *Journal of Biology, Agriculture and Healthcare* **5**, 1-6.

147. Khan, A. (2015b). Phenological charateristics of Brassica napus L. as influenced by biochar application and shoot cutting duration (days). *Civil and Environmental Research* **7**, 104-107.
148. Koçak, B., and Ortaş, İ. (2021). Short-Term Eucalyptus and Phragmites Biochar's Efficiency in Mineralization of Soil Carbon. *Journal of soil science and plant nutrition* **21**, 3346-3353.
149. Sahni, S., Prasad, B. D., Liu, Q., Grbic, V., Sharpe, A., Singh, S. P., and Krishna, P. (2016). Overexpression of the brassinosteroid biosynthetic gene DWF4 in Brassica napus simultaneously increases seed yield and stress tolerance. *Scientific reports* **6**, 28298.